Earth and Environmental Science

THE PRELIMINARY COURSE

Iain Imlay-Gillespie

Chris Huxley

PUBLISHED BY THE PRESS SYNDICATE OF THE UNIVERSITY OF CAMBRIDGE
The Pitt Building, Trumpington Street, Cambridge, United Kingdom

CAMBRIDGE UNIVERSITY PRESS
The Edinburgh Building, Cambridge CB2 2RU, UK
40 West 20th Street, New York, NY 10011–4211, USA
477 Williamstown Road, Port Melbourne 3207, Australia
Ruiz de Alarcón 13, 28014 Madrid, Spain
Dock House, The Waterfront, Cape Town 8001, South Africa

http://www.cambridge.edu.au

© Cambridge University Press 2002

Notice to teachers
It is illegal to reproduce any part of this work in material form
(including photocopying and electronic storage) except under the
following circumstances:
(i) where the text indicates that the student may copy or photocopy
a figure, the artwork may be photocopied for use within the school
or institution which purchases the publication. The material
remains the property of Cambridge University and photocopies
may not be distributed or used in any way outside the purchasing
institution;
(ii) where you are abiding by a licence granted to your school or
institution by the Copyright Agency Limited that permits the
copying of small parts of text, in limited quantities, within the
conditions set out in the licence;
(iii) where no such licence exists, or where you wish to exceed the
terms of the licence, and you have gained the written permission
of Cambridge University Press.

All activities in this book have been written with the current safety
regulations in schools in mind. However, it is recommended that
the experiments in this book be carried out in the presence of a
qualified teacher and after consultation with appropriate state
safety regulations for schools.

First published in 2002
Reprinted 2005

Printed in Singapore by Craft Print Pty Ltd
Typeface Garamond

National Library of Australia Cataloguing in Publication data
    Huxley, Chris
    Earth and environmental science: the preliminary course
    Includes index
    For senior secondary students
    ISBN 0 521 89020 9
    1. Earth sciences 2. Environmental Sciences. I. Imlay-Gillespie,
    Iain. II. Title.
550.712

ISBN 0 521 89020 9

Waiver
The publisher has used its best endeavours to ensure that the URLs for external
websites referred to in this book are correct and active at the time of going to press.
However, the publisher has no responsibility for the websites and can make no
guarantee that a site will remain live or that the content is or will remain appropriate.

Contents

Introduction — vi

SECTION 1
Planet Earth and its environment — 1

Chapter 1.1 Observing our universe — 2
What do we use to explore the universe? — 3
Electromagnetic radiation — 6
Continuous, emission and absorption spectra — 8
Classifying stellar spectra — 9
Summary — 11
Practical exercise: Flame tests — 11
Practical exercise: Spectroscopy: Measurement of spectra — 12
Practical exercise: Examining the spectra of the planetary nebula ZX1267 — 13

Chapter 1.2 The origin of the universe — 14
What is the universe? — 14
A brief history of cosmology — 14
Theories on the origin of the universe — 15
Dating the age of the universe — 23
Parts of the universe — 24
Summary — 28
Practical exercise: Measuring the luminosity of stars — 29
Practical exercise: Comparison of the characteristics of stars — 30

Chapter 1.3 The Sun and the solar system — 31
Solar nebular hypothesis — 31
Formation of the Sun — 32
Formation of the planets — 34
Processes shaping the terrestrial planets — 36
Structure of the Earth — 38
Density particle models — 39
Summary — 40
Practical exercise: Density of Earth materials — 41
Practical exercise: Locating meteorite impact sites in Australia — 41

Chapter 1.4 The early Earth and its atmosphere — 42
The early Earth — 42
Evidence for the Earth's age — 42
Earth's early atmospheres — 43
Formation of the oceans — 44
Banded iron formations — 45
The modern atmosphere — 45
The layers of the atmosphere — 47
Climate change — 48
Summary — 55
Practical exercise: Comparing climate change data — 56

Chapter 1.5 The early Earth and the start of life — 57
What is life? — 57
Formation of the first organic chemicals — 57
The first cells — 60
Fumaroles: The origins of life? — 61
The evolution of cells — 62
Photosynthesis and the atmosphere — 64
Summary — 65
Practical exercise: Anaerobic respiration and carbon dioxide — 66

Resources — 67

SECTION 2
Dynamic Earth — 69

Chapter 2.1 Dating rocks and other materials — 70
The geological time scale — 70
Methods of dating rocks — 72
Atomic structure — 74
Radioactivity and half-lives — 75
Dendrochronology — 78
Age of the Earth and Mount Narryer zircons — 79
Summary — 80
Practical exercise: Dating rocks — 81
Practical exercise: Drawing a simulated radioactive decay graph — 82

Chapter 2.2 Plate movements — 83
History of plate movements — 83
Crustal structure and movement — 87
Rejection of the continental drift hypothesis — 90
Evidence for the continental drift hypothesis — 90
Summary — 94
Practical exercise: Plate boundaries — 95
Practical exercise: Modelling convection currents — 95

Chapter 2.3
Magnetism and the wandering poles

The Earth's magnetic field	96
Palaeomagnetism	96
Magnetic inclination	97
Apparent polar wandering	97
Polarity reversals	98
Palaeomagnetic banding	98
Summary	100
Practical exercise: Investigating magnetic time scales	101

Chapter 2.4 Tectonic forces

Plate boundary movements	103
The importance of plate tectonics	110
Earthquakes and volcanoes	110
Summary	117
Practical exercise: Plotting earthquakes	119
Practical exercise: Extrusive rock chemistry	120

Resources 122

SECTION 3
The local environment 123

Chapter 3.1
The local environment as a system 124

Summary	126
Practical exercise: Analysing systems	126

Chapter 3.2
The elements of lithosphere and landscape 127

Talking about Earth materials	127
Characteristic properties of minerals	131
Describing rocks	133
Identifying rocks within your local area	139
Geology and landscape	141
Summary	143
Practical exercise: Observing and identifying a rock	144

Chapter 3.3
Soils: Features and formation 146

Characteristics of soils	146
Describing soils	148
Soil profiles	156
Soil fertility	158
Processes forming soils	159
Biological communities, climates and soils	162
Summary	165

Practical exercise: The biota in your area	166
Practical exercise: Areas of environmental importance in your area	167
Practical exercise: Analysing soil data	168

Chapter 3.4
Adaptations and survival in disturbed habitats 169

Adaptation and natural selection	169
Adaptations that allow survival in changing conditions	170
Successions	170
Describing plant communities	172
Summary	175
Practical exercise: Examining adaptations	176
Practical exercise: Describing a local environment	176

Chapter 3.5
Human impacts on the environment 177

Three factors affecting human impact on the environment	177
How we live and its effect on the environment	178
Habitat disturbance since 1788	179
Human impacts on ecosystems	184
Impacts of humans on water systems	185
Summary	188
Practical exercise: Dams and surface processes	189
Practical exercise: Current human impacts in your area	190

Chapter 3.6
Balancing development and conservation 191

When is an environmental change a problem?	191
Environmental and resource managers: Different priorities?	191
Environmental regulation	192
Summary	196
Practical exercise: Regulation of environmental impact	197

Chapter 3.7 Biodiversity 198

Types of biodiversity	198
The importance of biodiversity	200
The importance of refugia	201
Summary	202
Practical exercise: A field study report	202

Resources 203

SECTION 4
Water issues   205

Chapter 4.1
Water movement on the Earth   206
Earth systems and their distribution   206
Natural phenomena and the Earth's systems   207
The Earth's water budget   208
Factors affecting the movement of water on the Earth   209
Oceans   212
Summary   216
Practical exercise: Where is the water?   217

Chapter 4.2
Water and Australian environments   218
Features of the local environment   218
Evaporation and salinity   220
Water pollution   221
Summary   223
Practical exercise: Dissolved oxygen and water temperature   224
Practical exercise: Plants and salinity: Experimental design   224
Practical exercise: Water plants and pollution   225

Chapter 4.3   Water and weathering   226
The water cycle and rock breakdown   226
Types of weathering   227
Summary   230
Practical exercise: Weathering: A first-hand investigation   231

Chapter 4.4
Water resources: Past and present   232
Evidence of past water bodies in Australia   232
Tectonics, topography and drainage   236
Water pollution and conservation   237
Ground water: Issues, regulation and strategies   239
Summary   243
Practical exercise: Past aquatic environments   244
Practical exercise: Preparing a case study on environmental change   245

Resources   245

SECTION 5
Effective research   247
Looking for information   247
Requesting information   248
Collating information   248
Open-ended investigations   248
Using the Internet as a research tool   249
Internet exercise   251

Answering and asking questions   252

Glossary   254

Index   257

Acknowledgments

Cover image © Scenic World; *Fig. 1.1.1* Photo by Peter Hurford; *Figs 1.1.2, 1.2.6, 1.2.7, 1.2.9* and *1.2.10* Material created with support to AURA/STScI from NASA contract NAS5-26555; *Plate 2* © John M. Sarkissian; *Figs 1.1.6, 1.2.1, 1.2.4* and *1.2.5* and *Plates 6, 7* and *9* © NASA; *Plate 5* © Macmillan; *Fig. 1.2.8* © 1993 Bradford Technology Limited and Mayfield Consultants; *Fig. 1.5.2* © Department of Biological Sciences, University of Cincinnati; *Plate 11* Australian Picture Library/Corbis; *Fig. 1.5.3* © Colleen Cavanaugh; *Fig. 1.5.6* University of California Museum of Palaeontology; *Plate 12* Photo by Joseph Deuel; *Figs 2.2.5, 4.1.4* and *4.1.11* © Prentice Hall Inc.; *Figs 2.3.3, 2.3.5* and *2.3.6* © American Geophysical Union (AGU); *Plate 13* Aerial Photography Courtesy of US Geological Survey; *Fig. 2.4.6* © Redrawn from Sutherland F.L., Published by the Geological Society of Australia, NSW Div 1, *Fig.1*, p.29; *Figs 3.2.2, 4.1.2, 4.1.12* and *4.3.1* © Blackwell Science; *Figs 3.2.5, 3.3.15* and *4.3.2* © Open University Press; *Figs 3.2.7, 3.2.8, 3.2.9* and *4.4.8* © Australian Academy of Science; *Figs 3.2.13* and *4.4.2* © New South Wales Department of Mineral Resources, Sydney; *Figs 3.3.1* and *3.3.11* © Routledge; *Figs 3.3.6* and *3.3.10* © Pergamon Press; *Fig. 3.3.13* TAFE NSW (2000); *Figs 3.4.3* and *3.5.2* © Reed Books; *Plate 19* Copyright © Commonwealth of Australia, Geoscience Australia. All rights reserved. Reproduced by permission of the General Manager, National Mapping Division, Geoscience Australia, Canberra ACT. Apart from any use as permitted under the Copyright Act 1968, no part may be reproduced by any process without prior written permission from Geoscience Australia. Requests and inquiries concerning reproduction and rights should be addressed to the Copyright Officer, Geoscience Australia, PO Box 2, Belconnen, ACT, 2616, or by email to copyright@auslig.gov.au; *Fig. 3.5.3* Photo: Dusko Marie; *Fig. 3.5.4* © Blackwell Publishers Ltd; *Figs 4.1.1* and *4.4.10* © Oxford University Press; *Figs 4.1.7, 4.1.8* and *4.4.4* and *Plate 18* © CSIRO Publishers; *Fig. 4.4.11* © Redrawn from "Listen our Land is Crying" by Mary White, published by Kangaroo Press, Simon & Schuster (Aust.) Pty Ltd.

Every effort has been made to trace and acknowledge copyright. The publishers apologise for any accidental infringement and welcome information that would rectify any error or omission in subsequent editions.

Introduction

In this book we take a ride through time, starting at the beginning with the big bang. As we progress towards the present day we find out about the formation of our universe and its galaxies and stars. We travel forward to the creation of our local star—the Sun—and the solar system. We find out about the birth of our own planet and how it grew into a possibly unique object; one that not only supports life but also has been shaped by life itself.

In sections 3 and 4 we look at how some aspects of the Australian continent have been shaped by human influence. We discover how rocks are formed, which influences the development of the local environment, and examine the role of rocks in the formation of soils. We will also see how soils influence the flora and fauna that live in an area. By understanding processes in the local environment it is possible to understand human impacts on it and understand the necessity of legislation to protect it against exploitation and degradation. Our journey of understanding concludes with a global look at the Earth's water resources, their distribution and the importance of water for the planet and organisms living on it. Additionally, we will examine how water physically shapes our world and how, in turn, we as humans impact on its quality and availability.

The book is designed to allow both the school and independent student to follow the NSW Preliminary course in Earth and Environmental Science. It provides the students with regular review activities, which test their understanding of the text. Extension activities allow able students to take their understanding to a higher level. In addition, each section provides the students with a variety of practical exercises.

Planet Earth and its environment

In this section we start by examining how we are able to obtain information from outer space. We examine the theories explaining how our universe was possibly formed. We look at the different types of galaxies and stars that are found in the universe. We then focus on how our solar system formed and the structure of the different planets within the solar system. We finally arrive at the formation of our own planet. We discover how it was formed, then see how the composition of the planet's atmosphere changed as the Earth slowly cooled. We develop an understanding of how the seas formed and how the first organisms may have evolved in the seas.

CONTENTS

Chapter 1.1	Observing our universe	2
Chapter 1.2	The origin of the universe	14
Chapter 1.3	The Sun and the solar system	31
Chapter 1.4	The early Earth and its atmosphere	42
Chapter 1.5	The early Earth and the start of life	57
Resources		67

1.1 Observing our universe

OUTCOMES

At the end of this chapter you should be able to:

- identify that some types of electromagnetic radiation are used to provide information about the universe

- discuss inferences about the relationship between emission spectra of elements and spectral analysis and the composition of stars

- identify data and perform first-hand investigations using a spectroscope and appropriate light sources to observe the spectral lines of some elements

- gather, process and analyse information from secondary sources to match the spectral signature of elements with emission spectra of a star, in order to determine the elements present in that star.

We are inhabitants of a small planet circling an average-sized, main sequence (or middle-aged) star called the Sun. Our solar system is located in one of the outer arms of the Milky Way galaxy. There are 100–200 billion stars in the Milky Way and as many **galaxies** in the universe. The universe is the sum total of all space and its contents. Those contents include galaxies, **nebulae**, stars, planetary systems, **asteroids**, **comets**, meteorites and gases. Most of the universe is empty space.

Many thousands of years ago humans started their observations of space using visible light. It is only since the beginning of the twentieth century that we have been able to observe the universe by using other forms of radiation (collectively known as electromagnetic radiation). Each type of radiation gives astronomers different and unique information about the universe. Whether it is radio waves, visible light or X-ray, infra-red or ultraviolet radiation, they each give the astronomer different and unique information about the universe.

For as long as humans have existed, we have been looking at the stars both in wonderment and as a source of answers to our questions, such as how was the Earth formed and how old is the universe? It is only in living memory, with Neil Armstrong in 1969 setting foot on lunar soil, that we have been able to travel to our nearest neighbour—the Moon—and start to unravel some of the mysteries about Earth. Since early humans first built primitive observatories, such as Stonehenge, we have been measuring, examining and predicting cosmic phenomena. We have tried to relate these findings to our own solar system and, in particular, planet Earth. The gathered information allows us to understand the planet's past as well as what may occur to the planet in the future.

---

definitions

galaxies
large collections of stars, dust and gas in space; systems of millions or billions of stars held together by gravitation

nebulae
gas and/or dust clouds

asteroids
this word means 'starlike'. When viewed through a telescope, asteroids look like faint stars. They are made mostly of rock and maybe ice and are usually less than 2 km across but they can be over 100 km across.

comets
celestial bodies from space, usually of small mass that circle in elliptical orbits around the Sun. As a comet approaches the Sun it becomes visible because the surface of the centre, or nucleus, begins to warm and volatile gases evaporate. The evaporated molecules boil off and carry small solid particles with them, forming the comet's tail, or coma, of gas and dust. The coma absorbs ultraviolet radiation and begins to fluoresce. When the nucleus is frozen it can only be seen by reflected sunlight.

WHAT DO WE USE TO EXPLORE THE UNIVERSE?

Stars give off electromagnetic radiation and it is this radiation that allows us to obtain information about the universe. (Electromagnetic radiation is discussed on page 6.)

Humans have studied the stars for many millennia. Prior to the invention of the telescope, the earliest evidence of space exploration dates from Stone Age times. This exploration of the night sky could only be done with the naked eye. The results of this early exploration of space allowed early astronomers to create devices that could be used to track the movement of the stars, the Sun and our Moon. Stonehenge, in the south of England, is an example of one of these devices. It is a stone circle built around 3000 BC and is an astronomical instrument that was used to predict eclipses and the passage of the Sun and the Moon. (See Figure 1.1.1.)

Ground-based observations

A major step forward took place with the invention of the first type of telescope—the refracting telescope—in the seventeenth century by Galileo. It allowed astronomers to observe the heavenly bodies more closely. This was followed at the end of the seventeenth century by Newton's invention of an improved reflecting telescope. Galileo's telescope used lenses to bend light and magnify the image being viewed, while the Newtonian reflecting telescope used a combination of lenses and mirrors to both reflect and magnify light, resulting in higher magnification and clearer images.

Modern-day astronomers rarely look directly through telescopes. Instead they use more sophisticated, visible light sensitive devices to capture images. These devices include photographic films, photoelectric chips and photomultiplier tubes and they allow scientists to capture very small amounts of electromagnetic radiation. When seen with the naked eye, these small amounts of electromagnetic radiation normally appear to be very small and/or faint. Larger and clearer images can be produced by using visible light sensitive devices.

Optical astronomy remains one of the main ways of observing space. Scientists use optical telescopes (such as the Anglo-Australian telescope at Coonabarabran, New South Wales) to measure the temperature, chemical composition, distance and velocity of stars and galaxies. Other instruments can be attached to optical telescopes in order to obtain more detailed data, for example the 2-degree field instrument. This instrument is part of the Anglo-Australian telescope and analyses light that can come from up to 400 astronomical objects at the same time. This helps astronomers to accurately measure the position of these objects in relation to each other.

It is important that astronomers use different types of telescopes to look at the universe as each type gives us unique information. Some electromagnetic radiation wavelengths are able to penetrate the atmosphere. An example of these are radio wavelengths. Therefore, radio telescopes, which observe radio wavelengths, can be ground based. However, other wavelengths may be partly or totally absorbed by the atmosphere, resulting in a complete or partial loss of information. These include visible light, infra-red and ultraviolet wavelengths. While ground-based telescopes are of great use, space platforms have vastly improved our examination of these wavelengths.

A limitation of earth-based observations of space is that the atmosphere distorts or obscures our view of the universe. Using optical instruments from within the atmospheric confines of the planet to observe extraterrestrial phenomena presents a number of problems. These include light pollution, weather sensitivity, distortion due to atmosphere and limitation of the optics.

Figure 1.1.1 Stonehenge, a stone circle in the south of England used for predicting eclipses and the passage of the Sun and Moon.

REVIEW ACTIVITIES

1 Outline the importance of the electromagnetic spectrum in the ability of scientists to observe the universe.

2 Explain why ground-based telescopes do not give as good a view as space-based telescopes.

EXTENSION ACTIVITIES

3 Find a detailed map or photo of the Milky Way galaxy and locate the position of our solar system in it.

4 There are three types of optical telescopes: the Galilean (refracting) telescope, the Newtonian (reflecting) telescope and the catadioptric telescope. Compare the three types and explain their advantages and disadvantages.

Space-based observations

Space travel is limited to travelling within the Earth's solar system, especially the inner solar system. Most astronomical observation platforms are put into orbit around the planet (see Plate 1), with only a limited number being sent out further. There are a number of reasons why space travel is limited. They include the:

- type of propulsion unit, that is engine or rocket
- limited amount of fuel that can be carried
- time taken
- expense
- distance
- lack of technology.

Space travel is slow because the propulsion units only allow a relatively low speed. Building a vehicle for space travel is very expensive: tens or hundreds of millions of dollars. The distance that can be travelled is very limited in comparison to the size of the universe. As yet, we do not have the technology to travel anywhere fast or to sustain life during the long periods of time required for space travel between planets.

To avoid the problems of distortion, NASA built the Hubble Space Telescope (HST), an optical telescope, and placed it in Earth's orbit above the limits of the atmosphere. (See Figure 1.1.2.) As there is no distortion from the atmosphere, clearer images are gained. (See Figure 1.1.3.) However, our view is limited by optics and technology. One major problem in placing optical telescopes in space, orbiting the Earth, is cost. Satellites are very expensive to build and maintain. They are difficult to repair because they are in space and they require a large infrastructure to support them.

Satellites, such as the Cosmic Background Explorer satellite (COBE), are playing an increasing role in gathering data. This satellite is used to measure the background radiation of the universe and especially any variations in that radiation. It has discovered ripples in the cosmic background, which is a result of the 'big bang'.

Figure 1.1.2 The Hubble Space Telescope.

Radio telescopes

Radio telescopes, such as the one shown in Plate 2, are used to collect a wide range of radio signals coming in from outer space. Many objects in the universe emit relatively large amounts of radio waves. Nearly all types of astronomical objects give off some radio radiation, but the strongest sources of such emissions include pulsars, certain nebulae, quasars and radio galaxies. By the mid 1980s about 100 000 of these radio sources had been catalogued.

Radio astronomy was discovered accidentally. In the 1930s an engineer, Karl Jansky, was sorting out problems that affected transatlantic telephone calls, when he picked up radio signals coming from space.

Radio telescopes are shaped like a dish or mast and can be grouped (arrayed) to form a more sensitive instrument. Plans have been drawn for a large radio telescope array to be built in Western Australia and cover an area of 100 km² in size.

A number of plans have been put forward by institutions and governments around the world to build large radio telescope arrays ranging from 1 km² in size to much larger arrays. Radio astronomers feel there is a need for these large arrays because, as yet, what has been learnt has come from very weak signals, or 'whispers'. Larger arrays allow more whispers to be gathered and increase the possibility of gathering even weaker whispers.

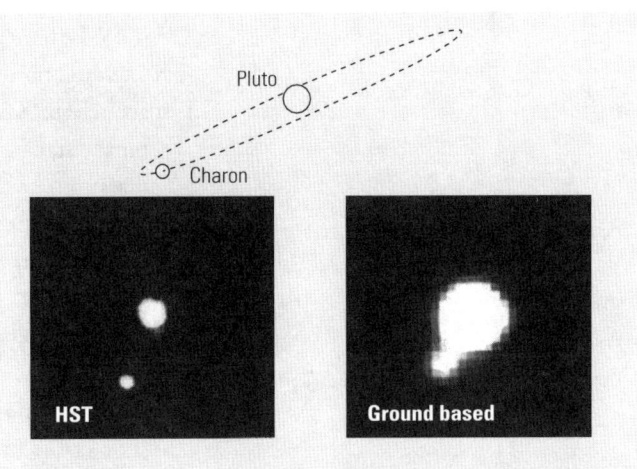

Figure 1.1.3 When Pluto and its moon, Charon, are viewed through a ground-based telescope they are seen as one mass. However, when viewed through the Hubble Space Telescope it is clear they are two separate bodies.

Microwave and radar astronomy

Microwave astronomy is a type of radio astronomy. Space is a source of microwaves in every direction. This is referred to as the microwave background. This background is believed to be the remnant from the big bang, which is the possible origin of the universe. COBE is making very precise measurements of the temperature of the microwave background.

Another use of microwave radiation is in radar astronomy. This is the application of radar to the determination of distances and planetary features within the solar system. A short burst of microwaves is transmitted in the direction of the object under study. The object reflects the microwaves back to Earth, where they are detected by the same antenna that sent the signal. The time between the signal being sent and received is easily measured. The object being measured does not propagate the signal, but acts as a passive reflector. Radar astronomy makes it possible to measure accurately the distance of the object from the Earth and the rate and direction of the object's spin. It also allows the object's surface details to be mapped precisely.

Infra-red telescopes

Various types of celestial objects—including the planets of the solar system, stars, nebulae and galaxies—give off wavelengths in the infra-red region of the electromagnetic spectrum.

Infra-red telescopes collect infra-red wavelength radiation released from hot bodies in the universe. All objects with a temperature higher than absolute zero or zero Kelvin (–273°C) emit infra-red waves. However, it is the objects that are hotter than their surroundings that interest astronomers. The molecules forming these objects have more energy than their surroundings and the energy is released from the molecules.

These objects emit visible light, but this is blocked by intervening dust particles. This prevents the objects from being seen from Earth. However, they can be examined using the techniques of infra-red astronomy. This form of astronomy has achieved better results since infra-red detectors have been placed on satellites above the atmosphere.

X-ray telescopy

X-ray radiation is a type of electromagnetic radiation. A large number of objects in space release X-rays. These include:

- binary stars—pairs of stars that orbit around a common point called the centre of mass
- remnant supernovas—the remains of a star that has exploded
- neutron stars—stars that are very dense and small (10–15 km in radius) and spin very quickly
- galaxy clusters—groups of galaxies found close to each other.

The Earth's atmosphere absorbs X-rays very efficiently. For accurate observation and measurement of objects that produce X-rays, it is necessary to carry telescopes and detectors by spacecraft high above the atmosphere.

REVIEW ACTIVITIES

1 Describe some of the possible problems that stop space travel being a realistic option in the exploration of deep space.

2 Other than visual information, what information can scientists gain from looking at objects in space?

3 Explain one of the main reasons NASA built the Hubble Space Telescope.

4 Explain why scientists prefer to build large radio telescopes rather than small ones.

5 Infra-red telescopy is not suitable for viewing all objects in space. Explain why.

6 Describe why the most efficient type of X-ray detectors are found on space satellites.

EXTENSION ACTIVITIES

7 A number of satellites have either collected or are collecting data from space. They include Galileo, Giotto, Magellan, Surveyor, COBE and Voyager. Choose one satellite and find out what data it was seeking and, if possible, what instruments were used to collect it.

8 Recently a number of spacecraft have been sent to Mars. Describe what information these missions were seeking to find out.

definitions

photon
a small bundle of electromagnetic energy

frequency
the number of peaks or troughs to pass a particular point every second

amplitude
the height of a wave, from the midpoint of the wave to the top of the peak. The maximum displacement of the medium from its equilibrium position.

wavelength
the distance between two adjacent troughs or crests on a wave

ELECTROMAGNETIC RADIATION

Electromagnetic radiation is made up of oscillating electric and magnetic fields, and propagates through space at 300 000 km/s. It includes gamma rays, X-rays, ultraviolet radiation, visible light, infra-red radiation, microwaves and radio waves. The distribution of these types of electromagnetic radiation forms an electromagnetic spectrum according to the amount of energy the types have. (See Figure 1.1.4.) The spectrum is continuous, so the different types of radiation merge. This means there are no defined boundaries between each part of the spectrum.

Electromagnetic radiation is a form of energy released by matter. Small quantities of energy are called **photons**, and they move in the form of a wave. Each type of electromagnetic wave (such as cosmic radiation, ultraviolet radiation and visible light) has a specific **frequency**, **amplitude** and **wavelength**. Electromagnetic waves all travel at the speed of light. They are non-mechanical waves because they do not require a material medium in which to travel.

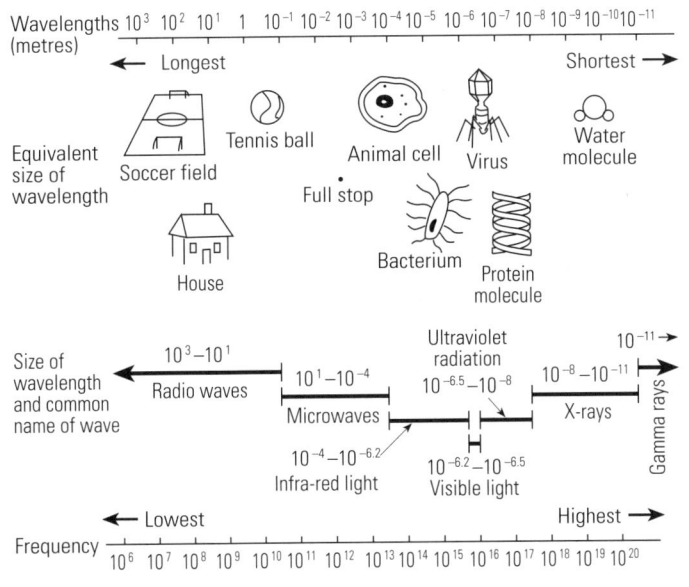

Figure 1.1.4 Diagrammatic representation of the electromagnetic spectrum.

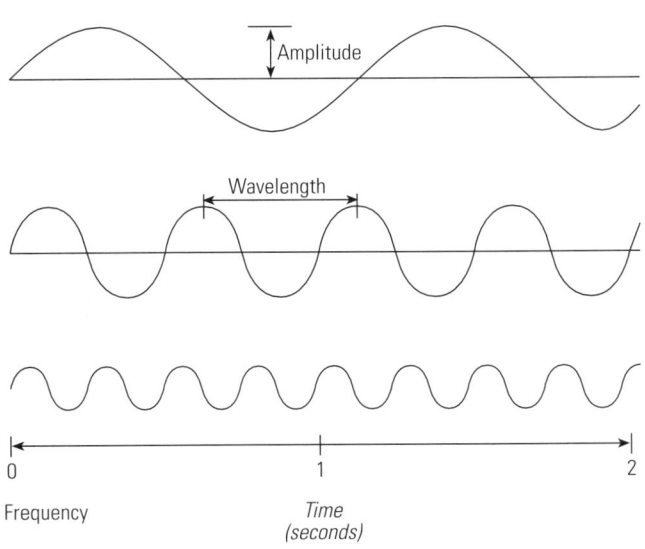

Figure 1.1.5 Electromagnetic radiation: transverse wavelengths.

Types of electromagnetic radiation

Gamma rays (or radiation) are composed of waves with very short wavelengths, but with very high energy. (See Figure 1.1.5.) These rays are emitted by radioactive atoms, which possess extra levels of energy that is released in the form of radioactivity (or radiation). Gamma rays are highly penetrating, being able to penetrate several metres of concrete and metal.

X-rays are a type of high-energy radiation that are emitted from fast-moving electrons. X-rays are also highly penetrating, being able to penetrate a variety of substances. However, they are absorbed by bones and elements such as lead and other elements with a higher atomic number than lead.

Ultraviolet radiation has a wavelength that lies between that of visible light and X-rays. Ultraviolet light is highly dangerous to living organisms as it may cause cancers or changes in genetic material. The Earth's surface is protected from the majority of its harmful effects by an ozone layer that absorbs the ultraviolet rays.

At about 700°C, the waves with shorter wavelengths become visible. This is visible light. These waves produce coloured light: red, orange, yellow, green, blue, indigo and violet. When these different coloured light waves combine they form white light.

Infra-red radiation is emitted by a hot object—one with a temperature higher than absolute zero or zero Kelvin. The wavelengths become shorter as the object heats up. Most of the infra-red radiation entering the atmosphere is absorbed by the ozone layer and dust particles.

Microwaves are short high-frequency radio waves that can travel at the speed of light. Microwave exposure can cause burns because it causes water and fat molecules to vibrate faster, producing heat. It has also been linked to cancers.

Radio waves form waves with long wavelengths ranging from a few millimetres up to 10 km. They are produced by oscillating electrons. Radio frequencies of over 100 Mhz up to UV frequencies are able to pass through the atmosphere, but AM radio frequencies around 1 Mhz are reflected by the ionosphere, a layer of the atmosphere.

Table 1.1.1 Wavelengths of different star colours

Colour of star	Wavelength (mm)
Blue	400–450
White	450–500
Yellow	500–550
Orange	550–650
Red	650–700

REVIEW ACTIVITIES

1
Refer to Figure 1.1.5 (page 7) and the text.
a Which wavelengths of the electromagnetic spectrum are absorbed by the Earth's atmosphere?
b Which has the longer wavelengths: X-rays or radio waves?
c Describe what happens to the frequency as the wavelength increases.

2
List the different coloured light waves that combine to form white light.

3
Distinguish between the terms 'frequency', 'amplitude' and 'wavelength'.

EXTENSION ACTIVITIES

4
Astronauts on the Moon communicated with Earth via radio waves. The Moon is 400 000 km from Earth. How long does it take a radio signal to cover this distance?

5
The space probe Pioneer 10 left Earth in March 1972 to travel to Jupiter, a trip of twenty-one months travelling at a speed of 12.24 km/s. How far is Jupiter from Earth, and how long does it take for radio messages to reach Jupiter?

6
Refer to Table 1.1.1 (page 7). Star A gives off light of a wavelength of 480 mm while star B produces light of a wavelength of 625 mm. What are the colours of the stars?

CONTINUOUS, EMISSION AND ABSORPTION SPECTRA

In 1666 Sir Isaac Newton discovered that white light could be split into its constituent colours of red, orange, yellow, green, blue, indigo and violet. This range of colours is called the visible spectrum of white light. White light can be split into its **visible spectrum** by passing the light through a glass prism.

During the nineteenth century Father Angelo Secchi S. J., the papal astronomer, invented the stellar spectroscope, which could be used to examine the light coming from the stars. He examined over 4000 stars and discovered many differences between individual stars. But the similarities he discovered were far more significant.

He knew that each chemical **element** emits its own characteristic radiation when heated in a gaseous form. When heated, the light from certain elements produces a line spectrum where only certain colours are present. These colours form narrow lines when viewed through a spectroscope. He discovered that stars are composed of a limited number of elements, including hydrogen and helium. This showed there is a commonality, possibly of origin, between stars.

Three types of spectra can be formed by **atoms** and **molecules**. They are continuous, emission and absorption spectra. For examples of each type of spectrum look at Plate 3. A **continuous spectrum** is formed when a hot, opaque gas, solid or liquid, emits light when under high pressure. In a very hot, opaque gas the atoms have high kinetic energy and collisions between them are very frequent. Within the atoms, electrons move to higher energy levels because they have gained energy from the collisions. They lose the energy rapidly, which causes them to drop back down to lower energy levels as they emit light. When viewed through a spectroscope the light is seen as emission lines.

An **emission**, or bright line, **spectrum** occurs when a hot gas is under low pressure. In this state it will emit a series of bright lines on a dark background. The emission spectrum of a gas occurs when energy is being released. It results in electrons dropping down from one energy level to another within the structure of the atom. In dropping, a pulse of light in the form of a photon is released at a wavelength between the two energy levels. Emission spectra occur when a hot material is emitting photons into a colder region.

An **absorption**, or dark line, **spectrum** occurs when light from a source that has a continuous spectrum is shone through a gas with a lower temperature and pressure than the source. The continuous spectrum will be observed to have a series of dark lines superimposed on it. These dark lines are the absorption spectrum, which represents the energy absorbed by atoms. In absorption spectra, photons are absorbed by atoms or molecules, resulting in one or more electrons moving to a higher energy level in the atom. Sometimes it may just result in an increase in the rotational energy of the atom or molecule.

The importance of stellar spectroscopy

Spectroscopy gives astronomers information about the composition of stellar bodies. Each element has its own unique arrangement of emission and absorption lines; they are the fingerprints of the element. In attempting to gain an understanding of celestial bodies, it is important to understand the information provided by all types of spectra.

The relationship between the structure of atoms and molecules was important in the development of the science of astrophysics. A change in the structure of an atom causes changes in the emission and absorption lines, or spectra, being produced. Therefore, when light from distant stars and galaxies is examined through a spectroscope, it is possible to say whether an element is present in, or absent from, that star or galaxy. As techniques have been refined it is also possible to tell the quantity of that element or compound in the distant object.

Spectroscopy allows scientists to determine how fast astronomical objects are moving. This can be done by measuring the change in the position of the lines that are produced by the chemical in the spectrogram. A spectrogram is a graphical representation of the spectrum produced by a hot gas or gases.

A celestial body moving away from the viewer will result in the absorption lines moving towards the red end of the spectrum, while a celestial body moving towards the viewer will result in the absorption lines moving towards the blue end of the spectrum. (See the discussion of the Doppler effect on page 17.) A red shift means a shift towards a lower frequency. This indicates that the light source is moving away from us. Conversely, a blue shift means a shift towards a higher frequency, indicating that the light source is moving towards us. The amount of change towards the red or blue wavelength depends on the object's speed.

definitions

visible spectrum
when white light can be split into the range of its constituent colours. The constituent colours are red, orange, yellow, green, blue, indigo and violet.

element
a substance that cannot be further divided by chemical methods. They are the basic substances that build up chemical compounds.

molecule
the smallest particle of matter; composed of two or more atoms

atom
the smallest part of an element that can exist as a stable entity. An atom is composed of a central nucleus containing protons possessing a positive charge and neutrons possessing no charge. Surrounding the nucleus are layers of electrons. Each layer possesses a different energy level.

continuous spectrum
a hot, opaque gas, solid or liquid, that under high pressure will produce a broad band of wavelengths of light, forming a continuous spectrum

emission spectrum
a hot gas that under low pressure will emit individual wavelengths of light. These form an emission spectrum, which is a series of bright lines on a dark background.

absorption spectrum
when light from a source that has a continuous spectrum is shone through a gas with a lower temperature and pressure, dark lines are formed on the spectrum. The lines are at the wavelengths of the atoms in the gas.

CLASSIFYING STELLAR SPECTRA

A spectral classification can be formulated by measuring the changes in intensity of the hydrogen lines on a spectrogram in relationship to the temperature of the gas. This was first done by Secchi in 1860. Having studied and catalogued over 4000 stars, he divided them into four spectral classes. In the 1890s the modern scheme started to evolve. In 1890 Edward Pickering and Willamina Fleming proposed a large number of stellar classes, labelling the classes A–Q. The classification was based on the maximum strength of the hydrogen absorption lines. The system continued to be refined and improved over the next century.

Figure 1.1.6 A distant galaxy composed of billions of stars.

The modern scheme is called the MK system. M and K are two of the classes of stars. (See Table 1.2.4, page 30.) The spectral class of a star is designated by one of seven letters (O, B, A, F, G, K and M), starting with the hottest type (O group) down to the coolest type (M group).

The width of a star's spectral lines determines how luminous it is. The narrower the lines, the more luminous it is. The sharpness of the lines can be used to further classify stars for each spectral type. The resulting luminosity classes are denoted by roman numerals and are divided into seven principal types. (See Table 1.2.3, page 30.)

REVIEW ACTIVITIES

1 Outline what happens to chemical elements when they are heated in a gaseous form.

2 Distinguish between continuous, emission and absorption spectra.

3 Explain why the line spectra of elements are so important.

4 Recount the two pieces of information that can be gained by viewing a distant galaxy using a spectroscope.

5
a Explain why electrons in an atom are found in different layers.
b Explain how electrons may move from one level to another.

6 Recount what sort of spectra has molecules or atoms found with increasing rotational energy.

7 Describe the two methods of star classification.

EXTENSION ACTIVITIES

8 Secchi catalogued 4000 stars, which he divided into four spectral groups. What were the characteristics of each of the four groups?

9 Research whether spectroscopy can be used on all celestial objects. If it can't, give the reasons why.

SUMMARY

- The universe is made up celestial bodies, such as galaxies, nebulae, stars and planets. Most of the universe is empty space.

- We observe the universe in a number of ways, including with the aid of optical, radio, X-ray and infra-red telescopes.

- We use ground-based and space-based platforms to observe space. Space-based platforms give us clearer views because, unlike ground-based platforms, they are not hampered by atmospheric distortion and the absorption of certain types of electromagnetic radiation.

- The electromagnetic radiation spectrum includes cosmic radiation, gamma rays, X-rays, ultraviolet radiation, visible light, microwaves and radio waves.

- White light can be split into different colours: red, orange, yellow, green, blue, indigo and violet.

- Each element has its own unique arrangement of emission and absorption lines; they are the fingerprints of the element.

- Spectroscopy gives astronomers information about the composition of stellar bodies and allows them to determine how fast astronomical objects are moving.

PRACTICAL EXERCISE
Flame tests

Equipment
- Samples of the following metal salts: potassium, barium, calcium, copper, lead, strontium, lithium and sodium
- Bunsen burner
- Nichrome wire
- Eight watch glasses
- 50 ml beaker
- 5M nitric acid
- Safety glasses

Procedure

1 Place each metal salt sample on a watch glass.

2 Place the nichrome wire into the beaker containing the nitric acid. Then place it into the flame. Repeat this step until the flame burns with a normal colour.

3 Once the nichrome wire is clean, dip the tip of the nichrome wire into one of the samples of metal salts on the watch glasses.

4 Place the tip of the wire into the hottest part of the flame. Observe and record the colour of the flame.

5 Record your results in a table.

6 Continue carrying out steps 2 to 5 until all salts have been tested.

ACTIVITIES

1 Why is nichrome wire used in preference to any other metal?

2 Why is it necessary to use nitric acid to decontaminate the nichrome wire?

3 Give a reason why it is preferable to use the hottest part of the Bunsen burner flame.

4 The metals tested were in the form of salts. Why did the non-metal part of the salt not produce a colour of its own when heated?

Caution: Care must be taken when carrying out this exercise. Some aspects of it are potentially dangerous. Be aware of the following:
- There is a danger of being burnt when using the Bunsen burner and heating the nichrome wire.
- When decontaminating the nichrome wire, avoid skin contact with the concentrated nitric acid, which is highly corrosive.
- Safety glasses must be worn while carrying out the experiment.

PRACTICAL EXERCISE
Spectroscopy: Measurement of spectra

In this exercise you will identify data and perform first-hand investigations using a spectroscope and appropriate light sources to observe the spectral lines of some elements.

Equipment
- Induction coil
- Transformer
- Light sources, such as a sodium vapour lamp, cadmium vapour lamp, mercury vapour lamp, oxygen vapour lamp, tungsten lamp and fluorescent light
- Hand-held spectroscope

Procedure

1 Set up the equipment as shown in Figure 1.1.7. The exercise needs to be performed in a darkened room.

2 Make copies of Table 1.1.2: one copy per light source. You will record your results in the tables.

3 Look through the eyepiece of the spectroscope at the light source.

4 For each of the observed lines, record the colours observed and their brightness (strong, medium or faint).

5 Draw the spectra for the light source.

6 Repeat steps 3 to 5 for each of the light sources.

ACTIVITIES

1 What was the main colour given off by each light source?

2 What type of spectra was given off by each of the light sources?

3 Did each light source produce the same number of lines?

4 Did the thickness of the coloured lines vary with the light source?

5 Research what the spectra of hydrogen and helium look like.

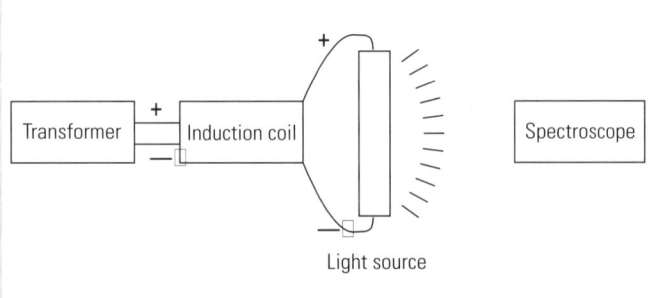

Figure 1.1.7 Set-up of equipment.

Table 1.1.2 Results

Type of light source

Line	Colour	Brightness
1		
2		
3		
4		
5		
6		
7		
8		
9		
10		

Drawing of spectra

PRACTICAL EXERCISE
Examining the spectra of the planetary nebula ZX1267

In this exercise you will gather, process and analyse information from secondary sources to match the spectral signature of elements with the emission spectra of a star, in order to determine the elements present in that star.

Background
NASA completed a mission using the Japanese Yohkoh soft X-ray and spectroscopic telescope to carry out a spectroscopic analysis of the planetary nebula ZX1267 in the region G134. Over a period of four weeks a large number of readings were taken of this nebula. A summary of the results was then tabulated. (See Table 1.1.3.)

ACTIVITIES

1 If the intensity is linked to abundance:
 a Which are the three most common elements?
 b Which two elements are the rarest in the nebula?

2 Explain why most of the elements in Table 1.1.3 are light gases and not metals.

3 Refer to Figure 1.1.8. Why do some of the elements have a number of different wavelengths?

4 Using Table 1.1.3 and Figure 1.1.8, identify which elements have the largest spikes on the graph.

Table 1.1.3 Spectroscopic readings from Yohkoh soft X-ray and spectroscopic telescope

Element	Wavelength (nm)	Typical intensity of G134 region*
Helium	402.6	4
Sulfur	406.9	2
Hydrogen	410.1	30
Hydrogen	434.0	50
Oxygen	436.3	2
Helium	447.1	5
Iron	465.8	1
Hydrogen	486.1	100
Helium	492.1	1
Oxygen	495.8	111
Oxygen	500.7	330
Nitrogen	519.9	1
Nitrogen	575.4	13
Helium	587.6	25
Oxygen	630.0	5
Sulfur	631.2	10
Oxygen	636.4	1
Nitrogen	654.8	12
Hydrogen	656.3	230
Nitrogen	658.4	36
Helium	667.8	16
Sulfur	671.6	4
Sulfur	673.0	5
Argon	713.6	10

*1 is the intensity of the Sun.

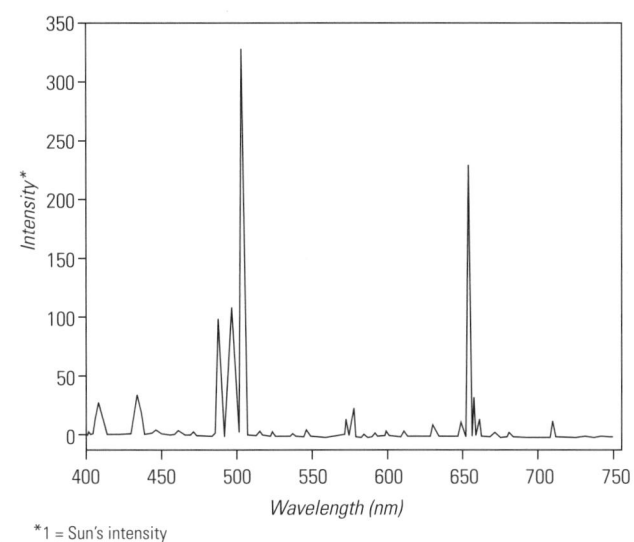

*1 = Sun's intensity

Figure 1.1.8 Graph showing results of spectrogram for planetary nebula ZX1267.

1.2 The origin of the universe

OUTCOMES

At the end of this chapter you should be able to:

- recall current scientific thinking about the origin of the universe

- compare two hypotheses developed to explain the existence of matter in the universe and describe the process of accretion of matter to form stars and planets

- recall the relationship between some major features of the universe and theories about the formation of the universe

- compare cultural beliefs with those of astronomers and other scientists that may arise in discussions of the origins of the Earth.

WHAT IS THE UNIVERSE?

The universe contains all the matter and energy in existence, including all the galaxies, stars and planets. The exact size of the universe is unknown. Its diameter is estimated at between 15 and 20 billion **light years**. Many scientists consider that the universe is still expanding. Some even believe that the rate of this expansion is actually increasing. The universe is generally thought to be between 10 and 20 billion years old. However, many scientists believe its age can be narrowed to between 12 and 15 billion years old.

A BRIEF HISTORY OF COSMOLOGY

Cosmology is the study of how the universe and time were created, the cosmological mechanisms in operation today, and what may possibly happen to the universe in the future. People have long discussed the origin of the universe. Today some people believe in the big bang theory. Others claim that everything happened by chance. Some believe the universe has been constant, not changing during the length of its existence. Still others believe that a supernatural being or beings created everything in the universe.

Ptolemy, a Greek philosopher in the second century BC, put forward the idea that the Earth was the centre of the universe and that everything orbited our planet. From the time of Ptolemy the ideas regarding the origin of the universe have changed as new evidence has come to light. In the early sixteenth century Copernicus put forward the theory that the planets revolved around the Sun and not the Earth. Copernicus's theory did not achieve wide support at the time because some argued that the relative positions of the stars should have changed when viewed from different parts of the Earth's orbit. This shift, or **parallax**, did not appear to occur. In addition the view accepted at the time allowed people to calculate the motion of planets with reasonably accuracy.

In the seventeenth century the imperial mathematician for the Holy Roman Empire, Johannes Kepler, showed that the planets orbited the sun in elliptical paths. Later that century Galileo was able to observe the orbiting of Jupiter's moons, as well as distant stars using the newly invented telescope. The work of both these scientists supported the Copernicus model and allowed it to gain acceptance.

In the nineteenth century Bessel measured the distance to the nearest star as being over 55 million, million km away. This was followed by the prediction of Kant (in 1755) and others that beyond our own Milky Way other galaxies had to exist. In the 1920s the astronomer Hubble finally established that other galaxies did exist beyond the boundaries of our galaxy. He also discovered that these galaxies were in fact moving away from us.

Einstein's work at the beginning of the twentieth century resulted in the general theory of relativity. (See page 18.) This theory assisted the Russian mathematician Aleksandr Friedman in the development of the theory of an expanding universe, which he proposed in 1922. This theory implied that the universe must have been born at one moment. British astronomer Sir Fred Hoyle in 1950 dismissively called this idea the 'big bang'. Sir James Jeans in 1928 first proposed the alternative theory called the steady state theory. It was revised by Hermann Bondi and Thomas Gold in 1948, and further developed by Hoyle. The steady state theory holds that the universe is always expanding, but maintaining a constant average density. Matter is being continuously created to form new stars and galaxies at the same rate as old ones become unobservable because of their increasing distance and velocity of recession. A steady state universe has no beginning or end in terms of time, and the average density and arrangement of galaxies is the same. As we shall see, this theory has little evidence supporting it, but a substantial amount of evidence disproving it.

Figure 1.2.1 The Milky Way is a spiral galaxy; so named because of its spiral shape. Galaxies are made up of billions of stars.

definitions

light year
the distance travelled by light in one Earth year
(1 light year = 9 500 000 000 000 km)

parallax
the relative positions of stars change when viewed from different positions of the Earth's orbit

REVIEW ACTIVITIES

1 Describe what the universe is.

2 Explain what cosmology is.

3 Draw a time line to show the development of our knowledge and understanding of the universe.

EXTENSION ACTIVITIES

4 Research the work of Copernicus and Galileo. In what way were their findings revolutionary and what was the response of society to these findings?

5 Johannes Kepler is often referred to as the 'founder of modern science'. Research why he has been given that title.

6 Stephen Hawking is possibly today's greatest cosmologist. What is his main interest and what was his bestselling book called?

THEORIES ON THE ORIGIN OF THE UNIVERSE

Creation theories

Some theories about the universe are based on beliefs and not scientific evidence. Different religions and nations have their own views. Creation theories or stories attempt to explain what occurred in a time prior to human records.

An example of a creation story is the Hindu belief that the four-headed god Brahma created the universe as well as humanity. Hindus believe that when Brahma sleeps the universe and everything in it is destroyed, and when he wakes the entire universe is created new, creating a perpetual cycle of creation and destruction.

The ancient Chinese believed in two opposing forces known as Yin and Yang. Yin is the power of darkness while Yang is the power of light and sunshine. Yin and Yang had a child named Pan Gu, who created the universe. His eyes became the Sun and the Moon, and his hair the trees and plants. His flesh became the Earth. His sweat became the rain, and the worms that left his corpse became people.

Muslims believe in the words of their holy book, the Koran, which says '…it was God who raised the heavens without visible pillars. He then ascended His throne and forced the sun and the moon into His service, each pursuing an appointed course. He ordains all things…It was He who spread out the earth and placed it upon the mountains and rivers.'

Many creation theories and beliefs contain elements of the steady state theory and see Earth as the 'centre' of the universe. Many Aboriginal stories provide examples of this. The one below is from the Karraur people:

Yhi lay asleep in the Dreamtime before this world's creation, in a world of bone-bare, windless mountains. Suddenly, a whistle startled the goddess. She took a deep breath and opened her eyes, flooding the world with light. The earth stirred under her warm rays. Yhi drifted down to this new land, walking north, south, east, west. As she did, plants sprang up from her footprints. She walked the world's surface until she had stepped everywhere, until every inch was covered with green. Then the goddess sat to rest on the treeless plain. As she glanced around, she realized that the new plants could not move, and she desired to see something dance. Seeking that dancing life, she descended beneath the earth, where she found evil spirits who tried to sing her to death. But they were not as powerful as Yhi. Her warmth melted the darkness, and tiny forms began to move there. The forms turned into butterflies and bees and insects that swarmed around her in a dancing mass. She led them forth into the sunny world. But there were still caves of ice, high in the mountains, in which other beings rested. Yhi spread her light into them, one at a time. She stared into the cave's black interiors until water formed. Then she saw something move—something, and another thing, and another. Fishes and lizards swam forth. Cave after cave she freed from its darkness, and birds and animals poured forth onto the face of the earth. Soon the entire world was dancing with life. Then, in her golden voice, Yhi spoke. She told her creatures she would return to her own world. She blessed them with changing seasons and with the knowledge that when they died they would join her in the sky. Then, turning herself into a ball of light, she sank below the horizon. As she disappeared, darkness fell upon the earth's surface. The new creatures were afraid. There was sorrow and mourning, and finally there was sleep. And, soon, there was the first dawn, for Yhi had never intended to abandon her creation. One by one the sleepy creatures woke to see light breaking in the east. A bird chorus greeted their mistress, and the lake and ocean waters that had been rising in mists, trying to reach her, sank down calmly. For eons of Dreamtime the animals lived in peace on Yhi's earth, but then a vague sadness began to fill them. They ceased to delight in what they were. She had planned never to return to earth, but she felt so sorry for her creatures that she said, "Just *once. Just this once."* So she slid down to the earth's surface and asked the creatures what was wrong. Wombat wanted to wiggle along the ground. Kangaroo wanted to fly. Bat wanted wings. Lizard wanted legs. Seal wanted to swim. And the confused Platypus wanted something of every other animal. And so Yhi gave them what they wanted. From the beautiful regular forms of the early creation came the strange creatures that now walk the earth. Yhi then swept herself up to the sky again. She had one other task yet to complete: the creation of woman. She had already embodied thought in male form and set him wandering the earth. But nothing—not the plants, not the insects, not the birds or beasts or fish seemed like him. He was lonely. Yhi went to him one morning as he slept near a grass tree. He slept fitfully, full of strange dreams. As he emerged from his dreaming he saw the flower stalk on the grass and trees shining with sunlight. He was drawn to the tree, as were all the earth's other creatures. Reverent and astonished, they watched as the power of Yhi concentrated itself on the flower stalk. The flower stalk began to move rhythmically—to breathe. Then it changed form, softened, became a woman. Slowly emerging into the light from which she was formed, the first woman gave her hand to the first man.*

Source: Georgia Southern University

REVIEW ACTIVITIES

1 Describe the two different bases for creation theories.

2 Explain why there are so many different creation stories.

EXTENSION ACTIVITIES

3 Research a creation story from the North American Indians or Eskimos.

4 Read chapter 1 of Genesis and recount the Christian story of the creation of the universe and Earth.

Big bang theory

The **big bang theory** was initially proposed in 1927 by a Belgian priest, George Lemaitre. He proposed the idea that the universe began with an explosive event that created matter. This matter is moving out from the site of the original explosion.

The proposal of the big bang theory led to the proposal of a number of other theories that start with the big bang but then try to explain the subsequent evolution of the universe.

> **definition**
>
> **big bang theory**
> the universe started with an explosive event, which created matter. The matter is expanding outwards from the point of the original explosion.

EVIDENCE FOR THE BIG BANG THEORY

The Doppler effect

We know the universe is expanding and getting bigger but is our planet, our solar system and our galaxy expanding? The answer is 'No'. This is because the gravitational force between atoms holds them in place. What does expand, in the scientific view, is space. The 'nothingness' between the galaxies is increasing, while the galaxies themselves stay the same and move away from each other. The evidence for this galactic movement is the change in the wavelength of the light we see coming from distant galaxies. This change in wavelength is caused by the Doppler effect.

The Doppler effect is the apparent change in the frequency and wavelength of a sound or light wave caused by a change in the distance between the source and the receiver. (See Plate 4.) The waves emitted by an object moving towards an observer will be compressed, causing a shift in wavelengths towards the blue end of the spectrum. This is called a blue shift. The waves emitted by an object moving away from an observer become extended. There is a move of the waves towards the red end of the spectrum. This is called a red shift. The colour of the light coming from distant galaxies appears slightly red and the greater the distance between a galaxy and the observer, the redder the light becomes.

The Doppler effect and the movement of the galaxies

The Doppler effect was seen in 1929 by Edwin Hubble. He found that the more distant galaxies were receding quicker than the closer galaxies. Therefore, their colour was slightly redder than the closer galaxies. In 1929 Hubble proposed a law that stated that the more distant the galaxy, the longer the wavelength (the greater the amount by which the wavelength of light is stretched) proportional to its distance. Hubble calculated that the universe is expanding uniformly at a rate of 75 km/s/megaparsec.

The discovery by Hubble that the universe is expanding was important because it supported Einstein's general theory of relativity. Einstein had shown mathematically that the universe was indeed either expanding or contracting. However, with no evidence supporting his model, Einstein manipulated his results by adding an extra factor called the cosmological constant: a sort of antigravity force that prevented the universe either expanding or contracting. With Hubble's discovery Einstein could now do away with the cosmological constant and show that the universe was indeed expanding or contracting.

Galaxy movement supports the big bang theory. Hubble's evidence showed that galaxies move and that the galaxies show a red shift, indicating that the galaxies are moving away from Earth. If the pathways of the galaxies are reversed it is found that all the galaxies started from a central point.

Astronomical measurements of distance

A number of measurements are used in astronomy. As distances are so great, it has been necessary to devise measurements suitable for these distances. They include the astronomical unit, light year, the parsec, kiloparsec and megaparsec.

- The distance to the individual planets is sometimes measured in astronomical units. The astronomical unit (AU) is the average distance between the Sun and the Earth: 149 597 870 km. The solar system is 80 AU in diameter.

- A light year (LY) is the distance that light travels in one Earth year (1 LY = 9 500 000 000 000 km). The Milky Way galaxy is 100 000 LY in diameter.

- A parsec is the distance at which a star shows a parallax, or apparent shift in position, when viewed from opposite sides of the Earth's orbit, of one second of arc (or 1/3600th of a degree). A parsec is equal to 3.26 LY. The nearest star, Proxima Centauri, is 1.25 parsecs away from Earth.

- A kiloparsec (kpc) corresponds to 1000 parsecs (3.26 x 10³ LY). Earth is approximately 8 kpc away from the centre of the Milky Way galaxy.

- A megaparsec is 1 000 000 parsecs (3.26 x 10⁶ LY), which is the average separation between galaxies in a galaxy cluster.

Background radiation

The discovery of background radiation provided further evidence for the big bang theory. The expanding universe began from an explosion and warmth from the explosion still fills the universe. This background warmth is called the cosmic background radiation and was discovered in 1965 by two US radioastronomers, Arno Penzias and Robert Wilson, working at Bell Laboratories. They discovered the radiation was completely uniform and came from every direction. This radiation, or warmth, has a temperature of three Kelvin (3K), which is three degrees above absolute zero (−273°C). Theoretically, space should be at absolute zero and this discrepancy can only be accounted for if there was once an event that produced high levels of energy.

The proponent of the steady state theory (see page 21), Fred Hoyle, on a radio broadcast in 1950 said, 'If the universe began with a hot big bang then such an explosion would have left a relic. Find me a fossil big bang.' Penzias and Wilson found this relic in the form of background radiation. The term 'big bang' stuck.

REVIEW ACTIVITIES

1 Evaluate the evidence supporting the big bang theory.

2 Explain how the Doppler effect occurs.

3 Explain how scientists have discovered that there is a difference between the speeds at which certain galaxies are receding from us.

4 Recount what Hubble's law states.

5 Describe why space, having a temperature of 3K, is so important to our knowledge of the creation of the universe.

EXTENSION ACTIVITIES

6 Explain what Einstein's general theory of relativity proposed and why is it important in our understanding of the universe.

7 Outline the major contributions made by Edwin Hubble in our understanding of the universe.

QUESTIONS UNANSWERED BY THE BIG BANG

A lot of evidence supports the big bang theory, and at the moment it appears to be the best theory for the formation of the universe. However, it is not a complete answer. Some of the questions that the theory cannot answer include:

- Where did the point singularity come from?
- What caused it to explode?
- Is the universe round, curved or flat?

As more evidence comes to light in support of the big bang, new questions arise. It appears that the theory of the big bang is only partially correct. Some of the questions that are thrown up by current evidence include:

- What is the actual age of the universe?
- Why do some stars appear to be older than the universe?
- What is the Great Attractor and why do galaxies in an area of the universe appear to be moving towards a single point?
- Why are some galaxies moving faster than others when, according to astrophysicists (such as Laurer and Postman), everything should be moving at the same speed?
- Why does the universe have a magnetic orientation? In April 1997, evidence was found that the universe has a common magnetic orientation or, in other words, a 'North Pole'. If there was a big bang and everything is moving out from a central point, there should be no magnetic orientation.

SEQUENCE OF EVENTS IN THE BIG BANG

Scientists generally solve problems by looking at cause and effect. For instance if the body suffers an illness (effect), doctors look for a cause. This cause could be bacterium or a virus, for example. Not all effects, however, are based on the premise that each effect has a cause. If we look at the decay of a radioactive element, it is found that the particles released do so in a totally random way. There is no cause for them to be released, it just occurs. In the same way, the start of the big bang did not have a cause, but the effects of it were universal. The sequence of the events in the big bang were as follows.

About 15 billion years ago, a tremendous explosion started the expansion of the universe. At this point all the matter and energy of space was contained at one point.

What existed prior to this event is completely unknown and is pure speculation. The start was not a conventional explosion, but rather an event filling all space with all the particles of the new universe rushing away from each other.

Fifteen billion years ago, all matter of all forms was concentrated in a dense point called a point singularity. This point contained various forces: charges, mass, energy, gravity, electrostatic forces and strong and weak nuclear forces. The changes in the inflating universe were initially very fast and only later did they slow down. Change continues today and will into the future.

At around 10^{-43} secs after the 'big bang', there existed almost equal amounts of **matter** and **antimatter** spreading out in all directions from the central point. As these particles of matter and antimatter moved outwards in the inflating universe, the particles collided with each other, in the process destroying each other and releasing large amounts of energy. Despite the large amounts of energy being released, the universe was starting to cool down.

At 10^{-33} secs the initial very fast inflation of the universe was over. Expansion of the universe continued, but at a slower rate. After the initial burst of collisions between matter and antimatter, the antimatter particles no longer existed and the only particles that remained were particles of matter.

At approximately 10^{-25} secs the universe continued to expand and cool. This cooling allowed particles to form. They included **photons**, **neutrinos**, **electrons** and **quarks**. At this stage, the temperature of the universe was still too high for the formation of **protons** and **neutrons**.

By about 10^{-6} secs, the universe had cooled to about 3000 billion K. At this temperature protons and neutrons formed from most of the photons, neutrinos, electrons and quarks.

Between one minute and 300 000 years, the temperature dropped to about 3000°C. At this temperature, protons and neutrons started to react together to form two types of heavy hydrogen: deuterium and tritium. The formation of hydrogen was rapidly followed by reactions between hydrogen and other protons, which resulted in the formation of helium. Evidence of primordial helium has recently been discovered in the outer fringes of the universe. These outer fringes are the oldest parts of the universe.

During this period of time the density of the universe decreased sufficiently that light could be perceived. Prior to this time, the photons that produce light were absorbed by the forming particles.

At 1 billion years after the big bang, the temperature of the universe had dropped to −270°C. Between 1 and 5 billion years from the start of the universe, elements heavier than helium began to form.

Five billion years from the start of the universe the galaxies (including the Milky Way galaxy) and, hence, stars have formed.

After 10 billion years our solar system formed. The temperature of space had dropped to 30K.

This time scale and events are only approximate, many scientists will give variations on the time and events. The problem is that nobody has any direct proof, as humans were not around at the time! The times and events are based on scientific and mathematical modelling and limited observations of the universe.

definitions

matter particles
fundamental particles. They have no known smaller parts. They are also called quarks.

antimatter
matter composed of the counterparts of ordinary matter, such as antiprotons instead of protons and positrons instead of electrons

photon
a discrete quantity of light energy, the energy being proportional to the frequency of radiation

neutrino
an elementary particle with zero electrical charge and a mass of zero when at rest

electron
an elementary particle that is a constituent of all atoms and has a minute mass of approximately 9.1×10^{-31} kg. It has a negative charge.

quarks
see matter particles

proton
an elementary particle present in every atomic nucleus, the number of protons being different for each element. A proton has an electric charge equal in magnitude to that of an electron but of opposite charge and has a mass of 1.7×10^{-27} kg.

neutron
an elementary particle that is a constituent of all atomic nuclei except normal hydrogen. It has a zero electrical charge and approximately the same mass as a proton.

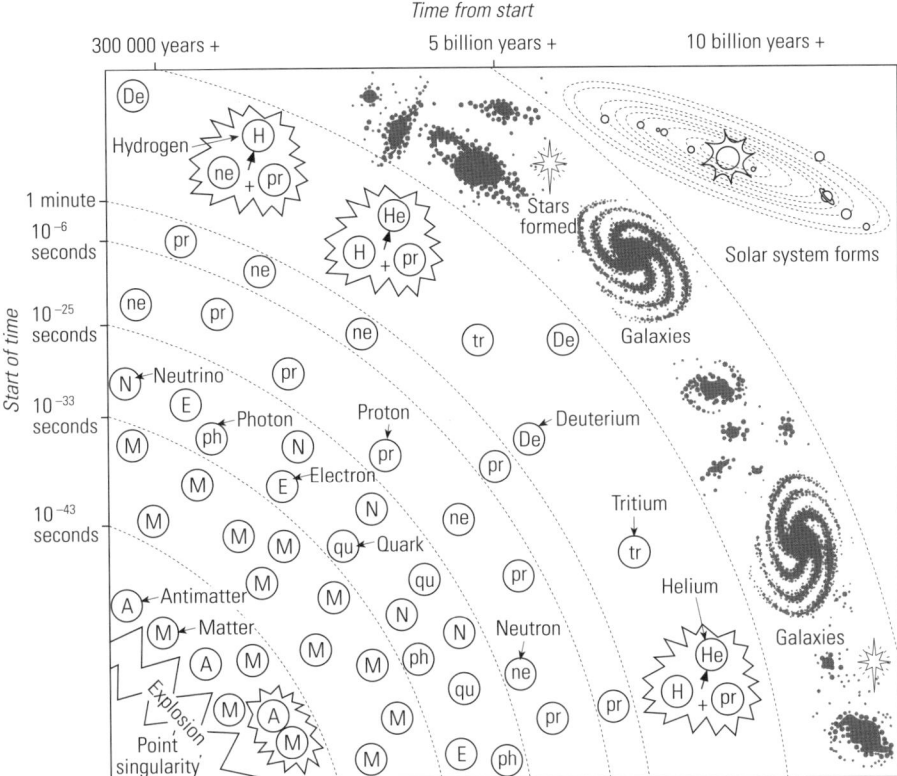

Figure 1.2.2 Diagrammatic representation of the history of the universe.

REVIEW ACTIVITIES

1 Outline the steps in the sequence of events in the formation of the universe.

2 Explain what happened when matter and antimatter collided.

3 Explain how helium was formed.

4 Describe why the first stages of the big bang would not have been visible.

5 Define the term 'point singularity'.

EXTENSION ACTIVITIES

6 There are differing theories as to the shape of the universe. Recount two alternative theories and explain their significance.

7 Research what cosmologists think may happen to the universe in the future.

The steady state theory

The **steady state theory** was originally proposed by Sir James Jeans in 1928 and was revised in 1948 by Hermann Bondi and Thomas Gold. It was further developed by Sir Fred Hoyle, the British astronomer and science fiction writer.

They did not agree with the idea that there was a sudden beginning to the universe. Their theory proposed that the universe did not have a beginning and will not end. Their view was that the universe is always expanding but maintaining a constant average density. It continuously creates new matter in the form of hydrogen. This formation of hydrogen appears to be derived from nothing and, according to the laws of physics, this is impossible. This theory also stated that the universe is eternal and has no edges.

> **definition**
>
> **steady state theory**
> a steady state universe has no beginning or end of time. The average density and arrangement of galaxies does not change.

PLANET EARTH AND ITS ENVIRONMENT 21

Hoyle proposed that the decrease in density of the universe, caused by its expansion, is exactly balanced by the continuous creation of matter condensing into galaxies. These newly created galaxies take the place of galaxies that have receded so far from the Milky Way that they are no longer observable. Thereby, the present appearance of the universe is maintained forever. The expansion and creation of matter work against each other and, therefore, a steady state of energy is maintained.

The steady state theory accommodates the idea of an expanding universe. The idea of an expanding universe has now been proved. A major difference from the big bang theory is that Hoyle moved away from the laws of modern physics by saying that new matter is being formed continuously. He proposed that matter was being created in the gaps of the expanding universe, at a rate of one hydrogen atom per year per 100 m. This was a baseless hypothesis, but because the amounts of hydrogen were so minute, it was impossible to disprove.

In support of the steady state theory, in 1957 Hoyle explained that some elements can be created by the action of supernovas. This creation of heavy elements is now accepted by both the steady state and big bang theories.

Another weakness of the steady state theory is the distribution of radio sources. The theory states that if the distribution of these radio sources is uniform then the fainter ones must be more distant. It has been found, however, that there are more bright radio sources at a greater distance than would be expected according to the steady state theory. The conclusion is that the universe is evolving or at least changing.

Quasars, or quasi stellar radio sources, were first discovered in the 1950s. At the time they were thought to be stars, but it was soon discovered that the radio sources originated from a wide variety of stellar objects. They included stars and supernova remnants. The main cause of quasars is the energy released by black holes at the centre of distant galaxies. Radioastronomy dealt a blow to the steady state theory. Scientists discovered that there was a greater abundance of radio-emitting objects lying at the edge of the visible universe. The discovery of quasars lent further support to the big bang.

The pulsating universe theory

The **pulsating universe theory** developed from the big bang theory. It proposes that the universe ends in a 'big crunch'. Gravity, this view argues, slows the expansion of the universe and then causes it to collapse towards its centre. When gravity exceeds the outward momentum it acts like a piece of elastic, which gets stretched and then returns to its original size. This process continues repeatedly.

pulsating universe theory
starting with the big bang, the universe expands. This is followed by a contraction of matter, called the 'big crunch'. This expansion and contraction occurs continuously.

REVIEW ACTIVITIES

1 Explain the reason for the steady state theory being proposed.

2 Summarise the evidence for and against the big bang theory and the steady state theory.

3 How does the big bang theory differ from the pulsating universe theory?

4 Discuss 'the big bang theory is the best theory in describing the creation of the universe'.

EXTENSION ACTIVITY

5 Fred Hoyle is remembered for the steady state theory, but he also won major scientific awards for his other astronomical work. Summarise some of his other major contributions.

DATING THE AGE OF THE UNIVERSE

There are a number of ways of measuring the age of the universe, including assessing radioactive decay and the speed of galaxies.

Radioactive decay

All living and non-living things possess some amount of radioactive material. The amount of the radioactive substance is fixed when the object forms. By measuring how much still remains it is possible to work out the object's age. This method has been used to age parts of the universe, such as stars and meteorites.

Elements such as plutonium and uranium are unstable. In order to become stable these elements decay, releasing radiation. There are three types of **radiation**: alpha, beta and gamma radiation. Alpha radiation is the emission of alpha particles or a positively charged particle composed of two protons and two neutrons (equivalent to helium nucleus) from the atom undergoing change. Beta radiation is the release of beta particles or an electron or positron spontaneously emitted by some radioactive nuclei of atoms undergoing change. Gamma radiation is high-frequency electromagnetic radiation released by radioactive atoms undergoing change. Gamma radiation resembles X-rays.

The emission of either alpha or beta particles from the nucleus of an atom is called **nuclear decay**. The emission of these particles results in a change from one isotopic form of an element to another or from one element to another. For example an atom of uranium undergoes a number of changes until it becomes an **isotope** of lead, which is a stable element. Each change in the uranium atom results in the formation of a different element. These changes, or stages of nuclear decay, may take a fraction of a second or millions of years to occur. As the decay takes place, it has been found that each of the radioactive elements loses mass at a definite rate. For example fluorine 20 loses 50% of its mass every eleven seconds. This loss is the result of radiation being released from the atoms. The rate of loss is called the element's half-life. It is the time taken for a radioactive element to lose half its mass. (See Table 1.2.1 and Figure 1.2.3.)

Table 1.2.1 Common isotopes and their half-lives

Isotope	Half-life
Carbon 14	5.7×10^3 years
Fluorine 20	11 seconds
Hydrogen 3	12.3 years
Iodine 131	8 days
Magnesium 27	9.5 minutes
Plutonium 239	2.4×10^4 years
Strontium 90	28 years
Uranium 238	4.5×10^9 years

radiation
in order to become stable some elements decay, releasing radiation. There are three types of radiation: alpha, beta and gamma radiation.

nuclear decay
the emission of particles from the nuclei of an element that results in a change from one isotopic form of an element to another or from one element to another

isotope
elements that have the same number of protons in their nucleus (atomic number) and similar chemical properties, but different atomic weights (their mass relative to ^{12}C)

definitions

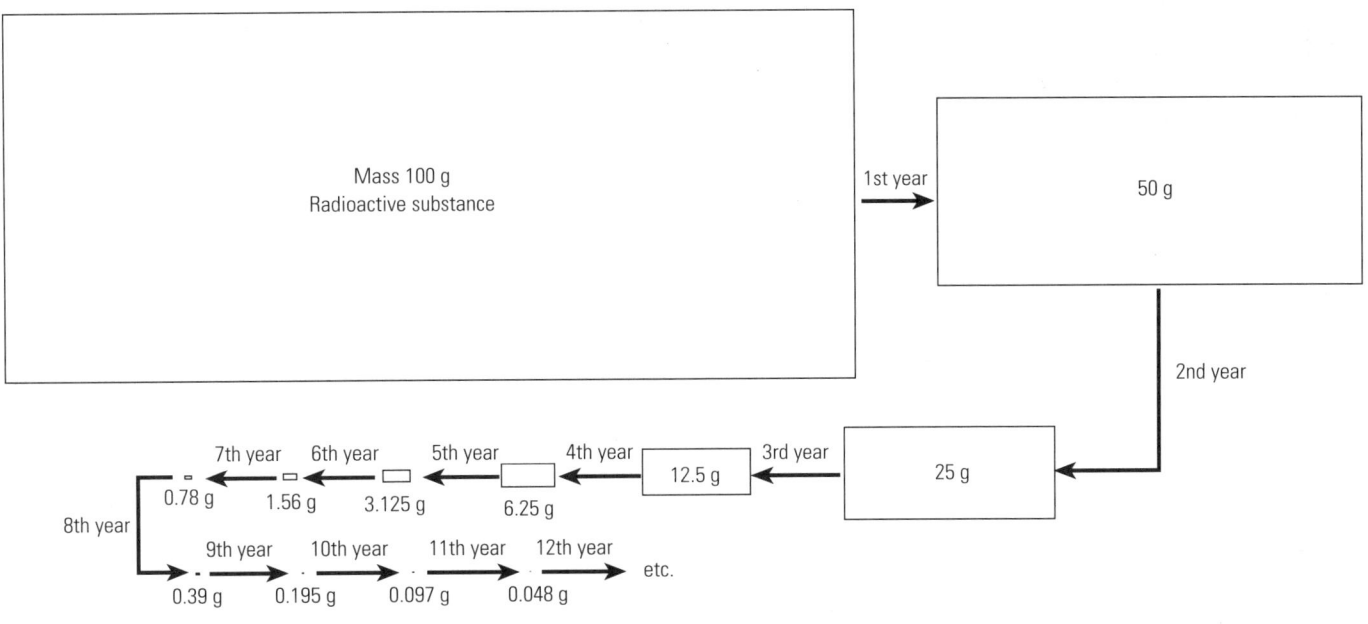

Figure 1.2.3 Diagrammatic representation of half-life.

Speed of galaxies

A second method of measuring the age of the universe is by working out how fast the galaxies are moving away from the centre of the universe. This can be done by using Einstein's general theory of relativity. (See page 18.) Once this is established, it is possible to work backwards to work out when the universe was formed. This method gives the origin of the universe at about 10–15 billion years.

REVIEW ACTIVITIES

1 Explain how we can date the universe.

2 Differentiate between the different types of radioactivity.

3 Define the term 'radioactive decay'.

4 Explain what is meant by the term 'half-life'.

EXTENSION ACTIVITY

5 Research how and why geologists use radioactivity as a tool.

Figure 1.2.4 A spiral galaxy.

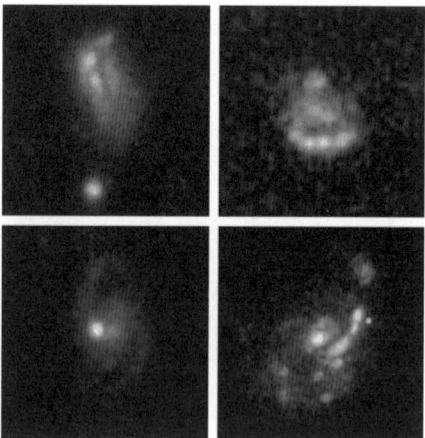

Figure 1.2.5 Irregular galaxies.

PARTS OF THE UNIVERSE

The whole concept of the universe is very difficult to grasp due to the immenseness of its structure, as well as the number of objects that are held within it. We are still trying to come to terms with what the universe contains. We are handicapped by using relatively crude equipment to observe the universe and limited by the locations from which we view the universe. The distance between the celestial bodies, as well as the number of bodies that can be observed, also limits our view.

Galaxies

Within the universe, there are many billions of galaxies and each galaxy has many millions or billions of stars. Galaxies are the largest structures in the universe. There are four types of galaxies and the types are named according to their shape: spiral, barred spiral, elliptical and irregular. Most of the older galaxies are spiral in shape, whereas the largest galaxies tend to be elliptical in shape.

Spiral galaxies usually consist of two major components: a flat, large disc, which often contains a lot of interstellar matter; and young star clusters, which extend outwards forming arms arranged in a spiral or barred spiral structure. (See Figure 1.2.4.) The spiral arms contain bright stars and the gaps between the arms may contain older, cooler stars. In barred spiral galaxies the stars are in a bar formation.

Elliptical galaxies are shaped like an egg. The stars are wrapped around the centre. Larger elliptical galaxies sometimes have a black hole in the centre.

Galaxies that are neither spiral nor elliptical are referred to as irregular. (See Figure 1.2.5.) They have no identifiable shape and are usually the smallest. Irregular galaxies may also be the building blocks for other types of galaxies.

There are two main theories as to how the galaxies originally formed. One theory states that the galaxies formed when clouds of dust and gas collapsed under their own gravitational pull. This collapse led to areas where the dust and gas were denser and in these areas millions or billions of stars formed, resulting in the formation of galaxies. The second theory is that lumps of matter clumped together to form galaxies. This process of clumping is still taking place today. Our galaxy, the Milky

Way, is currently absorbing a smaller galaxy. These mergers are thought to be common. The second theory is more popular nowadays, as there appears to be supporting evidence from Hubble Space Telescope photographs.

Stars

Stellar lifetimes range from around 40 000 years to longer than 10 billion years. It is therefore impossible for astronomers to watch a particular star go through all its life cycle phases. However, with the millions of stars around us it is possible to see stars at different points of their life cycles. By looking at thousands of pictures and using basic physics, astronomers have developed the sequence of a star's birth, life and death. (See Plate 5.)

> **protostar**
> a young star that is still forming ('proto' means 'before' or 'early')
>
> **red star**
> a small star of less than 0.1 solar masses
>
> **true star**
> a star in which nuclear fusion reactions are taking place
>
> **main sequence star**
> first and longest stage of a star's life
>
> **red giant**
> a late stage in a Sun-sized star's life. The outer layers expand and cool, and nuclear fusion is replaced with nuclear fission.
>
> **white dwarf**
> a late stage in a star's life. It is the result of the outer layers of the star dispersing and the core collapsing to form an extremely dense, small star. It may have half the mass of the Sun but it is only the size of the Earth.
>
> definitions

The birth of a star occurs when clouds of gas and dust up to two light years in size collapse into themselves. (See Figure 1.2.6.) Astronomers still do not fully understand the processes involved, but think that it is due to the star's own gravitational forces. As the collapse occurs, a **protostar** forms.

The life cycle of a star is strongly influenced by its mass. A protostar with a mass less than 0.1 solar masses will shrink. Despite the collisions between the atoms of the star it will never get hot enough for nuclear fusion reactions to begin. It will fade to form a small **red star** with a temperature of 3000°C before turning cold and dying. (See Figure 1.2.7.)

If the mass of the protostar is great enough, nuclear fusion will begin. This will result in an initial rise in temperature to 15 million °C followed by a cooling to 5000°C.

Nuclear fusion occurs due to the protons from the atoms travelling at very high speed and overcoming their natural repulsion from each other. The nuclei come together and nuclear fusion begins. Energy from fusion pours out of the core. Electromagnetic radiation is released and the cloud starts to shine. Once nuclear fusion occurs in a protostar it is a **true star**.

Figure 1.2.6 The birth of a star.

With newborn stars there is a link between colour and brightness: the whiter and hotter the star, the more brightly it glows; the redder and cooler it is, the dimmer its appearance.

When a new star is formed it is called a **main sequence star**. This stage of a star's life is the first and longest stage. A main sequence star, such as our Sun, will spend about 90% of its lifetime in this stage. During this time it is very stable. There are large reservoirs of hydrogen to power the nuclear fusion reactions. The star sends out energy at a steady rate and it is balanced with gravity pulling in and the pressure of its matter pushing out. It burns with a yellow colour and its surface temperature is about 5000°C.

Figure 1.2.7 A dying star.

When all the hydrogen is used up, the star's core shrinks. There is a decrease in nuclear fusion and an increase in nuclear fission, resulting in nuclear fission reactions replacing nuclear fusion reactions. As a result, the outer layers of the star expand and cool. The star forms a **red giant**, which can be up to 2 billion km in diameter. After this, the gases forming the outer layers drift into space and the remaining gases in the core collapse, forming a very small and dense Earth-sized star called a **white dwarf**. These stars are white and dim, but very hot. As they are small, they are hard to detect. The motion of binary star systems is used to detect them. Eventually the white dwarf will cool, leaving the gases to form a remnant cloud.

> **definitions**
>
> **supernova**
> a red supergiant star whose outer layers are blown off in a massive explosion. The remaining core collapses and forms a neutron star.
>
> **blue giants**
> massive stars with high temperatures, high luminosities and diameters ten to one hundred times that of the Sun

Stars that are two to six times bigger than our Sun have a short but spectacular life. Their life span is approximately 1 million years. The mass of these large stars creates enormous gravitational forces within the core of the star. The nuclear reactions occur more rapidly than in other Sun-sized stars. These stars are very hot—up to 25 000°C—and are blue in colour. When the fuel runs out there is a massive outflow of energy, which we call a **supernova**. Much of the star's matter is blown into space, and all that remains is the core of the star. These remains form a neutron star, which has a very large mass but a very small diameter—as little as 10 or 20 km—resulting in a very high density. The outer layers of the original star are blown into space, forming a cloud of gas and dust called a nebula.

CLASSES OF STARS

The different classes of stars can be arranged in a graph showing their luminosities plotted against their surface temperature. This graph is called a Hertzsprung-Russell diagram. (See Figure 1.2.8.) The majority of stars lie in a diagonal band across the graph. This band contains stars called the main sequence stars. There are other groups of stars that are not part of the main sequence. These include giants, supergiants and white dwarfs. Within these classes there are subclasses, such as the **blue giants**.

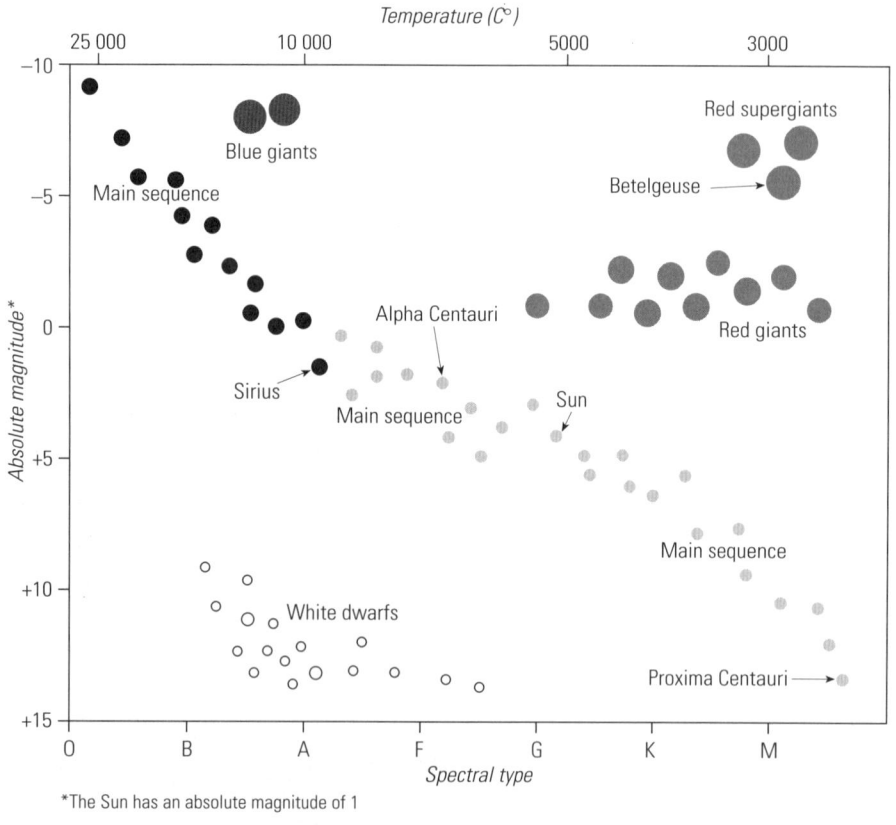

Figure 1.2.8 Hertzsprung-Russell diagram.

Nebulae

Nebulae are the gaseous and dusty materials expelled by a dying star just before its death. They are sometimes referred to as 'planetary nebulae', but they have nothing to do with planets. They orbit stars and that is where the name 'planetary nebula' comes from. Ninety-five per cent of all stars in the universe will create or end as nebulae. The other 5% of stars will end their lives as a supernova in one magnificent release of energy.

Many nebulae are given descriptive names, such as horsehead, crab and hourglass. (See Figures 1.2.9 and 1.2.10.) They are so named because they bear some resemblance to those shapes.

Asteroids, comets and meteorites

Apart from large bodies (such as galaxies, stars and nebulae) a number of smaller objects are found in the universe. These include asteroids, comets and meteorites.

Asteroids, or minor planets, are small bodies that orbit the Sun. At present more than 10 000 asteroids have been studied. The largest asteroid, Ceres, has a diameter of 1000 km. The orbit of most asteroids lies partially between the orbits of Mars and Jupiter in an area called the asteroid belt.

The origin of the asteroid belt is unknown, but two main theories have been proposed. The first theory is that they are the fragments of a planet that broke up long ago. This theory has lost support because there is far too little mass to constitute a planet and the marked chemical differences between individual asteroids show it could not have originated from a single planet. According to the second theory, the asteroid belt consists of the remains of the accretion disc of dust and gases that formed the Sun and our planets. (See page 32.) Astronomers believe that Jupiter's gravitational field strongly influences the belt and prevents further accretion, or fusing together, of the asteroids.

The origin of comets is still uncertain. Modern theories suggest that they were formed during the formation of the solar system. (See page 34.) Comets appear to originate in an area near the planet Neptune called the Kuiper belt. Far from the heat of the Sun, a comet has the appearance of a solid ball, which is formed from frozen gases and heavy substances. This frozen ball is called a nucleus. As it approaches the Sun, the surface of the nucleus warms and forms a tail or coma. The gases released by the Sun's heat absorb and reflect the Sun's radiation and become luminous.

There are two main sources of meteorites: the asteroid belt and comets. Large meteorites originate from material found in the asteroid belt and may be fragments produced from the collision of asteroids. Small meteorites are very fine dust-sized particles of rock that have been left behind by comets. They produce meteor showers, which burn up in the Earth's atmosphere.

Before meteorites enter the atmosphere and heat up due to friction caused by the Earth's atmosphere, these objects are classified as meteoroids. There are three main types of meteorites: stony; iron and nickel; and stony iron. As the name of the classes suggests, the composition of the meteorites varies.

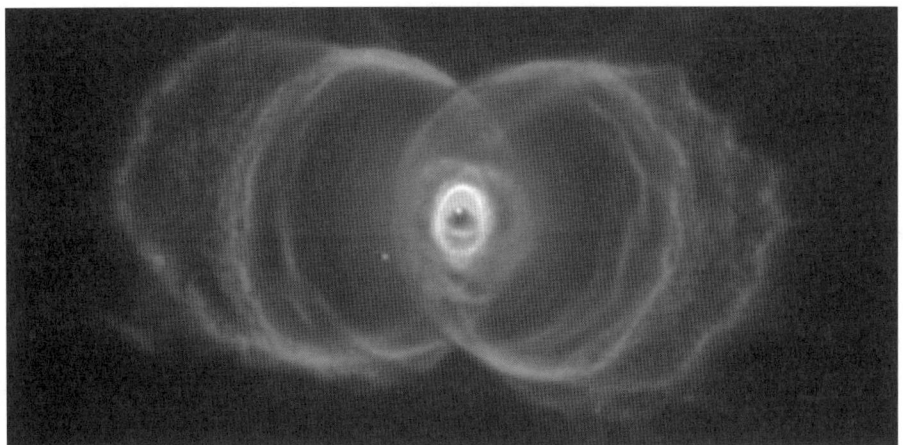

Figure 1.2.9 An hourglass nebula.

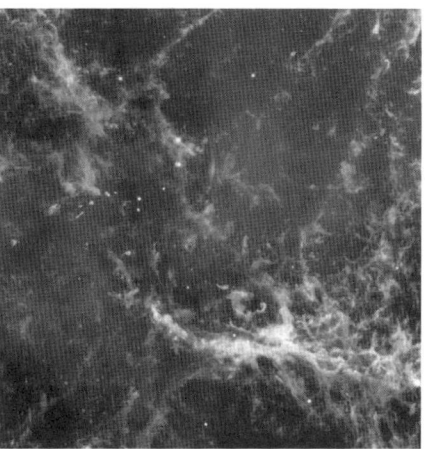

Figure 1.2.10 A crab nebula.

REVIEW ACTIVITIES

1 Describe the different types of galaxies.

2 Distinguish between the different classes of stars.

3 Explain what a red star is and how it differs from a main sequence star.

4 Define the term 'main sequence stars'.

5 Explain the difference between a red giant and a white dwarf.

6 Explain why very few meteorites hit the Earth's surface.

EXTENSION ACTIVITIES

7 Outline the life cycle of a star with a mass greater than 4 solar masses.

8 Explain why supernovae are important in the life cycle of stars.

9 Explain why asteroids may pose a threat to the Earth.

SUMMARY

- Creation theories are based on beliefs, not scientific evidence.

- Theories to the origin of the universe are: big bang; steady state; and pulsating universe.

- The big bang theory states that the universe started from an explosion at a point singularity. Evidence for the big bang theory: the Doppler effect; galaxy movement; and background radiation.

- The Doppler effect is the change in the observed wavelength and frequency of a sound or light wave coming from something moving away from or towards the observer. When the source is moving away from the observer, a reddish colour is produced. When the source is moving towards the observer, a blue colour is produced.

- Some of the questions left unanswered by the big bang: What is the age of the universe? Why are some stars older than the universe? What is the Great Attractor? Why are some galaxies moving faster than others? Why does the universe have a magnetic orientation?

- Steps in the big bang: the explosion of the point singularity; production of matter and antimatter; destruction of antimatter; formation of photons, neutrinos, electrons and quarks, which in turn formed particles called protons and neutrons; deuterium and tritium (types of hydrogen) formed, as did helium; heavier elements formed; the galaxies formed; and, finally, the solar system formed.

- The steady state theory states that the universe has no beginning or end of time, and that the average density and arrangement of galaxies remains the same.

- The pulsating universe theory states that when gravity exceeds the outward momentum of the universe's expansion, the universe collapses towards the centre and then re-expands.

- The age of the universe can be measured using radioactive half-lives, and by working out how fast the galaxies are moving and working backwards.

- The half-life of an element is the time taken for a radio-active element to lose half its mass.

- The four main types of galaxies are spiral, barred spiral, elliptical and irregular.

- The classes of stars include main sequence stars, giant stars, supergiants and white dwarfs. They can be arranged in a graph called a Hertzsprung-Russell diagram.

- Dying stars may produce black holes, supernovas or white dwarfs.

- Nebulae are the gaseous and dusty materials expelled by a dying star just before its death and are also the birthplace of stars.

- Asteroids are minor planets. Two theories for their origin have been proposed. The first theory proposes they are the remains of a planet, and the second proposes they are the remains of the accretionary disc.

- Comets are formed in the Kuiper belt. They are formed from frozen gases and particles of heavy substances, which are heated by the Sun.

- There are two main sources of meteorites: comets and the asteroid belt.

- The three main types of meteorites are stony; iron and nickel; and stony iron.

PRACTICAL EXERCISE
Measuring the luminosity of stars

This exercise will show that light intensity can vary with distance.

Equipment
- Masking tape
- Lamp holder
- A pearl and clear lightbulb of the following wattages: 20, 40, 60, 75 and 100
- Light meter
- Tape measure
- Darkened room
- Copy of Table 1.2.2

Procedure

1. The exercise needs to be completed in a darkened room. Put a 200 cm strip of masking tape on a table and mark off 25 cm intervals.

2. Place the lamp holder at the end of the strip, at the 200 cm mark.

3. Put the light meter on the zero mark and measure the light intensity of the lightbulb with the lowest wattage.

4. Repeat the measurement using a lightbulb with the same wattage but a different coating.

5. Record your results in the copy of Table 1.2.2.

6. Repeat steps 3 to 5 for all the different wattages and coatings.

7. Move the light meter to the 25 cm mark and repeat steps 3 to 6.

8. Repeat steps 3 to 6 for all intervals up to and including 200 cm.

9. Graph your results.

ACTIVITIES

1. What conclusions can you draw from this experiment?

2. Using the graph, was the drop in luminosity the same for each lightbulb? If not, explain why there was a difference.

3. Using the table of results, which lightbulb(s) lost the least amount of luminosity over the entire distance? Give a reason for this.

4. Comparing pearl and clear lightbulbs, is there any variation in their luminosities at the different distances?

5. If you had to repeat the experiment, what improvements would you make to it?

Table 1.2.2 Table of luminosity

Lightbulb type and wattage	Luminosity — Distance from lightbulb (cm)								
	0	25	50	75	100	125	150	175	200
Pearl									
20									
40									
60									
75									
100									
Clear									
20									
40									
60									
75									
100									

PRACTICAL EXERCISE
Comparison of the characteristics of stars

Table 1.2.3 Luminosity classes of stars

Luminosity class	Star type
I	Supergiant stars
II	Bright giant stars
III	Giant stars
IV	Subgiant stars
V	Main sequence dwarf stars
VI	Subdwarf stars
VII	White dwarf stars

Table 1.2.4 Spectral type of stars

Spectral group	Surface temperature (°K)	Spectral features
O	>20 000	Ionised helium visible
B	20 000–10 000	Neutral helium visible Hydrogen lines start to appear
A	10 000–7000	Helium lines weaker Strong neutral hydrogen lines visible
F	7000–6000	Hydrogen lines weaker Ionised calcium lines visible
G	6000–5000	Much weaker neutral hydrogen lines Ionised calcium lines very prominent Other metallic lines, such as iron, are visible
K	5000–3500	Neutral metallic lines are prominent Molecular bands are visible Most hydrogen lines are invisible
M	3500–2000	Molecular bands very visible

ACTIVITIES

Using Figure 1.2.8 (page 26) and Tables 1.2.3 and 1.2.4, complete the following activities.

1 What are the approximate absolute magnitudes of Sirius and Alpha Centauri?

2 Write a generalisation linking the absolute magnitude of the main sequence stars to their temperature and colour.

3 Describe the difference in the chemical compositions of a blue giant and the Sun.

4 Alpha Centauri, Sirius and the Sun are different ages. Which would be the oldest? Explain your reasons.

5 Compare and contrast the luminosity and temperatures of Proxima Centauri and Betelgeuse.

6 Two stars are compared with each other. Star A appears blue and star B appears red. Explain what can be deduced from this information.

7 Explain what happens to the position of a star on the Hertzsprung-Russell diagram if the temperature of the star is increased, but its size remains constant.

8 Explain what must be known when plotting stars onto a Hertzsprung-Russell diagram.

9 Two stars of the same spectral type are plotted on a Hertzsprung-Russell diagram. Star A is more luminous than star B. What would be the position on the graph of star A in comparison to star B?

1.3 The Sun and the solar system

OUTCOMES

At the end of this chapter you should be able to:

- identify the sequence of events described by scientists to outline the formation of the solar system
- explain the role of gravity in the formation of the Earth
- recall the explanation of density using a simple particle model
- describe the relationship between the density of Earth materials and the layered structure of the Earth
- gather, process and present information that outlines the sequence of events that led to the formation of the solar system.

SOLAR NEBULAR HYPOTHESIS

According to the currently accepted view, the Sun formed about 4.7×10^9 years ago from a cloud of gas and dust. The collapse of this cloud was triggered by a supernova explosion from a second-generation star. This second-generation star had itself been formed from the death of a previous star, which had resulted in an earlier supernova. The later supernova produced a gas and dust cloud called a solar nebula. The nature of the dust in the nebula depended on its position in the disc. As accretion of material in the disc slowed, it allowed the disc to cool. In the inner disc, however, there were vast numbers of collisions, which caused it to heat up. Scientists believe the **solar system**, composed of our own local star and its accompanying planetary system, was formed from this solar nebula. This idea is called the **solar nebular hypothesis**.

Evidence supporting the solar nebular hypothesis

There is no absolute evidence to definitely prove or disprove the solar nebular hypothesis. However, it is supported by a great deal of circumstantial evidence, such as:
- The ages of the members of the solar system are similar. All the planets within the solar system were formed 4.6 billion years ago.
- The orbits of the planets are in the same plane.
- The direction of motion and orbit of the planets are the same.
- The composition of the planets show similarities. There are two types of planetary bodies: terrestrial planets and Jovian (gas) planets. All terrestrial planets have a similar rocky composition. All Jovian planets also have a common composition.
- The Hubble Space Telescope (HST) has recorded images of nebulae containing areas in which protostars are forming.
- The HST has recorded images of gas and dust discs surrounding stars in the Orion Nebula, which is the part of the Orion constellation 450 parsecs or 1500 light years from Earth.
- Direct evidence has arisen from studying the spectra of light coming from nearby stars and looking for a 'wobble'. It is caused by planets orbiting one or more stars at the centre of the mass of a star system similar to our solar system.

solar system
a group of nine planets orbiting a main sequence star called the Sun

solar nebular hypothesis
the solar system is formed from a solar nebula

periodic table
a table of elements arranged in order of atomic number, the arrangement emphasising the chemical relationship between the elements

proton
the positively charged subatomic particle found in the nucleus of an atom

definitions

- Occasionally the planets in a star system form a line as they orbit their central, or parent, star(s). This line is called a transit. As the line of planets passes between the Earth and the parent star a small black shadow crosses the surface of the star and the amount of light being released from the star appears to reduce. This shadow is caused by a planetary body.

Supernova, solar nebulae and the formation of elements

The collapse of a large star causes a supernova, which results in the formation of a neutron star. When a large star collapses, the temperature reaches billions of degrees and a supernova is formed. The gas and dust cloud, or solar nebula, released by the supernova not only contains the original dust and gas elements, but also new elements formed by the supernova. Supernovas are thought to be responsible for the creation of many of the elements that are heavier than hydrogen and helium.

The known elements have been listed in a table called the **periodic table**. (See Figure 1.3.1.) The elements have been arranged in order of increasing atomic number. The atomic number of an element is the number of **protons** in the nucleus of the atom of that element.

Elements with an atomic number beyond number ninety-three on the periodic table either split into lighter elements or survive as heavy radioactive elements, such as uranium and thorium. The solar nebula contains supernovian elements as well as other elements that originated from interstellar space.

Key: Atomic number (1), Symbol (H), Atomic weight (1.008)

	1	2	3	4	5	6	7	8	9	10	11	12	13	14	15	16	17	18
1	1 H 1.008																	2 He 4.003
2	3 Li 6.941	4 Be 9.012											5 B 10.81	6 C 12.01	7 N 14.01	8 O 16.00	9 F 19.00	10 Ne 20.18
3	11 Na 22.99	12 Mg 24.31											13 Al 26.98	14 Si 28.09	15 P 30.97	16 S 32.07	17 Cl 35.45	18 Ar 39.95
4	19 K 39.10	20 Ca 40.08	21 Sc 44.96	22 Ti 47.88	23 V 50.94	24 Cr 52.00	25 Mn 54.94	26 Fe 55.85	27 Co 58.93	28 Ni 58.69	29 Cu 63.55	30 Zn 65.39	31 Ga 69.72	32 Ge 72.61	33 As 74.92	34 Se 78.96	35 Br 79.90	36 Kr 83.80
5	37 Rb 85.47	38 Sr 87.62	39 Y 88.91	40 Zr 91.22	41 Nb 92.91	42 Mo 95.94	43 Tc 98.91	44 Ru 101.1	45 Rh 102.9	46 Pd 106.4	47 Ag 107.9	48 Cd 112.4	49 In 114.8	50 Sn 118.7	51 Sb 121.8	52 Te 127.6	53 I 126.9	54 Xe 131.3
6	55 Cs 132.9	56 Ba 137.3	71 Lu 175.0	72 Hf 178.5	73 Ta 180.9	74 W 183.8	75 Re 186.2	76 Os 190.2	77 Ir 192.2	78 Pt 195.1	79 Au 197.0	80 Hg 200.6	81 Tl 204.4	82 Pb 207.2	83 Bi 209.0	84 Po 209.0	85 At 210.0	86 Rn 222.0
7	87 Fr 223.0	88 Ra 226.0	103 Lr 262.1	104 Rf 261.1	105 Db 262.1	106 Sg 263.1	107 Bh 264.1	108 Hs 265.1	109 Mt 268	110 Uun	111 Uuu	112 Uub	113 Une	114 Uuq	115 Unp	116 Uuh	117 Uns	118 Uuo

6	57 La 138.9	58 Ce 140.1	59 Pr 140.9	60 Nd 144.2	61 Pm 146.9	62 Sm 150.4	63 Eu 152.0	64 Gd 157.3	65 Tb 158.9	66 Dy 162.5	67 Ho 164.9	68 Er 167.3	69 Tm 168.9	70 Yb 173.0
7	89 Ac 227.0	90 Th 232.0	91 Pa 231.0	92 U 238.0	93 Np 237.0	94 Pu 244.1	95 Am 243.1	96 Cm 247.1	97 Bk 247.1	98 Cf 251.1	99 Es 252.0	100 Fm 257.1	101 Md 258.1	102 No 259.1

Figure 1.3.1 The periodic table.

REVIEW ACTIVITIES

1
Explain what the solar nebular hypothesis is.

2
Summarise the evidence supporting the solar nebular hypothesis.

3
Explain how an element's atomic number is determined.

EXTENSION ACTIVITY

4
Apart from an increasing atomic number, describe in what ways elements are arranged in the periodic table.

FORMATION OF THE SUN

Our Sun is currently in its long, stable mid-life and is classified as an average, main sequence star. It is thought that at least two generations of stars occupied the site of the present-day solar system and that the supernova formed from the second generation of those stars produced a solar nebula, which resulted in our Sun. The material from the supernova formed the building blocks for the new solar system.

The solar nebula rotated slowly due to forces formed in the explosion of the stellar body. As the cloud of gas rotated, the gases formed a thicker cloud as the gravitational forces pulled the gases towards the centre of the cloud. As the cloud became thicker and denser, the atoms moved closer together. The gas became hotter and denser due to this compression. The hydrogen atoms in the centre were very hot and densely packed and were subject to collisions with other atoms. These conditions resulted in nuclear fusion taking place in the core of this highly thick and dense cloud, which could now be called a protostar. The nuclear fusion reactions taking place inside the protostar produced immense amounts of energy. (See Figure 1.3.2.)

All stars get their energy through thermonuclear fusion of light elements into heavy elements. As nuclear fusion takes place in these stellar bodies new elements and isotopes of existing ones are formed. High temperatures are required so that the protons in each fusing atomic nucleus is overcome. For example the minimum temperature required for the fusion of hydrogen is 5 million °C. Elements with more protons in their nuclei require higher temperatures. For example the formation of carbon requires a temperature of about 1 billion °C. Hydrogen is the source of fuel for these fusion reactions.

The outward release of energy resulting from the reactions was balanced by the inward pressure of the colliding gases, and a protosun was formed. (See Figure 1.3.3.) The protosun started to spin faster and faster, at the same time collecting more gas and dust and becoming hotter and hotter. At this point the Sun was born. Even though a star is a massive object, its supply of hydrogen is not unlimited. Eventually, the hydrogen becomes depleted, the core contracts, the temperature rises and the star dies.

a. Nuclear fusion of deuterium atoms

$$^{2}_{1}H + ^{2}_{1}H \longrightarrow ^{3}_{2}He + ^{1}_{0}n + Energy$$

Deuterium + Deuterium ⟶ Helium + Neutron + Energy

b. Deuterium and tritium nuclear fusion equation

$$^{2}_{1}H + ^{3}_{1}H \longrightarrow ^{4}_{2}He + ^{1}_{0}n + Energy$$

Deuterium + Tritium ⟶ Helium + Neutron + Energy

Figure 1.3.2 Stellar fusion equations of different types of hydrogen.

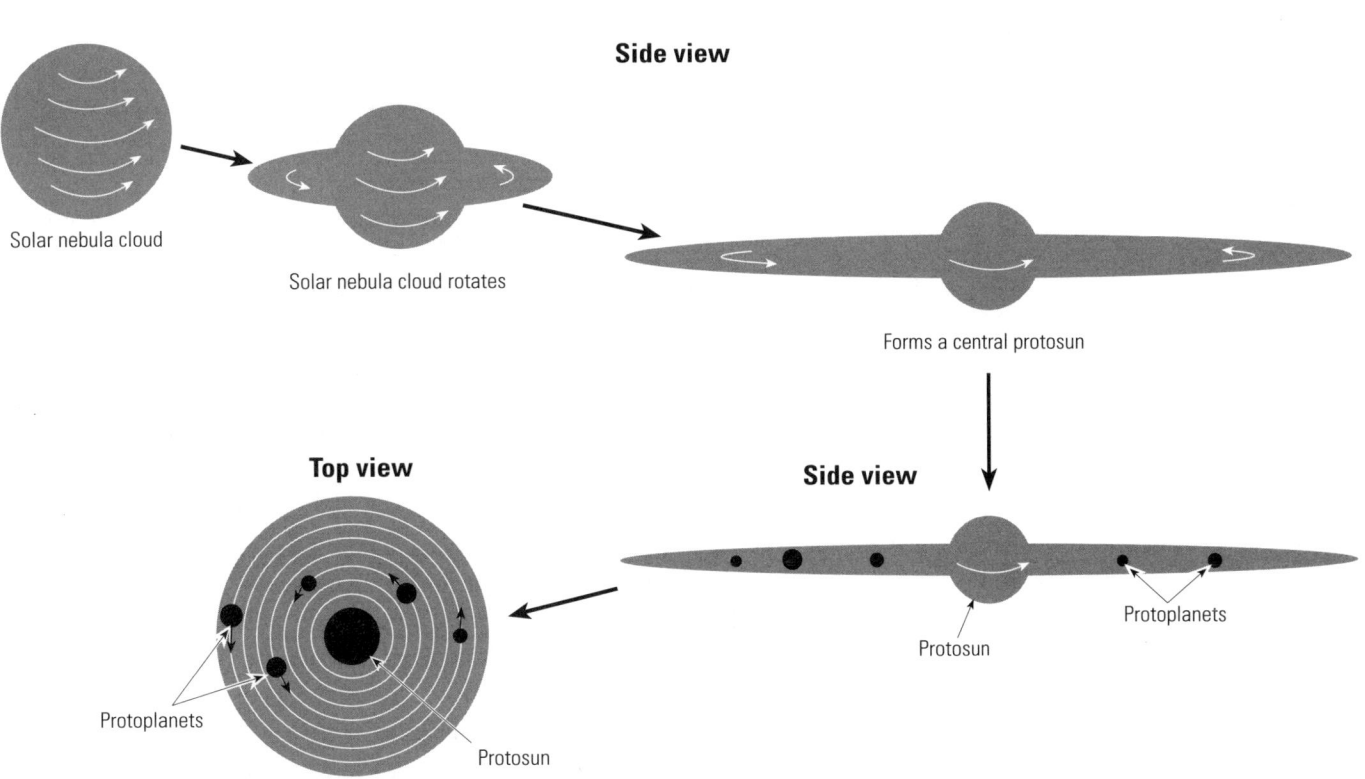

Figure 1.3.3 Diagrammatic representation of the formation of the solar system from a solar nebula.

REVIEW ACTIVITIES

1 Explain how and why nuclear fusion occurs in stars.

2 Explain what happens when the hydrogen supply runs out.

EXTENSION ACTIVITIES

3 Find out how deuterium and tritium differ from normal hydrogen.

4 In a Sun-type star, normally fusion reactions occur. However, as some Sun-type stars age, nuclear fission takes place. What is the difference between fission and fusion reactions?

definitions

protoplanetary disc
an early stage in a planet's formation

accretion
when material collides and sticks together

terrestrial planets
the inner planets formed from solid material by accretion: Mercury, Venus, Earth and Mars

Jovian planets
the outer planets composed mainly of gas: Jupiter, Saturn, Uranus and Neptune

planetesimal
an early stage in the formation of a planet

FORMATION OF THE PLANETS

Space missions have provided evidence to support the theory that all the objects in the solar system formed at the same time. As the sun formed, a disc of gas and dust surrounded it. This flattened rotating disc of gas and dust is referred to as the solar nebula. This solar nebula was the **protoplanetary disc**, the source of planetary formation. In the disc, there was turbulent motion of gas and dust. As the gas and dust moved in the disc some of it collided. Usually the material crashed into the surface of the new Sun. Sometimes, however, the material stuck together and remained in the disc. Collisions continued and by this process of sticking together, or **accretion**, the lumps of material increased in size. Eventually this resulted in the formation of the planets. To achieve an Earth-sized lump could have taken tens of millions of years.

Over a period of millions of years the process of accretion continued and eventually the planets of the solar system were formed. The solar system is made up of nine planetary bodies. The nine planets can be classified into two groups: **terrestrial planets** and **Jovian planets**. The terrestrial, or rocky, planets are composed primarily of rock and metal and have relatively high densities, slow rotation, solid surfaces, no rings and few satellites. The Jovian, or gas, planets are composed primarily of hydrogen and helium and generally have low densities, rapid rotation, deep atmospheres, rings and many satellites. On average, the Jovian planets have a diameter that is ten times greater than that of the terrestrial planets.

The planets and moons near the Sun consist of substances that are able to condense at high temperatures. These are mainly compounds containing oxygen, silicon, calcium, iron and magnesium. The Jovian planets exist further from the Sun, where temperatures are lower, allowing sulfur compounds, water and methane to condense. Due to their composition, the Jovian planets have a lower density than the terrestrial planets.

The Jovian planets

The cooler, outer portion of gas in the solar nebula condensed into solid particles, eventually resulting in the formation of the outermost planets. The outermost planets of Jupiter, Saturn, Uranus and Neptune are classified as Jovian planets. Their formation started with the accretion of dust, rock and ice. This accretionary mass may then have gravitationally pulled in the gases from the solar nebula. Having lower temperatures than the terrestrial planets and more abundant volatile chemicals, the Jovian planets could have grown by accretion to be ten or twenty times the mass of the Earth. Once a body had grown that large it had a stronger gravitational pull and was able to draw in large amounts of uncondensed gaseous material directly from the nebula. These gases caused an increase in size of the Jovian **planetesimal**. In turn, this increased the gravitational force, pulling in more gases. Because of their gaseous composition, these gas giants formed massive planets possessing a lower mean density. Their composition would also have resulted in the planets forming more rapidly than the terrestrial planets; only taking 3–10 million years to form.

The very outermost planets had a higher mean density because of the much lower temperatures, being so far from the Sun. This allowed carbon to form carbon monoxide gas, which was blown away. This loss of carbon prevented less dense methane from forming.

The terrestrial planets

If the elemental structure of stellar bodies, such as the Sun, is compared with the terrestrial planets of Mercury, Venus, Earth and Mars, we find they are different. Stellar bodies are composed of the elements hydrogen and helium, while the terrestrial planets are composed mainly of the elements carbon, oxygen, silicon, iron and a little hydrogen and helium. It is thought that most of the water and other volatile chemicals on the Earth's surface were delivered from the outer solar system by comets early in the history of the Earth.

A recent hypothesis states that the elements carbon, oxygen, silicon and iron are made from hydrogen and helium atoms during nuclear fusion within stellar bodies.

As the nebula disc shrank with the formation of a central protostar, thermal energy was released from the protostar, increasing the temperature to 2000°K. As the cloud changed in to a disc, with the protostar in the middle, it also began to cool. The temperature of the nebula disc decreased as the distance from the Sun increased and a variety of elements condensed. At around 1500K elements such as calcium, aluminium and titanium condensed, followed by iron, nickel and silicates. These elements make up what we call rocky material. Eventually, at low temperatures, water, ammonia and methane condensed to form ices. These ices were the most abundant chemicals within the nebula disc. The inner solar system never got cold enough to prevent volatile chemicals (such as ammonia, water and methane) being blown away before they could condense and form ices. Therefore, the inner planets (Mercury, Venus, Earth and Mars) were made out of rocky material. As a result of being made out of rocky material, the density of the terrestrial planets is, on average, three times greater than that of the Jovian planets.

All the planets were formed by the process of accretion. However, due to the variation in the material in the nebula disc and in the temperature, the outer planets are mainly gaseous, while the inner planets were formed from solid, rocky particles. The terrestrial planets formed when tiny grains of condensed material clumped together to build up huge solid bodies. This process occurred in the following stages:

1 Particles of about 1 micron in size collided and stuck together due to electrostatic charges, forming particles of up to 1 cm in size.
2 Particles of 1 cm in size collided with each other, forming bodies of around 1 km in size.
3 As the bodies became larger, smaller material was drawn to them due to their attraction to the larger objects through the pull of gravity. These accreting bodies formed distinct zones in the disc-shaped nebula cloud. The largest accretionary bodies reached a size large enough to be called planetesimal bodies, which in turn would form the planets of Mercury, Venus, Earth and Mars. These bodies began to dominate, collecting the remaining small bodies as they collided with them.

REVIEW ACTIVITIES

1 Describe how the composition of the Sun differs from that of Venus.

2 Explain why the Jovian planets were able to grow to a much larger size than the terrestrial planets.

3 Describe the order in which elements condensed on a planet such as Mercury.

4 Why do the outer planets have a lower density than those closer to the Sun?

5 Recount the steps in the process of accretion of the inner planetary bodies.

6 Explain the difference between a planetisimal and a planet.

EXTENSION ACTIVITIES

7 How many isotopes of hydrogen are there and in what ways do they differ from each other?

8 Explain how the gases in space can be 'blown away'. What are the winds that blow away the gases and where do they originate?

PROCESSES SHAPING THE TERRESTRIAL PLANETS

Once the terrestrial planets had been formed by the process of accretion, several other processes and factors influenced their shape. These included melting, volcanism, Sun–planet distance and, related to the latter, the presence or absence of a hydrosphere and biosphere.

Melting

The process of melting resulted in the formation of the different layers within the planet. The densest substances formed the innermost layer and the least dense substances formed the outermost layer.

When moving bodies collide, the kinetic energy that is gained from the motion of the body is converted into heat energy. As the process of accretion in the protoplanets grew, the collisions became larger and more heat was released. The terrestrial protoplanets started to melt and substances in these early planets either sank or floated to the surface, depending on their density. Iron and nickel, having greater densities, sank through the lighter material until they came to rest in the centre of the planet, forming the layer called the core. The lighter liquids, rich in lighter elements essential for life, floated to the surface. They formed the mantle and crust. These lighter elements included potassium, sodium, phosphorus, aluminium, silicon and oxygen. During and after this melting phase the terrestrial planets and the Moon continued to be hit by meteorites. (See Plate 7.) The massive, continuous impacts finished about 4 billion years ago. However, the terrestrial planets and the Moon continue to be hit by the occasional meteorite or meteorite shower. With the reduction of meteorite impacts, less heat energy was produced. This resulted in a cooling of the planets, causing the melting process to cease.

Volcanism

Volcanism is the process by which molten rock, ash, and steam and other gases are brought to the surface of the planet from the hotter interior.

After partial melting, the interior of the terrestrial planets remained hot because the radioactive elements present released radiation. All four planets are cooling, but the rate of cooling is determined by the size of the planet. Venus and Earth are large and are therefore cooling more slowly. They are able to retain heat for longer than the smaller terrestrial planets of Mars and Mercury, which have already frozen. Also, being larger, Venus and Earth have bigger reservoirs of radioactive substances. As Venus and Earth are still slowly cooling, the centre of the planets are hot and so volcanism is still occurring. However, Mars and Mercury are volcanically dead.

Sun–planet distance

Whether or not water can exist on a planet is dependent on the planet's distance from the Sun. If water can exist, the Sun–planet distance also determines whether it exists as a solid, liquid or gas because the distance determines how much heat energy the planet receives from the Sun. If the Sun–planet distance is great enough, oceans could form; further away ice sheets could form; while even further away it would be too cold for liquid water and water vapour to exist and there would be an absence of water in any form. The planets of Mercury and Venus are too hot for water to exist as a solid, liquid or gas because they are close to the Sun. The average surface temperatures of these two planets are 260°C and 480°C, respectively. Venus has water vapour in the atmosphere, but the average surface temperature is too high for liquid water. Of the four terrestrial planets, Mars is the furthest from the Sun. At –60°C, it is too cold for liquid or gaseous water to exist. Its temperature variation allows it to have some ice, but no liquid water. Earth is the Goldilocks planet. Venus and Mercury are too hot, Mars is too cold, but Earth is just right!

PRESENCE OR ABSENCE OF A HYDROSPHERE AND BIOSPHERE

Of the four terrestrial planets, only Earth has water on its surface as a liquid. As a result, Earth is the only planet that has a **hydrosphere** and **biosphere**.

The hydrosphere and biosphere play essential roles in the biochemical cycles that control the composition of the Earth's atmosphere. On the Earth, plants and microbes (such as bacteria) make organic matter from carbon dioxide and water and release oxygen, through the process of photosynthesis. The burial of organic matter in sediment removes carbon from the atmosphere but adds oxygen to the atmosphere. Venus has no life or buried organic matter; all the carbon dioxide is still in the atmosphere. As a result, there is a huge greenhouse effect because the carbon dioxide traps heat, preventing it from escaping into space.

The Earth came into being a long time ago. The shape and structure of planet Earth and the diversity of the organisms living in the hydrosphere and biosphere are the results of the planet-shaping processes.

definitions

hydrosphere
the totality of the Earth's water, including oceans, lakes, rivers, ice, snow and underground water

biosphere
the sum total of the Earth's organisms and organic matter

REVIEW ACTIVITIES

1 Explain why the early stages of the terrestrial protoplanets were hot enough for melting to occur.

2 Why do radioactive elements have an effect on the temperature of the planet?

3 A planet's distance from the Sun is vital for life. Explain why only Earth, not the other terrestrial planets, appears to be suitable for life. What effect does the distance have on the possible availability of chemicals, such as water, oxygen and carbon dioxide?

EXTENSION ACTIVITIES

4 Explain what evidence there is to support the idea that volcanism has occurred on each of the terrestrial planets.

5 What features of Mars makes it a possibility that life could have existed on that planet?

Table 1.3.1 The characteristics of the spheres and layers of the Earth

Sphere	Layer	State of material	Characteristic of material	Approximate thickness (km)	
Atmosphere	Thermosphere	Gaseous	Gaseous	500	
	Mesosphere	Gaseous	Gaseous	22	
	Stratosphere	Gaseous	Gaseous	25	
	Troposphere	Gaseous	Gaseous	18	
Lithosphere	Crust	Solid	Strong, rigid rock	55	**100**
	Upper mantle	Solid		45	
Asthenosphere	Upper mantle	Semisolid	Weak, plastic rock	250	
Mesosphere	Lower mantle	Semisolid	Strong, plastic rock	2605	
Core	Outer core	Liquid	Molten iron and nickel	2257	**3488**
	Inner core	Solid	Solid iron and nickel	1231	

STRUCTURE OF THE EARTH

The Earth is divided into a number of layers and spheres with different physical properties. (See Plate 8 and Table 1.3.1.) Its outermost sphere is the atmosphere, followed by the lithosphere, the asthenosphere and the mesosphere. At the centre is the core.

The **atmosphere** is the sphere surrounding the Earth, and reaches approximately 560 km from the Earth's surface. It contains a number of gases and helps protect the surface of the planet from harmful radiation and meteorites. (The atmosphere and the other spheres will be discussed in more detail in chapter 1.4.)

Below the atmosphere lies the **lithosphere**, which is about 100 km thick. This is the strong, rigid, rocky outermost part of the 'solid' Earth. Within this sphere there are two very important zones: the hydrosphere and the biosphere. The hydrosphere contains water in all its forms: oceans, lakes, streams, underground water, snow and ice. The other major part of the lithosphere is the biosphere, which is the totality of living matter on the planet. (See Plate 6.)

Under the lithosphere is the **asthenosphere**. The asthenosphere is a weak layer. Here the rocks are close to melting point, with some small pockets of molten rock, and therefore they have little strength. The asthenosphere is 250 km thick.

Below the asthenosphere is the **mesosphere**, which is composed of the rest of the mantle. The temperature of the mesosphere is very high, due to its proximity to the core. The rocks in this sphere are very strong due to the compression forces from the overlying layers.

Within these spheres, there are a number of compositional layers. These layers make up the actual planet. For example the lithosphere—the solid outer shell of the Earth—includes the **crust** and the uppermost rigid layer of the mantle, which is the region just below the crust that extends all the way down to the Earth's core.

Figure 1.3.4 A view of Earth from space.

There are two types of crust: the oceanic crust and the continental crust. The continental crust has a lower density than oceanic crust and is made up of granitic rock. The continental crust is on average 45 km thick. Oceanic crust has a higher density than continental crust and is mainly composed of basaltic rock. This crust is approximately 8 km thick. The continental crust, having less density than the oceanic crust, 'floats' higher than the oceanic crust and forms the land.

Under the crust there is a layer called the **mantle**, which is up to 2900 km in thickness. It contains rock at very high temperatures. The upper rigid part of the mantle forms part of the lithosphere, the asthenosphere is below this uppermost part of the mantle, while the lowest part of the mantle is the mesosphere.

The rocks within the asthenosphere are solid despite the very high temperature, although there are small pockets of molten rock. The rocks are able to flow like liquid and behave like a semisolid despite being solid. This is due to the very high compression forces acting in this layer. The flowing takes place over very long periods of time.

The mesosphere is approximately 2530 km thick, making it the largest part of the mantle. Although the temperatures of the rock are extreme, the rocks are very strong because they are so highly compressed.

At the centre of the planet is the **core**. It has a radius of 3450 km; the outer core has a radius of 2200 km while the inner core has a radius of 1250 km. The outer core is thought to be composed of iron and nickel. The inner core is solid while the outer core is liquid.

The Earth's core was formed very early in Earth's evolution as heavier molten iron sank towards the centre of the planet. As the Earth cooled and dissipated its internal heat towards the surface, some molten iron began to solidify to create the dense, solid inner core. Enormous pressure and temperatures keep the outer core liquid. Fluid iron has continued to solidify at the boundary between the two cores, so that over 1 billion years the inner core has grown steadily to its present diameter.

The two cores are the source of the planet's magnetic field. The magnetic field is lines of force that extend between the Earth's poles. It helps protect the Earth from potentially hazardous solar winds.

definitions

atmosphere
the gaseous envelope surrounding the Earth

lithosphere
the solid outer shell of the Earth

asthenosphere
the layer or shell of the Earth that lies directly below the lithosphere and behaves like a semisolid

mesosphere
the shell below the asthenosphere composed of rocks at a very high temperature which are very strong and highly compressed

crust
the crust forms most of the lithosphere and is made up of solid rock

mantle
the semiliquid layer below the crust

core
the innermost shell composed of a liquid outer layer and a solid inner layer

REVIEW ACTIVITIES

1 Refer to Table 1.3.1. Which Earth sphere is the thickest?

2 Compare oceanic crust with continental crust.

3 Compare and contrast the characteristics of the material found in the asthenosphere and the lithosphere.

4 Explain why the outer core is liquid and the inner core is solid.

5 Apart from temperature, what factors have had a major influence on the structure of the planet?

EXTENSION ACTIVITIES

6 The biosphere is the sum total of all living biomass on the planet. How far into the atmosphere does it extend and to what depth are living things found?

7 The inner and outer cores of the Earth are responsible for the formation of the Earth's magnetic field. Research how the magnetic field is formed.

DENSITY PARTICLE MODELS

Atoms and molecules are so small that it is impossible to actually see their arrangement in a solid, liquid or gas. The density particle model is an attempt to explain the arrangement of atoms, or molecules, in each of the three states of matter: solids, liquids and gases. The arrangement of the atoms in each of the states varies depending on the level of energy in those atoms in the solid, liquid or gas. As the level of energy increases, it allows the atoms to move more freely and the distance between the atoms to increase. The **density** of a substance is dependent on how closely the atoms are packed together. Following are density particle models for each of the states of matter.

definitions

density
the mass per unit of volume of an object. Density is calculated as follows:

$$\text{Density} = \frac{\text{Mass}}{\text{Volume}}$$

weight
the force with which an object is attracted to the Earth. This attraction will vary depending on the gravitational pull on the object.

mass
the quantity of matter or atoms found in an object when the force of gravity is not acting upon it

Solids

In a solid the atoms are generally closely packed. Each atom oscillates with the energy it possesses. As the energy levels increase, the atoms oscillate more. This results in them being less closely packed, producing a substance or object with a lower density.

Liquids

As the energy increases, so do the oscillations of the atoms so that the compaction is broken, allowing the atoms to move around slightly and randomly. This movement is greater than the movement found in the atoms that form a solid. There the atoms are unable to move around, but they are able to oscillate slightly. The atoms at the edge of the liquid form a loose attraction to each other, creating a film.

Gases

In a gas, the atoms possess more energy than those in a solid or liquid. The compaction of atoms that occurred in a solid is broken and, as the atoms possess more energy and move around more, there is a greater distance between the atoms of a gas. There is a large amount of space between the individual atoms. The atoms of a gas are able to move around randomly.

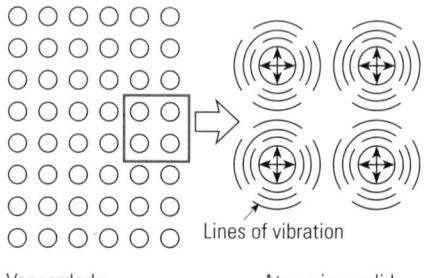

Figure 1.3.5 Arrangement of atoms in a solid.

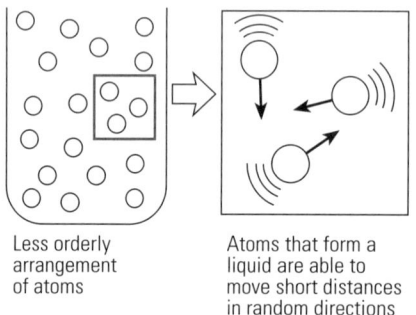

Figure 1.3.6 Arrangement of atoms in a liquid.

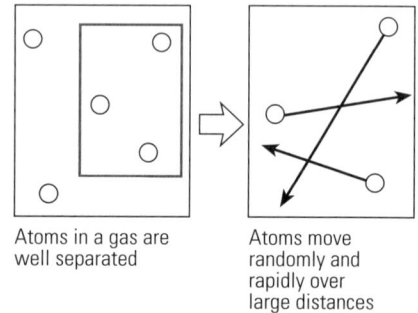

Figure 1.3.7 Arrangement of atoms in a gas.

SUMMARY

- The Sun formed about 4.7×10^9 years ago from a cloud of gas and dust, or nebula. The nebula was triggered by a supernova explosion from a second-generation star.

- The Sun released heat and light due to nuclear fusion reactions inside it.

- There are two types of planets: Jovian, or gas, planets; and terrestrial planets.

- The outer planets are the Jovian planets and they are formed from gases.

- The four inner planets are the terrestrial planets and they are formed from solid, rocky material.

- The Jovian planets have a higher mean density than the terrestrial planets.

- The outer planets are much colder than the inner planets.

- Planets are formed by the process of accretion of material from the dust and gas disc called the solar nebula.

- The Earth and the four other terrestrial planets became layered because they were hot and the material on the planets melted and then sank or floated according to their density.

- The planet forming processes for planet Earth were accretion, melting, volcanism, Sun–planet distance and formation of the biosphere and hydrosphere.

- Density is mass per unit of volume. It is the arrangement of the atoms inside matter.

PRACTICAL EXERCISE
Density of Earth materials

In this exercise you will measure the density of a selection of Earth materials representing three layers of the Earth: core, mantle and crust.

Equipment
- Electronic balance or triple beam balance
- Measuring cylinder
- Earth materials: iron nail or cube of iron; pieces of basalt, granite and serpentine; piece of limestone or marble or a cube of aluminium

Procedure
1 Design a method to measure the density of a selection of Earth materials using the equipment listed.

2 Once you have completed the experiment, complete the following activities.

ACTIVITIES

1 For each layer of the Earth, choose a material that represents it. List the layer, material and a reason for choosing it.

2 List the materials in order of density, from most to least dense.

3 Pumice is an Earth material released from volcanoes. It is very light. Design a method for measuring its density.

PRACTICAL EXERCISE
Locating meteorite impact sites in Australia

Photocopy the map of Australia (Figure 1.3.8). Plot the following meteor impact sites. The coordinates of each site are provided here to assist you.

1 Acracamp 32°S, 135°E
2 Tookoonooka 27°S, 143°E
3 Gosses Bluff 23°S, 132°E
4 Kelly West 20°S, 135°E
5 Teague 26°S, 121°E
6 Connolly Basin 23°S, 125°E
7 Goat Paddock 19°S, 127°E
8 Picaninny 18°S, 129°E
9 Spider 17°S, 125°E
10 Strangways 11°S, 134°E
11 Lawn Hill 19°S, 139°E

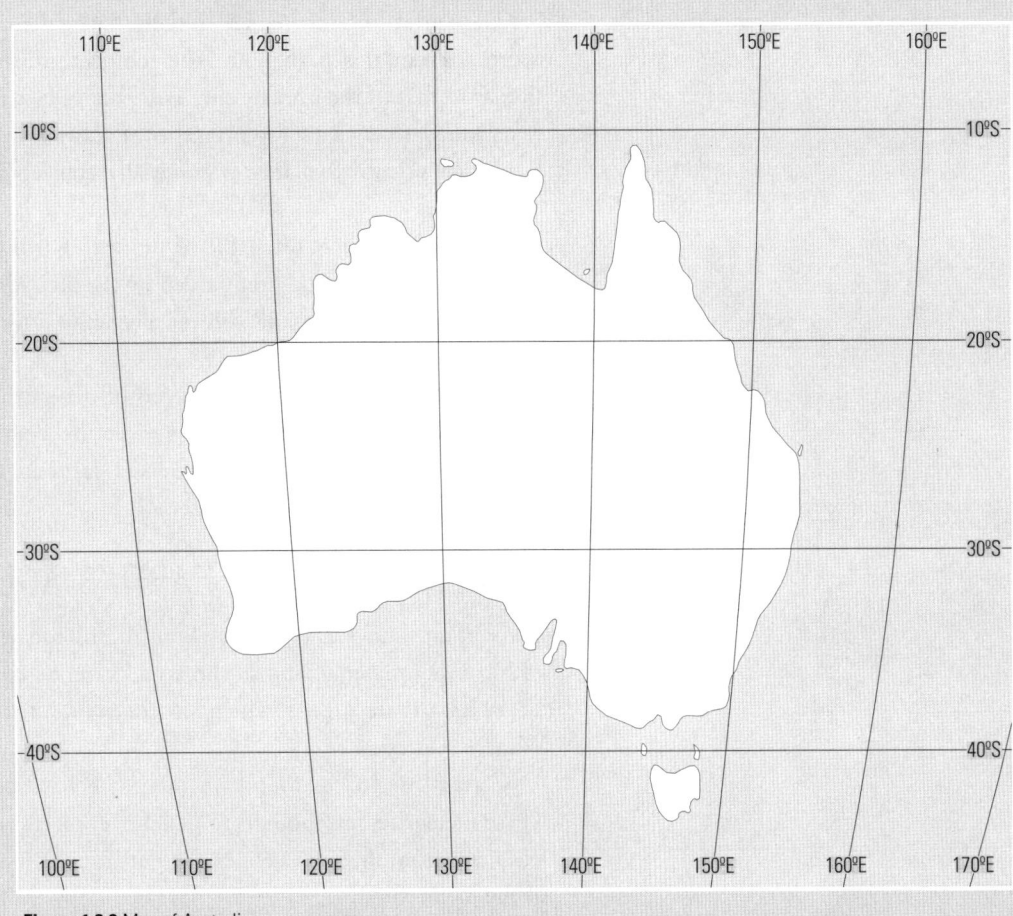

Figure 1.3.8 Map of Australia.

1.4 The early Earth and its atmosphere

OUTCOMES

At the end of this chapter you should be able to:

- describe the composition of the early atmosphere and compare it with the composition of the present atmosphere

- identify data sources, process and present information from secondary sources, to use available evidence about the Earth's earliest atmosphere and compare with the present atmosphere.

THE EARLY EARTH

The Earth formed about 4.5 billion years ago. However, it was not the Earth we would recognise today. The early Earth was a hostile, hot, and seismically and volcanically active planet. The soil-forming processes had not started and so the Earth's surface was just bare rock. If the early Sun had been as hot as it is today, it is likely that the planet would have remained unfit for life to have evolved. Instead, in the early stages of its life cycle the Sun was only 80% as hot as it is today.

The surface of the planet initially had a temperature of several thousands of degrees. From this high point the temperature dropped slowly to a temperature similar to that of the present day. This change in temperature took many millions of years and was very slow for a number of reasons:

- the newly formed planet was initially subject to continuous bombardment by meteorites
- the thin atmosphere offered little protection from the heat energy of the Sun
- the Earth's surface was bombarded by cosmic radiation coming from space
- the highly seismically and volcanically active crust released hot lava and gases onto the surface
- the radioactive elements in the layers of the Earth had a heating effect.

The surface of the planet was riven by countless volcanoes and shaken by numerous earthquakes. There was no soil, just bare rock. At this early stage there was a complete absence of surface water.

EVIDENCE FOR THE EARTH'S AGE

The earliest known crystals are about 4.2 billion years old. The crystals are located within rocky material, and so those rocks can be dated at more than 4 billion years. These crystals are zircons, which have been found at Mount Narryer in the Pilbara in Western Australia.

Certain meteorites found on the Earth have been dated to 4.56 billion years. These meteorites are thought to be leftover pieces of the solar nebula. As the planet was made from the solar nebula, the Earth's age cannot be very different from the age of the meteorites.

EARTH'S EARLY ATMOSPHERES

Like other important parts of the Earth's systems, the atmosphere we know today did not appear overnight. The very first atmosphere present during the accretion of the planet was a thin layer of silica-based compounds. This layer was blown away very quickly by the solar winds. This early atmosphere was rapidly replaced with another thin primary atmosphere of hydrogen, nitrogen, argon, neon and helium. This atmosphere was also blown away by the solar winds.

The Earth went through the process of differentiation, whereby the denser material sank towards the centre of the planet. As the planet cooled, gases and water vapour were released from the Earth's interior. This process was called **outgassing**. The gases that were released began to form a thicker atmosphere than previous atmospheres. The atmosphere was now thick enough to not be blown away by solar winds. The composition of this secondary atmosphere was largely water vapour, nitrogen, methane, hydrogen chloride, and some sulfur compounds, ammonia and carbon dioxide. Sampling the outgassing from modern volcanoes has shown that the gases released are approximately in the same proportions as the gases in the atmosphere. One of today's main atmospheric gases, which was almost completely missing, was oxygen.

Some scientists have put forward the idea that water vapour did not only come from the Earth's interior but possibly was also brought in by comets. Most scientists believe that the water vapour was the result of outgassing. Some elements are known to have been brought in by meteorites. It is thought that certain elements and compounds have their origin outside the planet. Meteorites have brought elements, such as silicon, iron and nickel. Compounds carried in by meteorites include simple amino acids and a variety of organic compounds. Some of these organic compounds were important constituents in the reactions required to produce the early stages of life on the planet.

definition

outgassing
the escape of gases from volcanic vents

REVIEW ACTIVITIES

1
Describe what the surface of the early Earth would have been like around 4.5 billion years ago.

2
Explain why there was no soil on the surface of the early Earth.

3
Explain why scientists consider that the planet was formed around 4.5 billion years ago.

4
Outline how the composition of Earth's early atmosphere changed.

5
Why is outgassing so important, and how do we know what gases were released?

EXTENSION ACTIVITIES

6
Explain the term 'solar winds'.

7
a Describe the origins of meteorites that hit the Earth.
b What are the main types of meteorites?

FORMATION OF THE OCEANS

As the Earth continued to cool, the temperature dropped sufficiently to allow water to condense and form bodies of open water. Water vapour also began to condense and it started to rain. Around 4 billion years ago the climate of the planet would have been similar to tropical monsoonal weather: very hot and humid. This rainfall may have continued for many thousands or millions of years and may have resulted in the formation of the oceans.

As the rain continued to fall, it caused a change in the climate. The atmosphere had high levels of carbon dioxide. (This gas, along with some of the other atmospheric gases such as methane, are known as 'greenhouse gases'.) This would have caused the atmosphere to act as an insulator, reflecting the heat back towards the Earth's surface, effectively trapping in the heat. This is called the 'greenhouse effect'. The surface temperature would have been relatively high because the Earth was still cooling down after its formation. The cooling would have been slowed by the insulating atmosphere. The rain would have washed some of the carbon dioxide out of the air. The net result was that the greenhouse effect was reduced and the planet's surface temperature dropped. The drop in surface temperature allowed more rain to fall, because colder air can hold less water vapour. The higher rainfall would have washed out more carbon dioxide, resulting in even higher rainfall. Eventually a balance was achieved.

The early rain was probably highly acidic. As the rainwater reacted with gases (such as carbon dioxide, sulfur compounds, nitrogen, and chlorine gases) it produced such acids as carbonic, sulfuric, nitric and hydrochloric. These acids would have reacted with the rocks of the crust, causing chemical weathering (breakdown of rocks by chemicals). As these reactions took place, the acids were neutralised. The products of neutralisation and of chemical and physical weathering processes (such as wind erosion, action of water and changes in temperature) led to the formation of sediments.

The Earth's oceans were probably formed around 4 billion years ago. The evidence for this is found in rocks of that age, which resemble rocks formed in the oceans today.

The formation of the oceans and atmosphere took place over many millions of years. During this time the process was fuelled by the outgassing from volcanoes.

REVIEW ACTIVITIES

1
The oceans were formed by continuous rain over many years. Describe what effect this rain would have had on the global climate.

2
Describe what occurs in the process of chemical weathering.

3
What acids are involved in chemical weathering?

4
Explain how geologists know that the oceans were formed around 4 billion years ago.

5
Explain why the oceans did not form directly after the formation of the planet.

EXTENSION ACTIVITIES

6
Explain what compounds are formed when an acid is neutralised.

7
Compare the main causes of greenhouse gases in the modern atmosphere to those of the atmospheric greenhouse gases of 4 billion years ago.

BANDED IRON FORMATIONS

Early banded iron formations may have formed when there was little or no atmospheric oxygen. Banded iron formations exist in rocks dating from 1.8–2.5 billion years ago. They are very large accumulations of iron-rich sedimentary rock that formed in shallow, elongated marine troughs or basins. These rocks are composed of alternating layers of iron-rich material (mainly magnetite) and silica.

A number of theories have been put forward to explain this phenomenon. The most widely accepted explanation states that iron (Fe^{2+}) and silica were in solution in the water. The iron was in its reduced state. Precipitation (formation of an insoluble mixture of the two elements) occurred. The amounts of iron varied on a seasonal basis but the silica levels remained constant. The variation in iron levels led to banding occurring.

A second explanation states that when the oceans first formed, the waters dissolved vast quantities of iron. The presence of the iron was due to the weathering of rocks in an environment that had no oxygen. As there was no oxygen, the iron was able to enter the water as iron ions and not combine with oxygen as happens in an oxygen-rich atmosphere. At the same time, primitive photosynthetic blue-green algae were beginning to proliferate in the surface waters. As the algae photosynthesised, waste oxygen was released. This free oxygen combined with the iron ions to form magnetite (Fe_3O_4), an iron oxide. The iron oxide precipitated out and sank to the bottom of the oceans, forming an iron-rich layer of iron oxide. The waste oxygen rose to toxic levels, killing the algae. The algae then sank to the bottom, forming an iron-poor silica layer. The silica came from the structure of the organisms. As the algal population declined it caused the oxygen levels to drop due to the lack of photosynthesis. The drop in oxygen levels allowed the algae population to increase again and the process repeated.

REVIEW ACTIVITIES

1 Recount the process by which banded iron formations were made.

2 Explain what magnetite is.

EXTENSION ACTIVITY

3 Are all banded iron formations found in very old rocks? If not, give an explanation for their formation.

THE MODERN ATMOSPHERE

The modern atmosphere developed from the early secondary atmosphere. The living processes, such as photosynthesis, allowed the amount of oxygen to increase. **Nitrogen-fixing bacteria** may have had an effect on nitrogen levels, while the decomposition of organic matter may also have had an effect on the proportions of gases in the atmosphere. (See Plate 9 and Tables 1.4.1 and 1.4.2 and Figure 1.4.1 on page 46.) Carbon dioxide levels in the atmosphere have varied throughout Earth's history. The fluctuations in the levels of carbon dioxide and other greenhouse gases have resulted in global warming (where greenhouse gases are at a high level) and global cooling (where greenhouse gases are at a low level). Increases in carbon dioxide levels have occurred throughout history and this is not a new change caused by humans. Large amounts of carbon dioxide have entered the atmosphere due to volcanic action, large-scale fires (such as major bushfires) and other natural processes.

nitrogen-fixing bacteria
a type of bacterium that is able to convert gaseous nitrogen to a more useable nitrate form

Table 1.4.1 Changes in atmospheric gases during Earth's history

Main atmospheric gases*	Earth's atmospheres and major events in their formation				
	First atmosphere		Second atmosphere	Third atmosphere	Present atmosphere
	Planetary formation	Early Earth cooling	Outgassing from volcanoes	Photosynthesis	Modern, relatively stable conditions
Nitrogen		X	X	X	X
Oxygen				X	X
Carbon dioxide			X	X	X
Hydrogen		X			
Hydrogen sulfide			X		
Hydrogen chloride			X		
Hydrogen bromide			X		
Hydrogen fluoride			X		
Ammonia			X		
Water			X	X	X
Carbon monoxide			X		
Methane		X	X		
Helium		X	X	X	X
Argon		X	X	X	X
Neon		X	X	X	X
Silicon-based compounds	X				

*Gas present in atmosphere

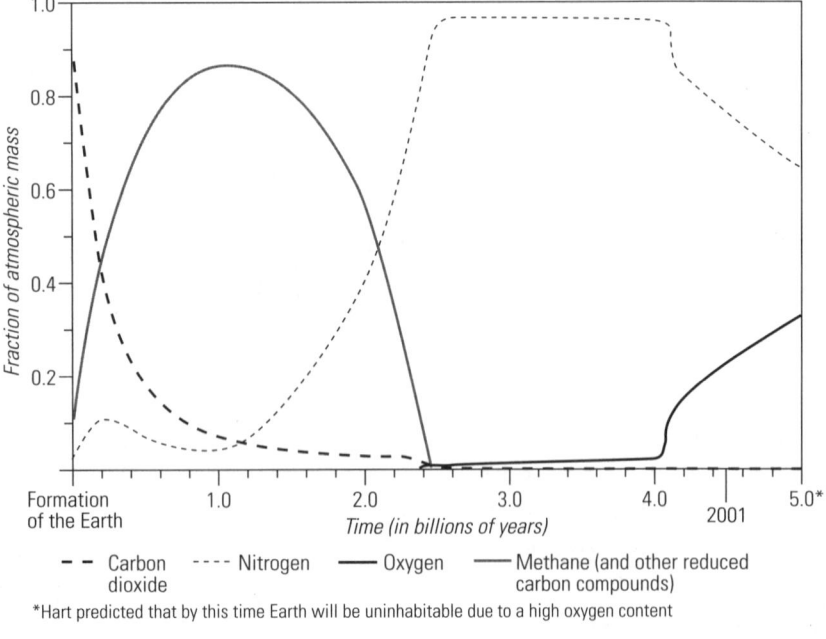

Figure 1.4.1 Hart's model of atmospheric gases during Earth's history.

*Hart predicted that by this time Earth will be uninhabitable due to a high oxygen content

Table 1.4.2 Percentages of gases in the modern atmosphere

Fixed gases in the atmosphere	Percentage of total atmospheric gases
Nitrogen	78.08
Oxygen	20.95
Argon	0.93
Carbon dioxide	0.035
Neon	0.0018
Helium	0.00052
Methane	0.00014
Krypton	0.00010
Nitrous oxide	0.00005
Hydrogen	0.00005
Ozone	0.000007
Other gases	0.002342
Additional variable substances in the atmosphere as an equivalent total percentage of atmospheric gases	
Water vapour	Between 1 and 4
Aerosols	Unknown quantity

At present scientists are gathering evidence showing how modern humans can affect the composition and function of the atmosphere. These effects can be seen by the changes in proportions of greenhouse gases, causing an increase in the greenhouse effect. This change is due to people's industrial, agricultural and domestic actions. People are also causing ozone depletion in the upper atmosphere, which is increasing the amount of ultraviolet radiation striking the Earth's surface. This has led to a higher incidence of skin cancers in humans and other organisms, as well as to changes in cellular genetic material, resulting in an increase in mutations.

There are three main types of constituents in the modern atmosphere: the gases, **aerosols** and water vapour. Aerosols are liquid droplets or solid particles that are so small that they remain suspended in the air. These aerosols include volcanic ash, sea salt crystals and smoke particles.

> **aerosols**
> liquid droplets or solid particles that remain suspended in the air

REVIEW ACTIVITIES

1 Explain the causes of the increase in greenhouse gases in the atmosphere.

2 What effects have greenhouse gases had on the planet?

3 Look at Hart's model of atmospheric gases (Figure 1.4.1). Explain why the carbon dioxide line seems to disappear.

4 Refer to Table 1.4.2. Recount the three main constituents of the atmosphere.

EXTENSION ACTIVITY

5 A number of gases are regarded as 'greenhouse gases'. Carry out research and find out the names of the most common greenhouse gases.

THE LAYERS OF THE ATMOSPHERE

The atmosphere is made up of four main layers. Each of these layers is separated from the next by thermal boundaries, which are called **pauses**. For example the boundary between the troposphere and the stratosphere is called the tropopause.

The **troposphere** is the layer closest to the Earth's surface. It is within the troposphere that life exists. This layer contains most of the heat-absorbing gases, or greenhouse gases, approximately 80% of the mass of the atmosphere and all the water vapour and clouds. The troposphere acts as both an insulator and a reflector. It traps heat energy but also reflects back 30% of the solar radiation. It is within this layer that most of the world's weather occurs.

Above the troposphere and tropopause lies the **stratosphere**. The stratosphere contains approximately 20% of the atmosphere's mass. It is here in the stratosphere that the **ozone** (O_3) **layer** exists. The ozone layer, which absorbs the most dangerous of the ultraviolet rays, appears to be under threat from ozone-depleting chemicals being used by humans. Depletion of ozone increases the bombardment of the planet's surface by ultraviolet rays. These rays cause cancers and also affect the chromosomes of organisms.

The next layer above is the **mesosphere**. There is less than 1% of the atmosphere's mass in this layer. There is a major temperature drop in this layer, from approximately

> **pause**
> a thermal boundary between each atmospheric layer
>
> **troposphere**
> the first layer of the atmosphere above the Earth's surface
>
> **stratosphere**
> the second of four layers of the atmosphere; found above the troposphere
>
> **ozone layer**
> a layer within the stratosphere containing ozone. It absorbs harmful ultraviolet radiation.
>
> **mesosphere**
> the third layer of the atmosphere; found above the stratosphere

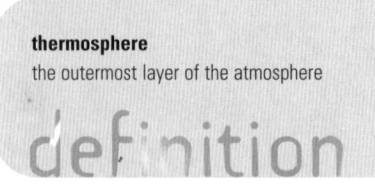

thermosphere
the outermost layer of the atmosphere

0°C down to –80°C. Its main function is to protect the planet from incoming meteorites. This layer is the first layer that is thick and dense enough to allow the effect of friction to destroy falling bodies, such as meteorites.

The outermost layer of the Earth's atmosphere is the **thermosphere**. This layer also contains less than 1% of the atmosphere's total mass. There is a major temperature increase, rising to 90°C. The thermosphere absorbs short wavelength ultraviolet rays.

REVIEW ACTIVITIES

1 Differentiate between a sphere and a pause.

2 Summarise the features of each of the atmospheric layers.

EXTENSION ACTIVITIES

3 Within the atmosphere there is found a layer called the Van Allen belt. Explain where it is located in the atmosphere and its function.

4 Explain the ways in which the atmosphere protects the planet.

CLIMATE CHANGE

How has the global climate changed?

The Earth was warmer 2 billion years ago and is now going through a cool period. Great ice ages occurred every 100 000 years, with small ice ages and cool periods punctuating the time between them. Conversely, there have also been hot periods. The exact cause of climate change at different times during Earth's history is unknown. However, some of the causes of climate change are discussed in the following section.

Causes of climate change

Ever since Pre-Cambrian times (up to 590 million years ago), large fluctuations in climate have occurred; ranging from periods characterised by ice ages to periods of global warmth. Ice ages have occurred at widely spaced intervals of geological time— approximately 200 million years apart. These ice ages lasted for millions, or tens of millions, of years. A range of factors have caused climate change over the Earth's history. These factors may be short term or long term, cyclical or random.

Some of these factors would have had a greater effect than others. Some changes in the climate may have been caused by a single factor, but it is more likely that most climate swings were caused by a combination of factors. These include:

- variations in solar output
- changes in the Earth's orbit (Milankovitch cycles)
- continental drift
- changes in the concentration of greenhouse gases in the atmosphere
- volcanic activity
- surface albedo, or reflectivity, feedback effect
- the amount of stellar dust in space between the Earth and the Sun, resulting in atmospheric albedo
- human activity
- sea level changes
- mountain building.

These factors will be discussed in more detail in the following pages.

Evidence has shown that global temperatures have gradually decreased over the last 60 million years and have fluctuated by between 5–10°C over the past 2 million years. These temperature fluctuations mean scientists can't definitely attribute the cause of the present change in global temperatures to modern, human-made greenhouse effects.

VARIATIONS IN SOLAR OUTPUT

Astronomers know that the Sun's output has not been constant throughout Earth's history. When the Earth first formed, the Sun's output was only 80% of what it is today. There is also some evidence that regular fluctuations in output do occur. Every eleven years the Sun undergoes a period of activity called the 'solar maximum', followed by a period of quiet called the 'solar minimum'. There appears to be a difference of 0.2°C in global temperatures between the solar minimum and the solar maximum. There also appears to be a correlation between Earth's temperature and sunspot cycles and a link between sunspot activity and seasonal changes. (See Figure 1.4.2.) Scientists are debating whether ice ages on Earth are related to fewer sunspots. In the late seventeenth century, sunspots were apparently absent for a number of decades. This coincided with the coldest phase of the Little Ice Age in Europe and North America.

CHANGES IN EARTH'S ORBIT

As the Earth orbits the Sun, there are subtle changes to the Earth's progress. This progress is affected by the Milankovitch cycles, which are **eccentricity**, axial tilt changes in the Earth's spin and **precession**.

The three variations in the Earth's orbit are thought to be partly responsible for past climatic changes. This is the basic idea proposed by the Serbian scientist Milutin Milankovitch in the 1930s. All three variations—eccentricity, polar axis and precession—affect the amount of solar radiation being received by the Earth, which in turn affects the Earth's climate, either heating or cooling the planet.

Eccentricity is the shape of the Earth's orbit around the Sun. The orbit gradually changes from being elliptical to nearly circular and then back to elliptical. This is referred to as the eccentricity cycle. This variation takes place over a period of 100 000 years. Currently we are in an orbit of low eccentricity; the orbit is nearly circular. Changes in the orbit results in a 20% variation in the amount of solar energy actually received by the Earth. Data analysis for the last 800 000 years of deep sea ocean sediments shows that ice coverage is at a maximum every 100 000 years, which matches the eccentricity cycle.

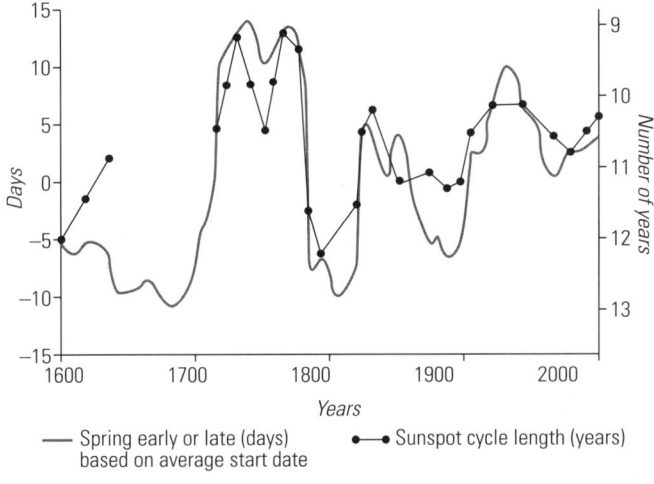

Figure 1.4.2 Sunspot cycle length and the effect on the start of spring.

The second orbital variation is due to changes in the Earth's axial tilt. The tilt, or roll, of the Earth's axis moves between 24.5° and 22.1°. The greater the tilt, the greater effect it has on the seasons. Polar regions receive less sunlight when the tilt is decreased and more when it is increased. This tilting is cyclical and it takes a period of 41 000 years to go through the whole cycle, that is from 24.5° to 22.1° and then back to 24.5°. At present the tilt is 23.5°. (See Figure 1.4.3, page 50.) When there is a small tilt, there is less climatic variation between summer and winter. Warmer winters result in more snow, because more water vapour can be held in the atmosphere. Cooler summers result in less of the snow and ice melting. Therefore, snow and ice accumulate and over a period of time there is a general cooling of the planet, resulting in an ice age occurring.

eccentricity
the shape of the Earth's orbit as it changes from an elliptical orbit to a circular orbit

precession
the wobble that the planet has as it spins on its axis

definitions

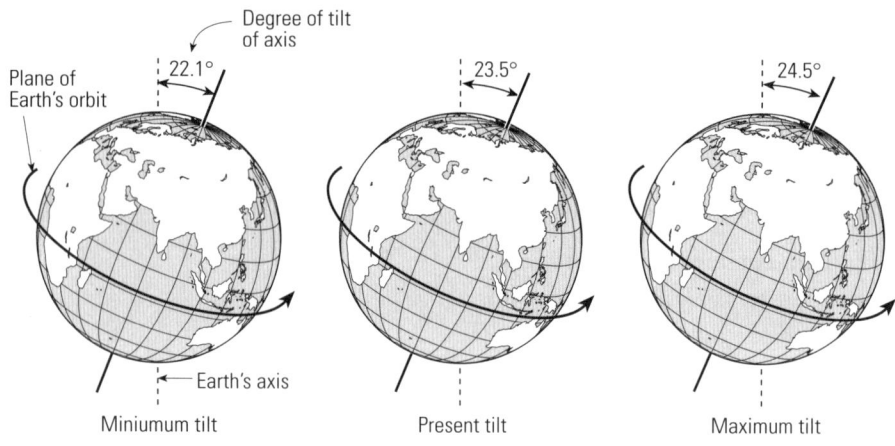

Figure 1.4.3 Minimum and maximum tilt of the Earth.

The third variation is related to the Earth's precession or wobble. (See Figure 1.4.4.) The planet wobbles in space like a slowly spinning top. Due to the wobble, the time of year at which the Earth is closest to the Sun varies. The point at which the Earth is closest to the Sun is called the **perihelion**. At present, the Southern Hemisphere's summer is in the perihelion, while its winter occurs at the furthest point in the orbit. This cycle takes 21 000 years to complete, and in approximately 10 500 years the position will be reversed; winter will be at the perihelion and summer will be at the furthest point.

As mentioned previously, the Earth's orbit is not perfectly round, but is slightly elongated. The Earth comes closest to the Sun in the first week of January. When the Northern Hemisphere experiences winter and receives the least amount of sunlight, the Earth as a whole receives the most; the swing is about 3° from peak to peak. This makes northern winters milder, since they occur when the Earth is most distant from the Sun. The opposite is true of the Southern Hemisphere. In January the Southern Hemisphere experiences summer. In the Southern Hemisphere, summers should be hotter and winters colder than in the Northern Hemisphere. However, the effect is weakened because most of the Southern Hemisphere is covered by oceans, which moderate the climate. At present, the northern winter occurs in the part of the Earth's orbit where the north end of the axis points away from the Sun. Since the axis moves around a cone, in 13 000 years it will point towards the Sun, putting it in midsummer just when the Earth is closest to the Sun. At that time the northern climate will be more extreme. More extreme winters produce more snow, which will reflect more heat and thus cause further cooling. The climate will change, leading to an ice age. Milankovitch believed that the cycles he had identified were not the only factors that may cause climate change because ice ages do not recur every 26 000 years. Instead, he believed the cycles were contributing factors.

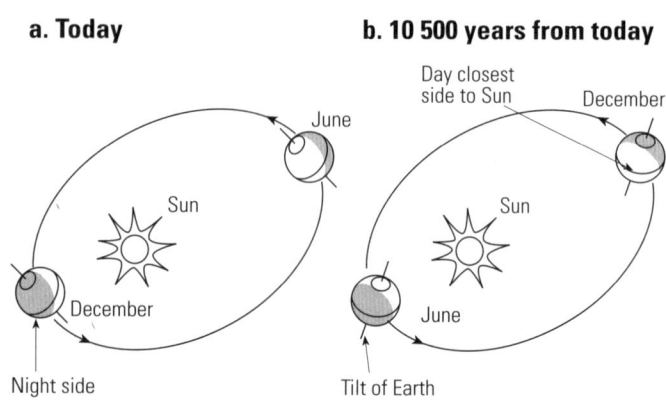

Figure 1.4.4 Precession, or wobble, of the Earth's orbit.

Acting independently of each other, the three factors change the amount and distribution of the Sun's energy over the surface of the planet at regular intervals. The consequences of this can be seen over the past 2.6 million years. The tilting of the planet dominated for the early part of this period, resulting in glaciation and interglaciation being of equal lengths. During the last 800 000 years the eccentricity of the Earth's orbit has dominated, resulting in the duration of glaciation becoming five to ten times that of the interglacial periods.

CONTINENTAL DRIFT

The continents are sitting on tectonic plates. As the plates move they take the continents with them. These travels have taken them from cold to tropical climes and, for some, back to cold climes. (See Plate 10.) Their own climate has changed depending on whether they are closer or further away from the poles or equator.

As the continents move, the pathways of the ocean currents change, the seas and oceans change shape, and passages open or close. This has had a significant effect on

definition

perihelion
the point on the Earth's orbit when it is closest to the Sun

the movement of hot and cold water currents. These changes affect the heat transfer around the world. As the ocean currents are one of the main means by which heat is moved around the surface of the planet, it results in climate change.

GREENHOUSE GASES

As mentioned previously, some causes of climate change are cyclical. An example of this is carbon. There are many forms and sources of carbon on Earth. Most carbon is locked up within molecules of various compounds. The sources of carbon are:
- organic carbon, which is found in material that has come from living or once living life forms
- inorganic carbon, which is found in rocks and minerals
- carbon atoms, which are found in gases such as carbon dioxide, methane and carbon monoxide.

As in most Earth systems, there is a connection between the various sources of carbon on the planet, which forms a cycle. (See Figure 1.4.5.) The carbon in one store is removed from it, transported to the next store and then converted to another form so that it can be stored again.

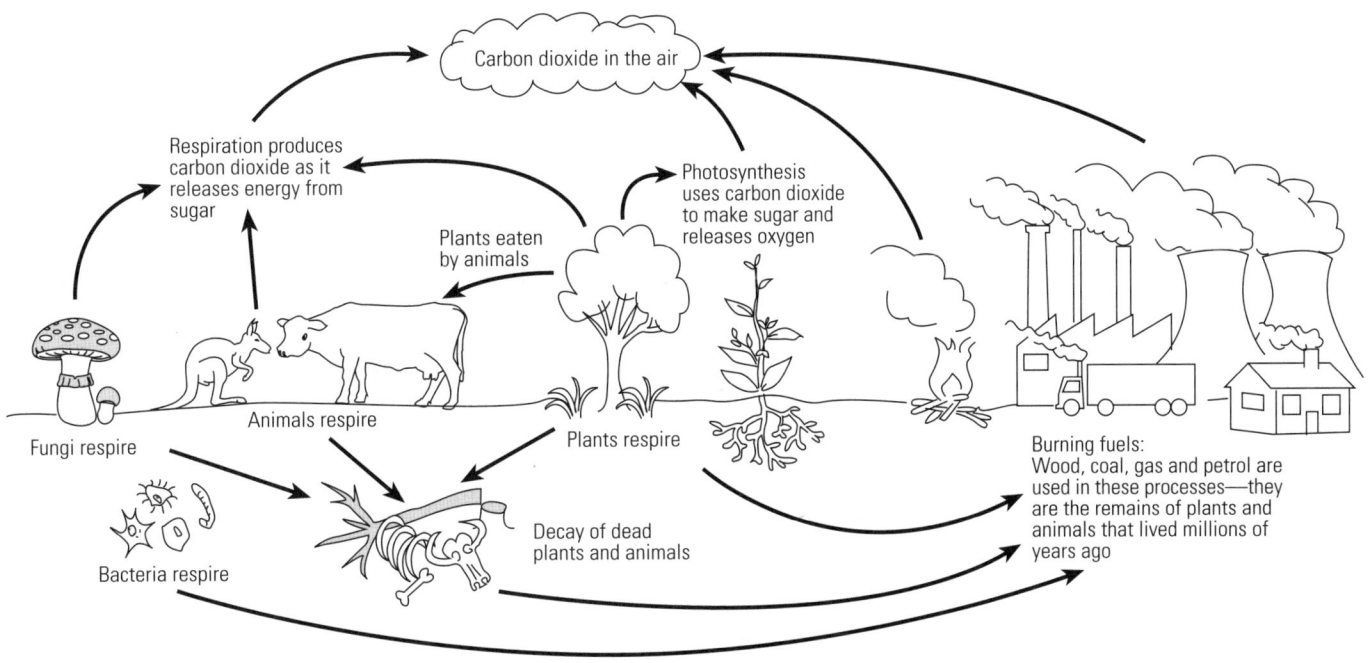

Figure 1.4.5 The carbon cycle.

Carbon was initially the atmospheric gas carbon dioxide. It was converted to carbohydrates and other organic compounds by plants. These organic compounds became fixed in rocks as fossil fuels, which are then burnt by humans. This releases the carbon to form carbon dioxide, which then re-enters the atmosphere. The cycle then repeats. Carbon dioxide, methane and carbon monoxide are all major greenhouse gases.

The greenhouse gases have an important effect on the climate of the Earth. The Earth's temperature is maintained by a balance between incoming radiation and outgoing radiation. (See Figure 1.4.6, page 52.) Trapping the radiation causes a natural greenhouse effect and makes the planet habitable.

The global temperature would be 33°C cooler than it is if not for the natural greenhouse effect. Humans are causing an increase in the greenhouse effect by deforestation, agricultural and industrial practices, the burning of fossil fuels and the release of CFCs. This increased greenhouse effect will produce higher surface temperatures and delay the onset of the next glacial period.

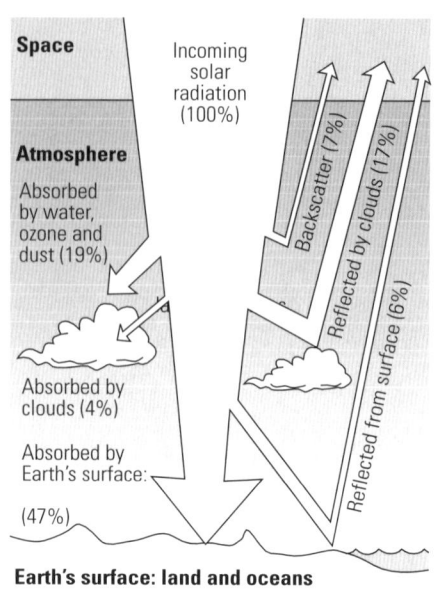

Figure 1.4.6 The balance between incoming and outgoing radiation.

Currently the planet seems to be entering a period of global warming. This episode of global warming appears to be caused by human activities. In the past, global warming occurred a number of times, and some episodes were far more severe than the present one.

VOLCANIC ACTIVITY

Volcanic activity has been slowing over the period of the planet's history. Therefore, over time its effects on the climate have decreased. Volcanoes, however, do still have an effect. When the Indonesian island of Krakatau erupted in 1883, it turned summer into winter in the Northern Hemisphere. The recent eruption of Mount Pinatubo in the Philippines caused a global drop in temperature of about 0.15°C. These changes only last for a short period, continuing for no longer than days, weeks, months or a few years.

SURFACE ALBEDO FEEDBACK EFFECT

The Earth's surface **albedo**, or reflectivity, is affected by its surface features. These can be ice sheets, water, aerosols, clouds, rock surfaces and vegetation. Land and ocean albedos are much lower than the albedos of snow and sea ice. Changes in snow or ice coverage have a huge effect on surface temperatures. Ice is white in colour and reflects solar radiation. The larger the ice sheet, the greater the amount of reflection of solar radiation. In winter, when the Antarctic ice sheet is at its greatest, so is the amount of reflection. This results in lower global temperatures.

Oceans and plants have a low albedo and absorb most of the radiation striking them. Increases in areas of ocean or vegetation lead to a decrease in albedo and a warmer Earth.

Over Earth's history there has been a fluctuation in surface albedo, which has resulted in temporary changes in the global climate. When warming occurs, more snow and ice melts and this lowers the surface albedo. The lower albedo causes more solar radiation to be absorbed by the surface, which in turn causes warmer conditions. Conversely, in colder conditions more ice and snow forms on the surface. This raises the surface albedo, which allows less solar radiation to be absorbed at the surface. This, in turn, causes cooler conditions.

ATMOSPHERIC ALBEDO

When fine dust enters the atmosphere it lowers the atmospheric albedo, which blocks the solar radiation reaching the Earth's surface. This causes a reduction in the Earth's temperature. The dust can come from volcanic action or dust storms. Atmospheric albedo is also affected by cloud cover and the amount of water vapour in the atmosphere.

HUMAN ACTIVITY

Recently the burning of forests in Indonesia and other countries caused an increase in smoke particles in the air, resulting in a temperature change. Ozone depletion and greenhouse emissions caused by human activity also have a direct effect on the climate because they increase the amount of greenhouse gases entering the atmosphere.

SEA LEVEL CHANGES

Throughout the Earth's history the planet has been subject to changes in sea levels due to the melting and freezing of the ice caps as well as Earth-forming processes. The global sea level and the Earth's climate are closely linked. Over the last one hundred years the global climate has increased by 1°C, resulting in the melting of glaciers. If warming continues, by the year 2100 Iceland will have lost 40% of its glaciers. If all the Earth's glaciers were to melt it would result in the sea levels rising by 50 cm. However, if the polar ice caps were to melt it would result in a rise of 80 m in sea

definitions

albedo
the reflectivity of a surface

rain shadow
the sheltered side, or downwind flank, of a mountain range, which receives less rainfall than the upwind flank

orogenesis
the process of mountain formation

levels. During cold climate events, such as ice ages, sea levels fall because there is a shift in the water cycle to storage of water in ice sheets. During the last glacial maximum, sea levels dropped by 125 m. During the interglacial intervals, sea levels have risen to as much as 20 m above present levels. Sea level changes have had a direct effect on surface albedo. When sea levels are low there is greater reflectivity of solar radiation because of the increase in the size of global ice sheets.

MOUNTAIN BUILDING

Mountains and high plateaus have a profound influence on climate. Rain falls on the upwind flank, which is the side of the mountain facing the oncoming clouds. The opposite side of the mountains, or downwind flank, receives very little rain and is described as being in the **rain shadow**. (See Figure 1.4.7.) An example of this occurs in the Blue Mountains of New South Wales, where the eastern slopes on the Sydney Basin side are the upwind flank and the western slopes are in the rain shadow and therefore receive less rain.

During mountain building, or **orogenesis**, changes to the direction of airflows can occur. As mountains rise, stratospheric airflows may be diverted or even blocked. In the formation of the Himalayas, there was a 2 km uplifting of the Tibetan Plateau. This blocked mid-latitude airflow, causing stronger monsoons. It also affected the whole of the Northern Hemisphere. The formation of mountain ranges in North America and Central Asia may have been one of the factors that led to the last major ice age in the Northern Hemisphere.

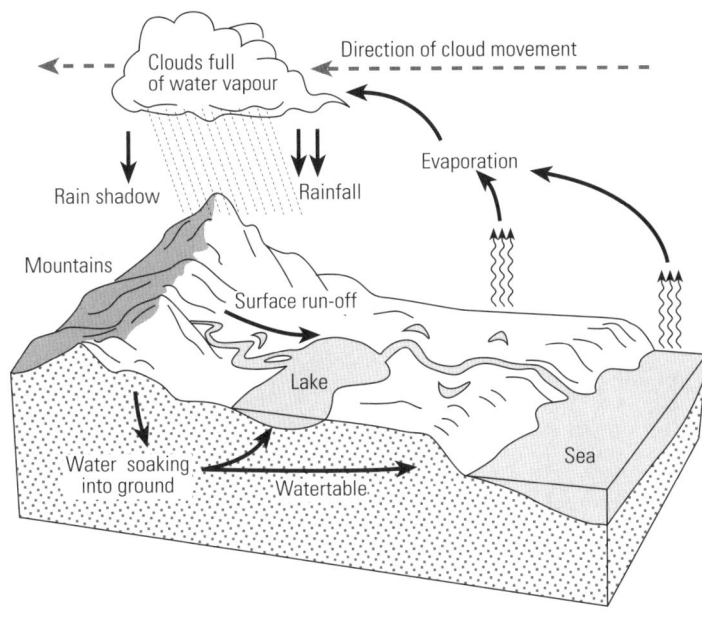

Figure 1.4.7 The effect of mountain building on rainfall patterns.

REVIEW ACTIVITIES

1 Describe the pathways taken by carbon within the carbon cycle.

2 Describe what effect sunspot activity has on the Earth's climate.

3 Explain the importance of the Milankovitch cycles in relation to climate change.

4 Define the terms 'orogenesis', 'perihelion' and 'albedo'.

5 Tabulate the factors that affect climate changes. Which of them are cyclical or random and which are short term or long term?

EXTENSION ACTIVITIES

6 Research which gases increase the greenhouse effect.

7 Describe what climate change took place as a result of the 1883 Krakatoa eruption. How long did these climate changes last?

Evidence of climate change

A number of sources can provide scientists with information about fluctuations in climate over long periods of the Earth's history. These sources include ice caps; layered sediments; ancient rock strata, or soil horizons; and tree rings.

It has taken many millions of years to achieve the current thickness of ice at the poles. As the ice formed, by the freezing of water, it trapped small bubbles of air. This air has since remained entombed in the ice. This has allowed scientists to gain information about conditions in earlier atmospheres, such as previous cycles of the greenhouse effect.

To sample the ice, scientists drill into the ice and take an ice core, in the same way as geologists take a rock core when drilling for oil. Once the core has been taken, isotopes of oxygen can be sampled from the core. Earth scientists measure the ratio of O^{16} to O^{18} radioisotopes in the sample cores. The normal ratio of O^{16} to O^{18} is 490:1. However, as the temperature drops there is an increase in the proportion of O^{18}. From the measurements of ice cores, an accurate picture can be drawn of fluctuations in concentration.

It is possible to construct a temperature gradient through the polar ice. This process can be done with the other gases, such as carbon dioxide or rarer gases such as methane. Temperature records from Antarctic, Greenland and Siberian ice cores have shown there are cycles, as predicted by the Milankovitch cycles.

Ice cores can give scientists a wide range of information, including levels of radioactivity in the water and air, and levels and types of atmospheric pollution. By the measurement of sodium levels in the ice cores, scientists can calculate the storminess of seas. The level and type of dust in the ice cores can be an indication of levels of volcanic activity or the direction and strength of air currents. Changes in the ice cores can also provide information regarding climate change and global warming.

Geologists can also use information from deep sea sediments to gain information about previous climate changes. It is possible to measure the amounts of oxygen isotopes in these sediments and relate them to temperature variation over a period of time. Marine sediments that have been deposited on the ancient floors of oceans contain microfossils of **plankton** in the form of foraminifera and photosynthetic coccoliths. There are many types of plankton microfossils and scientists can link each of these to a particular type of environment.

The marine sediments provide a detailed and continuous record of climate change. This record indicates that over Earth's history there have been decreasing deep sea water temperatures, along with a build up of continental ice sheets.

Sedimentary rocks can give us information about their distribution and **mineralogy** and the weathering processes that have affected them. They also preserve a record of short-term changes in acidity and changes in the chemical composition of water and air.

Ancient soil horizons, called paleosols, give us an indication of former land surfaces and climates. Palaeontologists can deduce previous climates by studying the fossils of ancient animals and plants. Fossilised pollen grains, in old bogs and lake sediments, give a good indication of climatic conditions.

The size of sediments and their composition in layered lake sediments can reveal seasonal variations. Lake sediments can also indicate changes in climatic conditions by different mineralogy and the sediment chemistry. Different climatic conditions will show up as changes in animal and plant compositions. Some sediments contain higher levels of salts and evaporites, which are an indication of warmer conditions.

Ancient fossilised coral reefs can give us indications of changes in the climate, as well as evidence of changes in sea levels.

Dendrochronology, the study of tree rings, can give very detailed information of the climate during the life of the tree. Each year of the tree's growth can be observed by cutting horizontally through the tree's trunk and comparing each growth ring. Most of the tree's growth takes place during the summer months. A poor summer—one that is either cold and/or with low rainfall—would mean little growth and therefore the ring produced in that year would be very narrow. A warm and wet growing season would result in a broad growth ring.

plankton
very small or microscopic animals and plants living in the oceans

mineralogy
the study of minerals, including their formation, composition, properties and classification

definitions

REVIEW ACTIVITIES

1 Recount how humans have affected the composition of the atmosphere.

2 Summarise the causes of climate change.

3 Which features in lake sediments give information about seasonal variations?

4 Explain the importance of coccoliths and foraminifera.

5 Explain how dendrochronology can inform scientists about ancient climates.

EXTENSION ACTIVITIES

7 Write a discussion on the topic 'the atmosphere is in a constant state of change'.

8 Describe how coral reefs can give information about climate changes.

SUMMARY

- When the Earth first formed it was very hot, and both seismically and volcanically active.

- Shortly after it formed, the Sun was only 80% as hot as it is today.

- Cooling of the planet's surface was slow because of bombardment by meteorites and cosmic radiation, a thin atmosphere, a seismically active crust and heating by radioactive elements.

- Evidence of the Earth's age comes from zircons in the Pilbara.

- Earth's first atmosphere was composed of hydrogen, neon, nitrogen, argon and helium.

- The first atmosphere was blown away by the solar winds.

- Most of the secondary atmospheric gases and water vapour were released by outgassing from volcanoes.

- The first atmosphere was replaced by a secondary atmosphere of mainly water vapour, nitrogen, methane, hydrogen chloride, some sulfur compounds, ammonia and carbon dioxide.

- The water making up the oceans originally was water vapour released in the form of rain.

- Rain washed out a lot of gases. As it fell, the rain turned acidic.

- Formation of the oceans and atmosphere took millions of years.

- Banded iron formations were formed by precipitation of dissolved iron and silica, then layers of dead algae rich in silica, followed by more iron oxide.

- Constituents of the modern atmosphere are gases, aerosols and water vapour.

- The layers of the atmosphere are the troposphere, stratosphere, mesosphere and thermosphere and each layer is separated by a thermal boundary called a pause.

- The ozone layer protects the Earth from the most dangerous ultraviolet radiation.

- Climate change has been affected by a combination of factors, including variations in solar output, changes in the Earth's orbit, continental drift, greenhouse gases, volcanic activity, surface albedo, atmospheric albedo, human activity, sea level changes and mountain building.

- Ice ages occur every 100 000 years.

- The greenhouse effect, caused by gases such as carbon dioxide and methane, forms an insulating layer that holds in the planet's heat.

- Evidence of climate change can be found in a number of sources, including ice caps, deep sea sediments, paleosols, fossil coral reefs and tree rings.

PRACTICAL EXERCISE
Comparing climate data change

ACTIVITIES

1 Refer to Figure 1.4.8.
  a What climatic change factors could account for the troughs and peaks in the temperature line?
  b As the temperature drops, the carbon dioxide level also drops. What producer of carbon dioxide is likely to be affected by temperature?

2 Refer to Figure 1.4.9.
  a Describe the likely trend of the oxygen line prior to the Cambrian period.
  b Compare the oxygen and carbon dioxide lines over the last 150 000 years.

3 Refer to Figures 1.4.9 and 1.4.10. Which of the two gases follows the temperature line most closely?

4 Refer to Figure 1.4.9.
  a During which period(s) were the temperature changes greatest?
  b Which period had the most unstable temperatures?

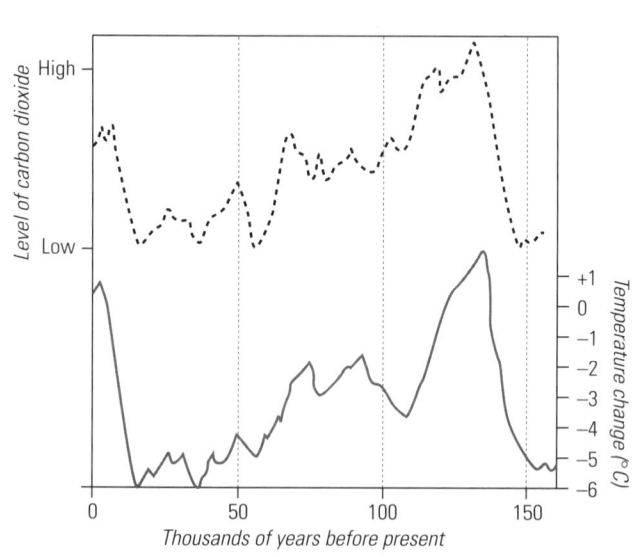

Figure 1.4.8 Comparison of levels of carbon dioxide and temperature.

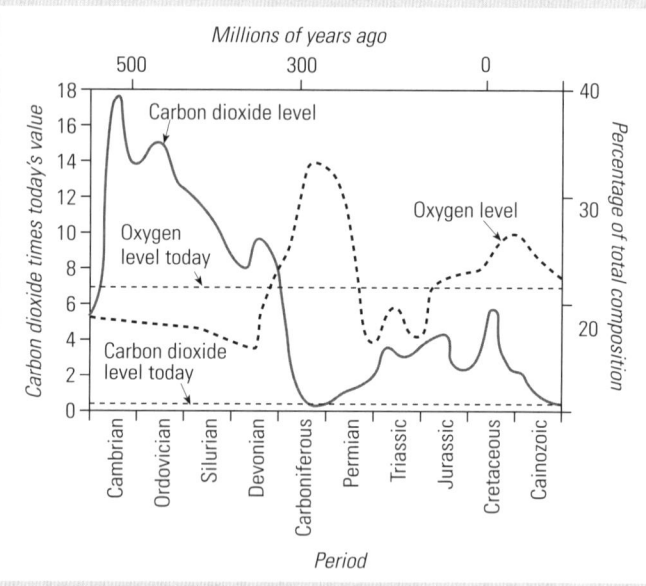

Figure 1.4.9 Oxygen and carbon dioxide levels over the Earth's history.

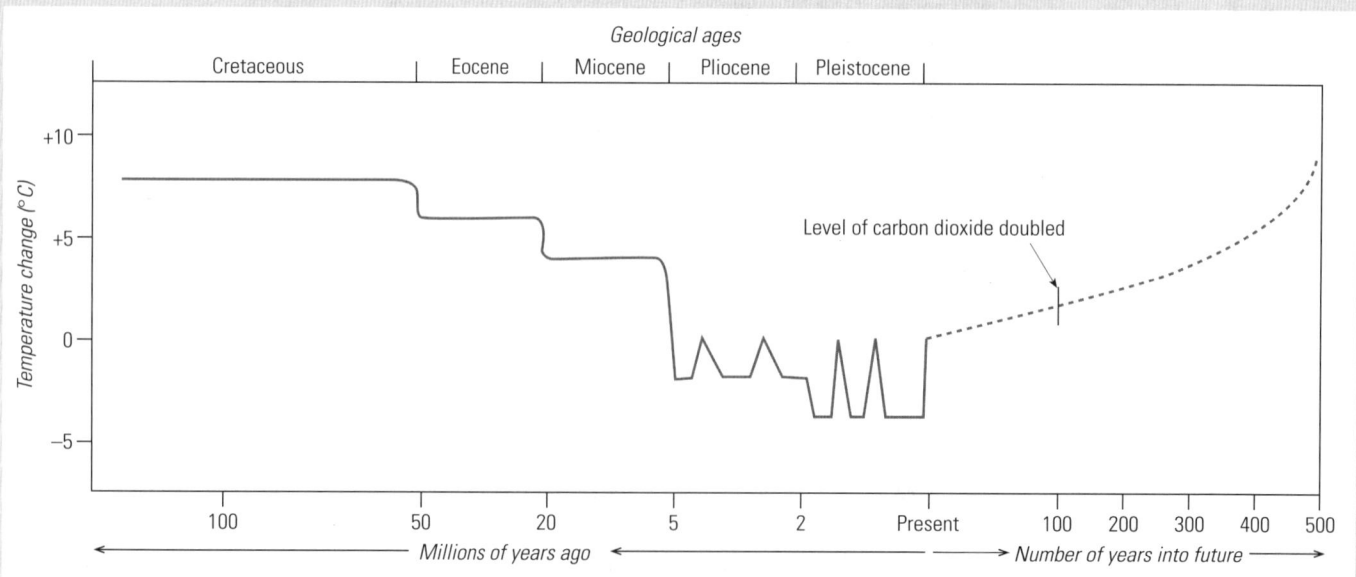

Figure 1.4.10 Temperature fluctuations based on the ratio of radio-isotope forms of oxygen in sea floor sediments.

1.5 The early Earth and the start of life

OUTCOMES

At the end of this chapter you should be able to:

- summarise the experiments of Urey and Miller and consider the importance of their findings to our understanding of how amino acids may have originated on Earth

- gather and process information from secondary sources about the synthesis of amino acids by discharging an electric spark in mixtures of methane, ammonia, hydrogen and water

- outline the evidence that indicates how the first cellular organisms (archaeobacteria) may have developed and describe their mode of respiration (anaerobic fermentation)

- gather and process secondhand information about both ancient archaeobacteria and present day archaeobacteria that live near fumaroles and submarine vents known as black smokers

- identify photosynthetic bacteria as the first organisms to release oxygen into the environment

- discuss the roles of precipitation and photosynthesis in the removal of carbon dioxide from the early atmosphere

- predict and explain the differences in composition of the oceans before and after the evolution of photosynthesis.

WHAT IS LIFE?

As our biological knowledge increases, it becomes increasingly difficult to give a simple definition of what life is. It is relatively easy to define what life is when examples include most members of the animal and plant kingdoms. However, when scientists attempt to apply the same criteria to such organisms as viruses, it creates great debate as to whether or not those organisms are living. A possible definition of life should include the fact that, during the course of its life, the organism would have the ability to carry out the basic processes of respiration, nutrition, growth, reproduction, excretion and have the ability to move and to respond to its environment. It would also have the ability to pass its characteristics on to its offspring.

All known life forms are composed of a limited number of compounds, and carry out chemical reactions.

FORMATION OF THE FIRST ORGANIC CHEMICALS

Early Earth was hot, humid and seismically and volcanically active. It was bombarded by cosmic rays and meteorites, hit by continuous lightning strikes and riven by volcanoes and earthquakes. The atmosphere contained large amounts of nitrogen, methane and ammonia. The surface rock was weathering, releasing salts into water and, when it started to rain, large amounts of different acids were formed. All these reactions released more chemicals into the environment.

Life, or the precursors of life, could not occur on the planet until certain criteria were met. Firstly, the surface temperature had to be low enough to allow water to condense. Secondly, bodies of surface water, for example oceans, had to form.

The precursors of life were in fact simple **amino acids** and simple nucleic acids in the form of **ribonucleic acid (RNA)**. There are twenty known amino acids. The importance of amino acids is that they join together into chains to form **proteins**. Some proteins are required as the building blocks of cells. Another group of proteins form enzymes, which are required in a large number of chemical reactions within the cell. The chains of amino acids can form millions of combinations of lengths and sequences. Each combination produces a different protein. The shape of the amino acid chain will also influence the function of the enzyme. Long chains, possibly with cross links, generally have structural functions, whereas chains forming globules are normally enzymes and are involved in chemical reactions.

RNA is one of the simplest forms of chemicals capable of copying and transferring information. RNA is a simple nucleic acid that is involved in making proteins. It reads the instructions for making new proteins from the **deoxyribonucleic acid (DNA)** molecule. DNA is the inherited material of all living things. It has the ability to pass on its structure to copies of itself. The structure of DNA is translated by the cell into the structure of protein molecules. The RNA organises the sequence of amino acids to produce the protein whose code was given by the DNA.

These simple chemicals—amino acids and RNA—were possibly formed by the action of ultraviolet radiation, heat and lightning in an aqueous solution.

The two main types of **polymers**, or long chain molecules, formed soon after the Earth's formation were proteins and nucleic acids. Amino acids have the opposite charge to clay particles, which causes them to be attracted to the surface of those particles. A high concentration of these amino acids collects on the surface of the clay particle. This leads to a certain amount of bonding together of the amino acids, and the forming of long chains. Amino acids could also concentrate as a surface film on top of water. Evaporation near volcanic centres due to the intense heat and spontaneous formation of the amino acids into long chains led to simple proteins forming. These same conditions may have also formed simple nucleic acids.

Some scientists believe that some simple organic chemicals were brought in on meteorites or comets. Other scientists believe life itself was brought to this planet from outer space. This is the theory of panspermia. From astronomical observations and previous space missions, comet dust grains are known to contain various minerals and some organic compounds. At present, their detailed chemical structures are unknown. Until scientists acquire proper samples of cometary material from a passing comet, it is impossible to either support or rule out comets as being the origins of life on Earth.

Evidence from stony meteorites shows that some simple organic compounds do indeed come from space. However, those organic compounds are unlikely to have been the precursors of Earth's organic compounds. It is unlikely because the amounts of these precursor compounds would have been insufficient to produce the abundance of organic compounds needed to create the earliest life. The idea that panspermia was the origin of the organic material required for life is incompatible with the findings of Oparin and other scientists, such as Miller.

definitions

amino acid
an organic compound, required in the formation of proteins

ribonucleic acid (RNA)
a nucleic acid needed for making proteins

proteins
very complex organic compounds, required for numerous processes in living things

deoxyribonucleic acid (DNA)
a nucleic acid which holds the blueprints of life, responsible for inheritance

polymer
formation of large molecules consisting of repeated structures

Chemosynthetic origin of life

In 1923, the Russian biochemist A. I. Oparin stated that before living cells could arise, living chemicals must be formed. His hypothesis stated that biological chemical evolution had to occur before the possibility of cellular evolution. He believed that the simple chemicals present in the early atmosphere could be converted into biological chemicals. J. Haldane, a British biologist, arrived independently at this idea in 1929.

It was not until the 1950s that Oparin's hypothesis was tested. Harold Urey believed in the hypothesis as proposed by Oparin and Haldane, and decided to test it. Urey and Stanley Miller, one of Urey's graduate students, set up an apparatus to simulate the conditions of the early Earth. (See Figure 1.5.1.) The apparatus contained a variety of gases: hydrogen, ammonia, methane and water vapour. The gases were circulated between two chambers: the upper one representing the atmosphere and the lower one representing the oceans. An electrical discharge was passed through the upper chamber—the atmosphere. This electrical discharge representing lightning—the energy in the early atmosphere.

Once the experiment was completed, the atmosphere and ocean chambers were tested for a variety of chemicals. It was found that the chambers contained amino acids as well as other organic chemicals. The experiment was repeated a number of times. During subsequent experiments, ultraviolet light was used instead of electricity. It was discovered that **nitrogenous bases** were produced along with **ribose** and **nucleotides**. These chemicals were the constituents of genetic material. The experiment proved that it is possible for non-living matter to produce living material. This is the chemosynthetic origin of life. Chemosynthetic evolution is the most popular of the hypotheses as to how life originated on Earth.

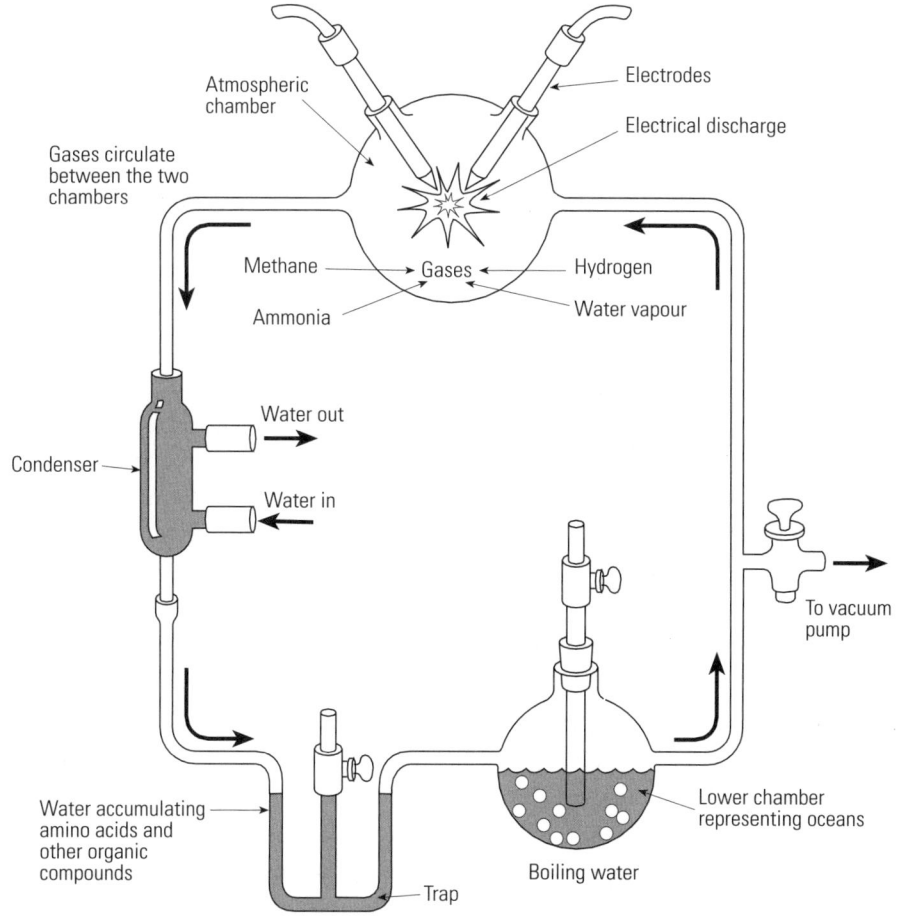

Figure 1.5.1 Apparatus used in the Urey-Miller experiment.

definitions

nitrogenous bases
part of the DNA and RNA molecules

ribose
a sugar required in the structure of nucleic acids

nucleotides
part of the nucleic acid molecule, formed from a ribose sugar and a nitrogenous base

REVIEW ACTIVITIES

1 At times it is difficult to recognise when something is a living entity. How can living things be distinguished from non-living things?

2 Explain why nucleic acids, such as RNA or DNA, are vital for life.

3 Describe the conditions that may have led to the formation of nucleic acids.

4 Outline the roles of Oparin, Haldane, and Urey and Miller in developing our understanding of the early stages of cellular evolution.

5 Summarise the reasons why the precursors of life could not have existed on the Earth when it first formed.

6 Recount how molecules, such as chains of amino acids and nucleic acids, may have been formed.

7 Explain the chemosynthetic origin of life.

8 Why was it important that the first organic compounds were amino acids and nucleic acids?

EXTENSION ACTIVITIES

9 Research the evidence for and against the theory of panspermia.

10 'Urey and Miller's experiment would not have worked if they had chosen gases from an earlier atmosphere.' Discuss this statement.

THE FIRST CELLS

Scientists now mainly accept that the first organic chemicals that existed were the amino acids and RNA. Which came first is difficult to know as it is rather like the chicken and egg problem. Today RNA is required for protein synthesis, but whether this was true originally is unknown. It is thought that as the concentrations of elements and compounds increased in the oceans they reacted with each other and formed increasingly complex molecules. Those that were stable survived; those that were not broke down. Some of the molecules were hydrophobic (water hating) and so they had a tendency to clump together and form a boundary between themselves and the surrounding aqueous solution; rather in the same way as a drop of oil forms in water. It has been found that proteins and lipids have parts that are hydrophobic. Experiments have shown that when a number of chemicals are added to water they clump together in small drops, or **microspheres**. (See Figure 1.5.2.) Microspheres are thought to be the precursors of cells. They are still a long way from being cells, and our knowledge of how this gap was bridged is very limited.

Scientists accept that the early Earth processes formed amino acids and RNA some

microsphere
a stage before the formation of primitive cells

definition

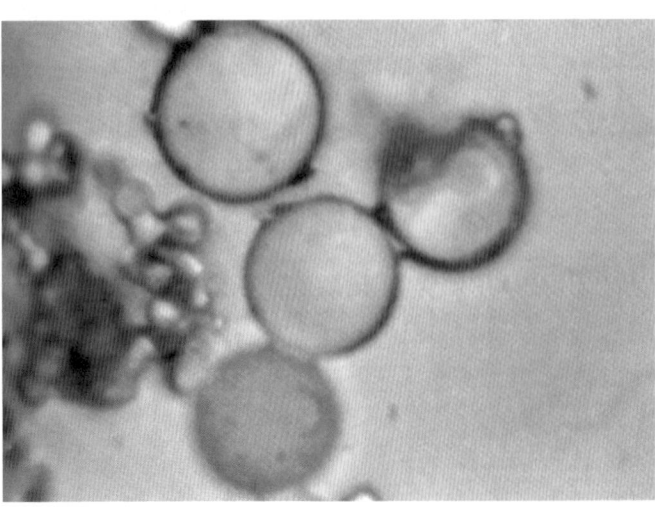

Figure 1.5.2 Simple biological microspheres.

time after 4.5 billion years and we now have evidence of the first cells existing around 3.8 billion years ago. What we do not know is the sequence of events during the intervening 700 million years.

The first evidence of cells comes from rocks in the Isua region of Greenland. Recently the rock there has been found to contain specks of carbon, and radio-isotopic analysis has identified the carbon as being a product of biological processes. The simplest organisms that are capable of carrying out biological processes are cells. What is absent from the Isua rocks are fossil cells. The carbon is the only evidence that cells existed during the formation of the rocks.

REVIEW ACTIVITIES

1 Explain how microspheres are formed.

2 The earliest evidence of cellular activity dates from around 3.8 billion years ago. Describe what the evidence is.

EXTENSION ACTIVITIES

3 What sort of chemical processes would produce compounds containing carbon?

4 Why were the carbon traces more likely to be preserved in the rocks of Greenland than other European or North American countries?

FUMAROLES: THE ORIGINS OF LIFE?

Since the late 1960s, scientists have known about the existence of **fumaroles**, or submarine vents. (See Plate 11.) These are deep sea volcanic vents that release sulfide-type compounds. These compounds are precipitating out, forming a black plume of chemicals. Hence, they have been nicknamed 'black smokers'. Other submarine vents produce a white or clear liquid and are referred to as 'white smokers'. They release cooler water and often contain compounds of barium, calcium and silicon. Deep sea volcanic vents are generally located in the area of sea floor spreading. The environment around these vents are subject to very high temperatures (up to 380°C in the vents themselves), very high pressure and very high levels of inorganic nutrients. The oceanic crust around these vents is only about 6 to 8 km thick and is tectonically active. Areas of sea floor spreading have a large number of these vents. The vents release vast amounts of chemical compounds into the oceans. The compounds include methane, sulfides and nitrates.

It has been found that deep sea vents support unique ecosystems containing bacteria, giant worms and molluscs, for example. The ecosystems rely both on the chemicals and heat to support their food web. The main energy chemical is hydrogen sulfide. The bacteria belong to a group called the **archaeobacteria**. (See Figure 1.5.3, page 62.) Some scientists think that life may have originated around these deep sea vents and that **photosynthesis** may have evolved within this type of ecosystem. Archaeobacteria exist around the vents because they are **methanogenic** or **sulfaphilic**. This means they are methane feeders or sulfur feeders. They also tend to be thermophilic, which means they can only live in very hot conditions. These bacteria provide the giant worms, molluscs and other organisms with organic compounds.

definitions

fumarole
deep sea volcanic vent

archaeobacteria
a very primitive type of bacteria

photosynthesis
the process by which plants make their own food from inorganic compounds

methanogenic
organisms that use methane as a source of food

sulfaphilic
organisms that use sulfur as a source of food

Figure 1.5.3 Archaeobacteria in tube worms from deep sea thermal vents.

The importance of these fumarole-centred ecosystems is that they exist in extreme conditions, possibly akin to some of the conditions of early Earth. The organisms living around fumaroles need to survive extreme pressure and temperature and high concentrations of possibly toxic chemicals. Until the discovery of these fumarole-centred ecosystems it was thought that it was only solar energy that could be used by food webs.

Scientists also have further evidence that archaeobacteria can live in other extreme conditions similar to early Earth. They live in hot water springs in temperatures higher than 100°C and they have been found in salt lakes. These conditions of high temperature and highly saline and acidic conditions may well have resembled some of the conditions of early Earth. Some of these simple types of early bacteria also live within ice and snow in glaciers; these are thermophobic, meaning they hate heat.

REVIEW ACTIVITIES

1 Assess the importance of deep sea fumaroles in extending our knowledge of early life.

2 When living in the vicinity of a fumarole, why do methanogenic or sulfaphilic organisms have an advantage?

EXTENSION ACTIVITIES

3 Explain why black smokers can support a specialist ecosystem but white smokers do not support a unique ecosystem.

4 Explain why sunlight cannot penetrate deep water.

definitions

prokaryote
a primitive type of cell, lacking membrane-bound organelles (structures with a specialised function). Examples are bacteria and blue-green algae.

eukaryote
advanced cells that have membrane-bound organelles. Examples are animal and plant cells.

aerobic respiration
release of energy using oxygen

anaerobic respiration
less efficient form of respiration by which energy is released in the absence of oxygen

ribosome
a cell organelle involved in the production of proteins

heterotrophs
organisms that are unable to make their own food

THE EVOLUTION OF CELLS

At the present time scientists have no evidence to show how the first cells formed or what they looked like. The first traces of cells are the products produced in the cells, not the actual cells. (See the previous section on the first cells, page 60.) What is known is that fossil evidence of early cells shows prokaryotic types that are very similar to some of those still living today. The first eukaryotic fossil cells appeared at a later date than prokaryotic fossils.

Today there are two main types of cell: the **prokaryote** and the **eukaryote**. The earliest cells were prokaryote cells. Evidence shows that they first appeared somewhere around 3.5–4.0 billion years ago.

As the first prokaryotes lived in an oxygen-free environment—the atmosphere contained no oxygen—they had to be able to respire without oxygen. Respiration is the process by which organisms release energy. There are two main types of respiration: aerobic and anaerobic. **Aerobic respiration** uses oxygen in the release of energy, while **anaerobic respiration**, or fermentation, is the release of energy in the absence of oxygen. (See Figures 1.5.4 and 1.5.5.) Most organisms carry out aerobic respiration. The early archaeobacteria, which lived in the absence of oxygen, had to rely on the less efficient anaerobic respiration, which releases smaller amounts of energy then aerobic respiration.

Prokaryote cells are very simple. They consist of a cell wall and a cell membrane inside of which is located protoplasm containing **ribosomes**, which are involved in the production of proteins, and a strand or ring of DNA. This type of cell dominated the early Earth for a period of about 2 billion years until their more complex cousin—

the eukaryote cell—became dominant. The eukaryote cell achieved dominance because of its specialised organelles, which made it more efficient. Also, the majority of eukaryotic cells were able to utilise oxygen through aerobic respiration.

Two major types of prokaryote are the bacteria and cyanobacteria (or blue-green algae). (See Figure 1.5.6.) The probable first type of prokaryote to evolve was the simple form of bacteria called archaeobacteria. The archaeobacteria were slightly simpler in structure then the cyanobacteria. Also, the archaeobacteria were **heterotrophs** (see Figure 1.5.7), meaning they were able to utilise the chemical compounds, such as methane or sulfur, in the water as a source of energy. This allowed them to flourish around the chemical-rich deep sea vents.

As the levels of dissolved chemicals in the oceans dropped, it put pressure on the heterotrophic prokaryotes and allowed the more complex cyanobacteria to become dominant. Some scientists consider that this type of prokaryote evolved around the deep sea vents and spread out from the vents to the rest of the oceans. The cyanobacteria, which are photosynthetic prokaryotes, had an evolutionary advantage over their cousins because they were autotrophs. This meant they were able to make their own foods from simple molecules using either solar or chemical energy by the process of photosynthesis. (See Figure 1.5.8.)

$$\text{Glucose} + \text{Oxygen} \longrightarrow \text{Carbon dioxide} + \text{Water} + \text{Energy}$$

Figure 1.5.4 Equation for aerobic respiration.

$$\text{Glucose} \longrightarrow \text{Alcohol} + \text{Carbon dioxide} + \text{Energy}$$

Figure 1.5.5 Equation for anaerobic respiration.

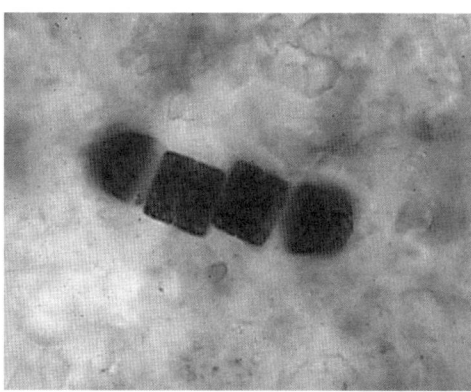

Figure 1.5.6 Fossil cyanobacteria in Bitter Springs, Australia found in chert (a dense, hard sedimentary rock).

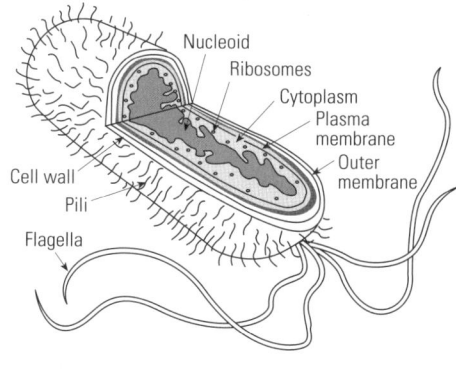

Figure 1.5.7 Diagram of a typical heterotroph prokaryote.

$$\text{Water} + \text{Carbon dioxide} \xrightarrow{\text{Solar or chemical energy}} \text{Glucose} + \text{Oxygen}$$

Figure 1.5.8 Equation for photosynthesis.

REVIEW ACTIVITIES

1 Describe the evidence that indicates the levels of oxygen in early seas.

2 Using the information in the text and Figures 1.5.4 and 1.5.5, differentiate between aerobic and anaerobic respiration.

3 Distinguish between heterotrophs and autotrophs.

4 Refer to Figure 1.5.8. Describe what happens in the process of photosynthesis.

5 Explain why it is likely that prokaryotic heterotrophs were the first organisms to develop.

6 What are the two main types of prokaryotes?

7 Explain why cyanobacteria had an advantage over the other types of cells.

EXTENSION ACTIVITY

8 Research the date of the first eukaryotic fossil cells and how this compares with the date of the first prokaryotic fossil cells.

PHOTOSYNTHESIS AND THE ATMOSPHERE

The evolution of photosynthetic prokaryotes had a major effect on the atmosphere. Prior to their evolution there was little or no atmospheric oxygen. What little oxygen there was in the atmosphere was due to the dissociation of water by ultraviolet radiation, producing free oxygen and hydrogen.

The arrival of photosynthetic organisms led to the release of oxygen into the water. Initially it reacted with iron (II) ions to produce iron oxide, which eventually resulted in banded iron formations. (See chapter 1.4, page 45.)

Oxygen levels took time to build up. If rocks formed from sea floor sediments are studied, it is found that sediments older than 1.8 billion years show signs of having been formed in an oxygen-poor (reduced) environment, while younger sediments show indications of an oxygen-rich environment.

Eventually enough oxygen was produced by photosynthesis and seismic activity to allow oxygen levels in the atmosphere to build up. Some of the oxygen was converted into ozone, which formed a layer in the atmosphere. This reduced the amount of ultraviolet B and infra-red radiation striking the Earth's surface, thereby reducing the heating effect of the radiation. Ultraviolet B radiation damages the chromosomes in cells, resulting in sudden changes in the DNA structure and leading to mutations in organisms. Thus, the reduction in the amount of ultraviolet B radiation striking the Earth decreased the amount of damage to organisms. Organisms were now able to colonise shallow waters. This invasion of shallow waters and formation of the ozone layer took place around 600 million years ago.

As the oxygen levels increased, the methane in the atmosphere was removed due to the reaction between methane and atmospheric oxygen. This resulted in the formation of carbon dioxide. There was also a decrease in the amount of carbon dioxide in the air, as the autotrophs required it for the process of photosynthesis.

As the prokaryotic autotrophs moved into shallow water, new forms emerged. One of these new forms were **stromatolites**. They are a colonial form of photosynthesising cyanobacteria. Most cyanobacteria live as separate single-celled organisms. A few forms, including stromatolites, live together in colonies. This may provide a better survival strategy for the species as it gives increased protection, a division of labour, a sharing of resources and a higher reproductive rate. The stromatolites are formed by layers of photosynthesising cyanobacteria. The cyanobacteria eventually die and are replaced by a new layer of cyanobacteria, which grow on top of the old layer. These stromatolites are some of the oldest fossils, dating back more than 3 billion years. (See Plate 12.) They thrive in warm, shallow aquatic environments. Ancient stromatolites built reefs in much the same way as coral does today. They are nearly extinct, living a precarious existence in only a few localities worldwide. (See Figure 1.5.9.)

definition

stromatolites
cyanobacteria that form columns in shallow, warm seas

Figure 1.5.9 Stromatolites.

REVIEW ACTIVITIES

1 Explain the importance of ultraviolet radiation in the early stages of evolution.

2 Assess the importance of photosynthetic prokaryotes on the Earth's early atmosphere.

3 Explain what a stromatolite is.

4 Describe the conditions and appearance of the early Earth.

5 Why did the cooling of the Earth take so long?

6 What are the main reasons as to why life was possible on Earth?

7 Why couldn't organisms originally colonise shallow waters?

8 List the steps in the formation of early life.

EXTENSION ACTIVITIES

9 Explain how the eukaryotic cell evolved.

10 Why is the Bitter Springs formation in central Australia an important fossil site?

SUMMARY

- Life is a collection of systems that possess the ability to work together within the environment in which it exists, to maintain its form and function and to pass on its characteristics to its offspring.

- The first organic compounds formed were amino acids and RNA.

- Amino acids are the building blocks of proteins, which in turn build cells.

- RNA is involved in copying information in order to make proteins.

- Work by Oparin, Haldane, and Urey and Miller showed how simple elements and compounds, methane, ammonia, water and hydrogen formed simple organic compounds in the form of amino acids and nucleic acids by the means of lightning.

- Microspheres were the precursors of cells.

- The first signs of life were found in rocks in the Isua region of Greenland—not cells but cellular activity.

- First life may have originated around deep sea vents, or fumaroles.

- The first cellular life were prokaryotes. The two major types are bacteria and cyanobacteria. The first bacteria were archaeobacteria.

- Photosynthetic prokaryotic cyanobacteria caused changes in the atmosphere: oxygen levels increased, the ozone layer formed and carbon dioxide levels reduced.

- With the formation of the ozone layer, the amount of harmful ultraviolet B radiation penetrating the atmosphere was reduced and life could enter the shallow seas.

PRACTICAL EXERCISE
Anaerobic respiration and carbon dioxide

In this exercise you will design an experiment to show that carbon dioxide is released in anaerobic respiration.

Background

Early life, such as prokaryotic organisms, had to carry out anaerobic respiration because there was no free oxygen. These early organisms were not able to carry out aerobic respiration due to the lack of oxygen. They still needed to release energy, so the process of anaerobic respiration evolved very early on.

Organisms require glucose in order to carry out respiration and thereby release energy, irrespective of whether they carry out aerobic or anaerobic respiration.

Procedure

1 Bubble carbon dioxide through limewater. Limewater will turn milky if carbon dioxide gas is bubbled through it.

2 Use the questions below to help you design an experiment to show that carbon dioxide is released during anaerobic respiration. Also refer to Figure 1.5.5 (page 63).

a What are the ingredients needed to allow anaerobic respiration to take place?
b Does anaerobic respiration take place in solid or liquid form?
c How detailed do we want the experiment?
d Should the experiment be carried out in a test tube, a small beaker or a large beaker?
e Does the temperature matter? If it does, what temperature should it be?
f What gas are we trying to collect or test?
g How are we going to collect the gas? Do we need to collect it or can we do something else with it?
h How can we bubble the carbon dioxide through the limewater?
i How does the limewater show that carbon dioxide gas is present?
j How are we going to show the results?
k Are there any variables that need to be taken into account, such as amounts of ingredients, temperature, amount of light, and length of time?
l Should there be a control with which we can compare our experiment? (If so, describe the control you would devise.)
m Are there any precautions that need to be taken in order to carry out the experiment safely?

RESOURCES

Archaeobacteria

Archaeobacteria *fig.cox.miami.edu/~161hon4/evolution.htm*

Archaeobacteria *volcano.und.nodak.edu/vwdocs/msh/p_a/p_aa/p_aawrpb.html*

Astronomy

Comets *whyfiles.news.wisc.edu/011comets/time_capsules.html*

Measuring the Universe *doug-pc.itp.ucsb.edu/online/plecture/kirshner/*

NASA *www.nasa.gov/gallery/photo/index.html*

Observing the Universe *www.windows.umich.edu/cgi-bin/tour_def/teacher_resources/space_astronomy/contents.html*

Orbimage, Global Imaging Information *www.orbimage.com*

Students for the Exploration and Development of Space *seds.lpl.arizona.edu/*

Welcome to the Planets *pds.jpl.nasa.gov/planets/*

Deep sea vents (fumaroles)

'Black Smokers' on the Sea Floor *walrus.wr.usgs.gov/pubinfo/smokers.html*

Expedition to Explore Black Smokers *www.pbs.org/wgbh/nova/abyss/*

Black Smokers *www.amnh.org/nationalcenter/expeditions/blacksmokers/black_smokers.html*

Dreaming stories

Dreamtime Legends *www.artistwd.com/joyzine/australia/dreaming*

Dreaming Stories *www.dreamtime.net.au/dreaming/storylist.htm*

Earth's interior

Earth's Early History *www.Ldolphin.org/Early.html*

Inside the Earth *pubs.usgs.gov/publications/text/inside*

History of the universe

The Evolution of the Universe *www.sciam.com/specialissues/0398cosmos/0398peebles.html*

Mysteries of Deep Space: Interactive Timeline *www.pbs.org/deepspace/timeline/index.html*

Observatories and telescopes

Canberra Deep Space Communication Complex *www.cdscc.nasa.gov*

Canberra Planetarium and Observatory, Dickson, Canberra. Tel. 02 6249 7817

Cowra Darby Falls Observatory, Observatory Rd, Darby Falls. Tel. 02 6345 1900

Central Coast Koolang Observatory, Bucketty. Tel. 02 4998 8216 *users.hunterlink.net.au/koolang/*

Coonabarabran Skywatch Observatory, Coonabarabran *www.lisp.com.au/~skywatch*

Deep Space Communications Complex, Tidbinbilla (40 km from Canberra). Tel. 02 6201 7838

Gilgandra Observatory, Gilgandra *www.gilobs.com.au*

Goulburn Magellan Observatory, Lake Bathurst. Tel. 02 4849 4489

Grove Creek Observatory, Trunkey (about 65 km from Bathurst) *www.gco.org.au/*

Linden Observatory, Blue Mountains *www.physics.usyd.edu.au/~ptitze/aa/linden4.html*

Macquarie University Observatory, North Ryde. Observatory operated by Southern Skies Mobile Observatory *www.southernskies.com.au/macu.htm*

Mount Stromlo Observatory, Weston Creek, Canberra *msowww.anu.edu.au/exploratory*

Australia Telescope compact array located at the Paul Wild Observatory, Narrabri *www.narrabri.atnf.csiro.au/*

Parkes Australia Telescope National Facility, Visitors' Centre, Parkes Observatory, Parkes *www.parkes.atnf.csiro.au/*

Siding Spring Observatory, near Coonabarabran *www.sidingspringexploratory.com.au/intro.htm*

Sydney Observatory, Observatory Hill, The Rocks *www.phm.gov.au/observe/*

University of New South Wales, Kensington, Sydney. Sky and Space Nights plus other programs. Tel. 02 9385 5752

University of Western Sydney, Nepean: Nepean Observatory, Werrington North Campus *www.uws.edu.au/astronomy/observatory/*

Wollongong University Science Centre and Planetarium, Fairy Meadow *www.uow.edu.au/science_centre*

The solar system

Solar System: Formation
www.physics.gmu.edu/classinfo/astr103/CourseNotes/sls_form.htm

Spectroscopy

How to Build a Simple Spectroscope
geo.arc.nasa.gov/sge/jskiles/top-down/mystery_solved/myst_solved_by_R_and_D.html

Simple Spectroscope
asd-www.larc.nasa.gov/edu_act/simple_spec.html

Spectral Lines
www.colorado.edu/physics/PhysicsInitiative/Physics2000/quantumzone/

Spectroscope *kids.infoplease.com/ce5/CE049035.html*

The Spectroscope
chemscape.santafe.cc.fl.us/chemscape/catofp/measurep/spectro/fspectrs.htm

Spectroscopy *phys.educ.ksu.edu/vqm/*

Stars and galaxies

The Spectral Types of Stars
www.skypub.com/tips/basics/spectra.html

Star Colours *www.nova.org/~sol/chview/ch5.html*

Stars and Galaxies *www.eia.brad.ac.uk/btl/sg.html*

Taking a Measure of the Universe
doug-pc.itp.ucsb.edu/online/Plecture/Kirshner

CD-ROMs

'The Inner Planets', Marcom Projects

'The Outer Planets', Marcom Projects

'The Universe', Marcom Projects

Magazines

Astronomy

Astronomy Now

New Scientist

Newton Graphic Science, Australian Geographic publication, vol. 1–3

Scientific American

Sky and Space

Space Illustrated

Videos

'Beyond the Infinite: Looking to the Edge of the Cosmos', VEA (Video Education Australia)

'The Big Picture: The Universe', available at ABC Shops

'Destination Cosmos', VEA

'The Elastic Universe: Can the Big Bang Survive?', VEA

'EMR and the Stars', Learning Essentials

'The Hydrogen Spectrum', Learning Essentials

'Origins', Learning Essentials

'Our Rocky Neighbours: The Inner Planets', Marcom Projects

'Our Solar System', VEA

'The Outer Planets: The Gas Giants', Marcom Projects

'Radioactive Decay', Learning Essentials

'Radioactivity', Learning Essentials

'The Stars', BBC Productions, available at ABC Shops

'The Universe', BBC Productions, available at ABC Shops

'Waves: Energy in Motion', Learning Essentials

Dynamic Earth

In this section we will be looking at some of the most important discoveries and theories that have shaped modern geology. Over the last one hundred years techniques have been developed that allow the age of rocks and other materials to be measured accurately. This has enabled us to date rocks, which has which led to a more accurate geological time scale on which to hang the Earth's history. During the last one hundred years knowledge of how rocks are laid down in crustal plates, and how these plates have moved and shaped the world we live in, has increased dramatically. Geological studies of the plates have given us an increased knowledge about earthquakes and volcanoes. As yet we are unable to predict these catastrophic events with great accuracy, but are moving towards that eventuality.

CONTENTS

Chapter 2.1	Dating rocks and other materials	70
Chapter 2.2	Plate movements	83
Chapter 2.3	Magnetism and the wandering poles	96
Chapter 2.4	Tectonic forces	103
Resources		122

2.1 Dating rocks and other materials

OUTCOMES

At the end of this chapter you should be able to:

- recall an appropriate model that has been developed to describe atomic structure

- explain radioactivity in terms of the decomposition of atomic nuclei

- outline the conditions under which an atomic nucleus is unstable and decomposes

- explain how the relative percentage of remnant radio-isotopes can be used to measure absolute ages of materials, including rocks

- identify the age and explain the significance of the oldest mineral grains in Australia (Mount Narryer zircons of the Pilbara).

This chapter covers the dating of rocks and other materials and explains the significance of atomic structure and radioactivity to this process. The atomic structure of an element decides whether or not that element is stable. If it is not, it may achieve stability by releasing radiation. This release of radiation is uniform and can be measured precisely. If rocks or other materials are releasing radiation, it is possible to calculate mathematically how much radioactive material it contained originally, and how long ago that was. Thereby, the age of the rock or material can be ascertained.

THE GEOLOGICAL TIME SCALE

Scientists divide the Earth into a number of eons called the 'geological time scale', or 'geological column'. The eons are arranged according to the rock types and sort of fossils found in each one. The divisions are arbitrary, as with any human-made division. However, they do serve as useful labels.

There are a wide variety of geological time scales. Many countries have included their own names for some of the divisions. In the USA, for example, the divisions of the Upper and Lower Carboniferous have been replaced with the Pennsylvanian and Mississipian divisions.

A typical geological time scale has four major divisions. (See Table 2.1.1.) The first division comprises eons, each of which covers millions or even billions of years. Eras are smaller divisions of eons. The era divisions are associated with a particular form of life. The eras are followed by even smaller units called periods and epochs. Some geological time scales also include an even smaller division called ages.

The divisions of the geological time scales have been given names in either Greek or Latin (or given a geographical name to commemorate a significant find) to establish or characterise the division. The English versions of these names are given in brackets in Table 2.1.1.

Table 2.1.1 A geological time scale

Eon	Era	Period	Epoch	Start of event (millions of years ago)
Phanerozoic (visible life)	Cainozoic (recent life)	Quaternary (see Tertiary)	Holocene (wholly recent)	Recent 0.01
			Pleistocene (most recent)	2.0
		Tertiary (derived from the eighteenth and nineteenth century geological time scale that separated crustal rocks into fourfold divisions, quaternary and tertiary being the only terms to survive)	Pliocene (more recent)	5.1
			Miocene (less recent)	24.6
			Oligocene (slightly recent)	38.0
			Eocene (dawn of the recent)	54.9
			Paleocene (early dawn of the recent)	65.0
	Mesozoic (middle life)	Cretaceous (Latin for 'chalk', after the chalk cliffs of southern England and France)	Late Early	144
		Jurassic (named after the Jura Mountains)	Late Middle Early	213
		Triassic (a threefold division of rocks in Germany)	Late Middle Early	248
	Palaeozoic (ancient life)	Permian (province of Perm in Russia)	Late Early	286
		Upper Carboniferous (see Lower Carboniferous)		320
		Lower Carboniferous (named after the Welsh coal deposits)		360
		Devonian (named after Devonshire, a county in England)	Late Middle Early	408
		Silurian (named after Silures, an ancient Welsh Celtic tribe)		438
		Ordovician (named after Ordovices, another ancient Welsh Celtic tribe)		505
		Cambrian (named after 'Cambria', the Latin word for Wales)	Late Middle Early	545
Precambrian	Proterozoic (early life)	Late Middle Early		2500
	Archaean (ancient)			3900
	Hadean (beneath the Earth)			4560

REVIEW ACTIVITIES

1 Which is the older epoch: Eocene or Pliocene?

2 List the periods from youngest to oldest.

3 How long did the Ordovician period last?

4 The Devonian period directly preceded which period?

5 What does the term 'Archaen' mean?

6 What does the term 'Palaeozoic' mean?

EXTENSION ACTIVITIES

7 In the US geological time scale why are there two divisions called the Pennsylvanian and Mississippian divisions?

8 Explain why, in the majority of geological time scales, the Archaen era has fewer divisions than the post-Archaean eras.

METHODS OF DATING ROCKS

Geologists deal with two types of time: relative and absolute. Relative time is the order, or sequence, of past events. Absolute time is the time, in years ago, when a specific event happened. By using the two types of dating, geologists have been able to construct a geological time scale. As our knowledge increases, the dates of each epoch become more accurate and more subdivisions of the epochs have to be formed.

A rock can be dated based on its composition and its placement in relationship to the surrounding rock layers.

Dating methods can be classified as either relative or absolute. The absolute methods give the rock ages in years before the present, commonly in millions of years. Relative dating is the chronological arrangement of fossils, rocks or events with respect to the geological time scale.

Relative dating

Rocks in one locality or region are compared with or correlated to rocks from other regions. This allows geologists to build up a chronological order of geological events in the Earth's history. This form of dating is called **relative dating**. Relative dating methods are dependent on stratigraphic order. An individual layer of sediment is called a stratum (plural = strata) from the Latin word for 'layer'. The layering that results from a pile of layers deposited one on top of another is called stratification.

The study of strata is called **stratigraphy**. In a group of sedimentary or volcanic rocks that has not been greatly disturbed by folding or faulting, each layer is younger than the layer below it but older than the layers above it. This is called the **law of superposition**. Relative dating also relies on the **law of horizontality**. This states that sediments are originally deposited in layers parallel to the Earth's surface.

Geologists also use another principle when using relative dating and that is the principle of cross-cutting relationships. It states that a rock unit must always be older than any feature that cuts or disrupts it. If a rock has a fracture, the rock will be older than the fracture or the material in the fracture.

Many sedimentary rocks, siltstone or limestone contain fossils, which are the remains (usually the hard parts) of dead animals or plants that were buried by deposited sediment. It is common for geologists to obtain an approximate age of rocks from fossils in the rocks. This system works well for sedimentary rocks deposited during the last 500 million years. Rocks older than that are more difficult or impossible to date using fossils, due to the scarcity or absence of fossils.

During the eighteenth century the English canal engineer William Smith made careful notes of the shells and other fossil material at various levels in the rock strata he examined in the course of his work. From his observations he proposed the **principle of faunal succession**. This principle states that each rock formation contains a unique assemblage of fossil flora and fauna. The fossils of the flora and fauna succeed one another in a definite and recognisable order. Smith correlated fossil assemblages from different rocks over a wide area, and thereby was able to identify strata of the same relative age.

Some fossils are classified as index fossils, which are fossils that identify and date those strata in which they are found. They can be broadly distributed (perhaps even worldwide) and confined to a narrow stratigraphic range. The dates of these fossils are known and thus the position of other fossils in relation to these (before, with or above) allows the relative age to be determined.

Absolute dating

Absolute (or radiometric) **dating** methods give the age of rocks in number of years before the present, and are based on the atomic breakdown of certain unstable elements. The rate of decay of a large number of atoms of a radioactive element is constant and unaffected by the precise physical or chemical conditions in which the material occurs. It does not matter whether the material is on the top of a mountain or deep within the Earth; the rate is always the same. Nor can humans affect the rate.

Absolute dating is mainly used on igneous rocks but can be used for sedimentary and metamorphic rocks. Both sedimentary and metamorphic rocks have undergone some form of transformation, which may affect the original radio-isotopes in the rock material and make absolute dating inaccurate. Sedimentary rocks are formed from the accumulation of discrete mineral or rock particles derived from weathering and the erosion of pre-existing rock. The newly formed rock may contain radio-isotopes with differing half-lives, making it harder to date. Metamorphic rocks are formed as a result of partial or complete recrystallisation in the solid state of pre-existing rocks. Absolute dating presents problems when dating sedimentary rocks. Most sedimentary rocks consist largely of mineral grains that were formed long before the layers, or strata, that contain them were deposited. Absolute dating will tell the geologist the age of the crystals in the sedimentary rock but it will not tell when the layers were set down.

Atomic structure, radioactivity and why it can be used as a dating tool are discussed in the following sections.

definitions

relative dating
the dating of rocks by working out their place in a sequence of rocks in one locality and then comparing or correlating these rocks to those in other regions

stratigraphy
the study of rock layers or strata

law of superposition
each rock layer is younger than the layer below it if strata have not been overly disturbed

law of horizontality
sediments are deposited in layers parallel to the Earth's surface

principle of faunal succession
each rock formation contains a unique assemblage of fossils and each assemblage succeeds one another in an orderly and predictable way

absolute dating
gives the age of the rocks in years based on atomic breakdown of unstable elements

REVIEW ACTIVITIES

1 Explain the factors upon which the process of relative dating is dependent.

2 Explain what stratigraphy is.

3 Define the law of superposition.

4 Explain why absolute dating is more accurate on igneous rocks than other rocks.

5 Differentiate between relative and absolute dating methods.

EXTENSION ACTIVITY

6 Research what methods are used by geologists to date sedimentary rocks older than 500 million years and metamorphic rocks.

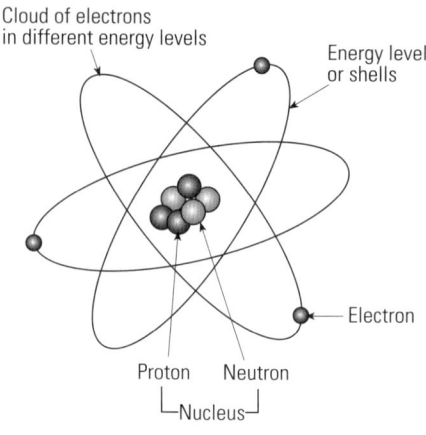

Figure 2.1.1 The structure of an atom.

ATOMIC STRUCTURE

Over the last 150 years a number of models for the structure of the atom have been proposed. Scientists have had difficulty in studying atoms and determining their structure due to their very small size, being no more than 10^{-11} m in size. As they are so small they have not been directly observed, but their properties have been studied. Current work on subatomic particles being undertaken by certain laboratories, including the European CERN laboratories, has helped in extending the modern atomic model.

Atoms are made up of three types of particles: neutrons, protons and electrons. The electrons are negatively charged, the protons are positively charged, and the neutrons possess no charge, that is they are neutral in charge. Two of the particles—the neutron and proton—are located in the centre of the atom, in an area called the nucleus. Each of these two particles has more mass than the third particle, the electron.

The proton and neutron possess a type of energy that is an 'attracting' force. The electron possesses a different type of energy called 'kinetic energy'. These forces balance one another.

The electrons form a cloud with different energy levels, or **shells**, orbiting the nucleus. (See Figure 2.1.1.) These layers are formed by electrons possessing the same amount of energy as each other. The electrons of one layer will possess a different level of energy than the electrons in another layer. If the electrons lose or gain energy they will move from one layer to another, and if they have enough energy they may break away completely from the atom.

The arrangement of the electrons in the layers is in a set pattern and form the **electron configuration**. The first shell contains two electrons; the second shell contains eight; the third shell contains eighteen; the fourth shell contains thirty-two; and the fifth shell contains fifty.

There are usually an equal number of protons and electrons in an atom and these can be calculated by looking at the atom's atomic number. The atomic number is equal to the number of protons in the nucleus of the atom. Helium has an atomic number of 2. Therefore, it possesses two protons and two electrons. Magnesium has an atomic number of 12. Therefore, it has twelve protons and twelve electrons. The number of electrons, protons and neutrons in the atom varies with each of the 108 different elements.

There are two types of atoms in the universe. One type is a stable form and the other type is unstable. The unstable type is one where the nucleus of the atom releases energy in the form of particles or waves. This release of energy from the nucleus is called **radioactivity** and results from the decomposition of the nucleus—the atom is trying to achieve a stable form. There are sixty known unstable atoms and they are called **radio-isotopes** or radionuclides. They are either pure elements (such as uranium) or unstable forms of common elements (such as oxygen or carbon). (See Table 2.1.2.) There may be a number of different radio-isotope forms of the same element. For example uranium 238 and uranium 235 both have ninety-two electrons and ninety-two protons, but uranium 238 has 146 neutrons and uranium 235 has 143 neutrons.

definitions

shells
the energy levels surrounding the nucleus of an atom. The shells are occupied by electrons.

electron configuration
the arrangement of electrons in the different shells

radioactivity
spontaneous emission of energy from unstable atom nuclei

radio-isotopes
elements that have the same atomic number and similar chemical properties but different atomic weights

Table 2.1.2 Radioactive decay of common isotopes

Radio-isotope	Half-life (x 10^8 years)	Daughter radio-isotope	Material in which parent is abundant
Potassium 40	13	Argon 40	Feldspar, mica, amphibole, volcanic rocks
Uranium 238	45	Lead 206	Uranium ores, zircon
Uranium 235	7.04	Lead 207	Uranium ores, zircon
Thorium 232	141	Lead 208	Zircon
Rubidium 87	488	Strontium 87	Potassium rich minerals, mica, feldspar, igneous and metamorphic rocks
Carbon 14	0.0000573	Nitrogen 14	Organic matter, such as wood and charcoal, dissolved limestone

REVIEW ACTIVITIES

1 Explain why our study of the structure of the atom has to rely on scientific models.

2 Describe the difference in the type of energy possessed by the three types of subatomic particles.

3 Describe how the electrons are arranged within the atom.

4 Differentiate between stable and unstable atoms.

5 What is radioactivity?

EXTENSION ACTIVITIES

6 Describe the development of the atomic model over the last 200 years.

7 One of the main tools used by the CERN laboratories in the study of the atom is an instrument called a 'doughnut'. Find out what it is and what is it used for.

RADIOACTIVITY AND HALF-LIVES

Radio-isotopes are constantly being formed in the nuclear reactions of stars. Other atoms formed at the birth of the universe are still radioactive, and will continue to be so for many more thousands of millions of years, until eventually they will end up as a stable form. The presence of radio-isotopes means that since the big bang, the Earth, all living things and everything around them are radioactive. Natural radioactivity is present inside the human body from the food, water and air we absorb. It is also common in the rocks and soils that make up our planet, in water and in building materials. There is nowhere on Earth that you cannot find natural radioactivity.

There are three sources of radioactivity:
- *Primordial*—from before the formation of the Earth.
- *Cosmogenic*—formed by cosmic rays, which is radiation of an extremely high penetrating power originating outside the Earth's atmosphere, consisting principally of charged particles moving at near the velocity of light.
- *Human produced*—enhanced or formed due to human actions.

The amounts of radioactive substances in materials can be measured by means of special instruments, such as Geiger Muller Tubes. These measurements can be used to work out ages of materials by calculating the half-life of the radioactive substances the materials contain.

Setting the radiometric clock

'The radiometric clock' is a name given to the time taken for all the radio-isotopic atoms in a radio-isotope to change to more stable atoms. (See Figure 2.1.2.) It is important to understand what causes the radiometric clock to start ticking.

When a new mineral grain has formed, it locks up all the atoms in that crystal, as if it was in a sealed atomic bottle. If some of the atoms are radioactive, and provided no daughter atoms were present in the mineral at the time of formation, we can determine how long ago the bottle was sealed. By measuring the number of remaining parent atoms (the original radio-isotope) and the number of daughter atoms (the intermediate products formed as a result of radioactive decay of a radio-isotope) it is possible to calculate when the crystal was formed. If some daughter atoms were trapped at the start, this has to be taken into account.

The temperature at which the radio-isotope is sealed in a crystal is called the **closure temperature**. For example in a potassium-rich mineral that has crystallised from molten rock, argon is trapped within the rigid crystal of the potassium-rich mineral. It is at the point when argon becomes trapped that it is sealed off, so that neither parent nor daughter radio-isotopes can enter or leave the crystal. Sometimes rocks are reheated when they are buried deeply. During this reheating, if the rock is taken past its closure temperature radio-isotopes may be lost, and the clock will reset, only to begin again when the temperature once again drops below the closure temperature. When rocks are eroded and transported away in fragments, the **radio-isotopic ratios** of the individual particles are not altered. This means that even a small fragment of rock can be dated using this method.

definitions

closure temperature
the temperature at which the radio-isotope is sealed in a crystal

radio-isotopic ratio
the ratio of parent to daughter radio-isotopes

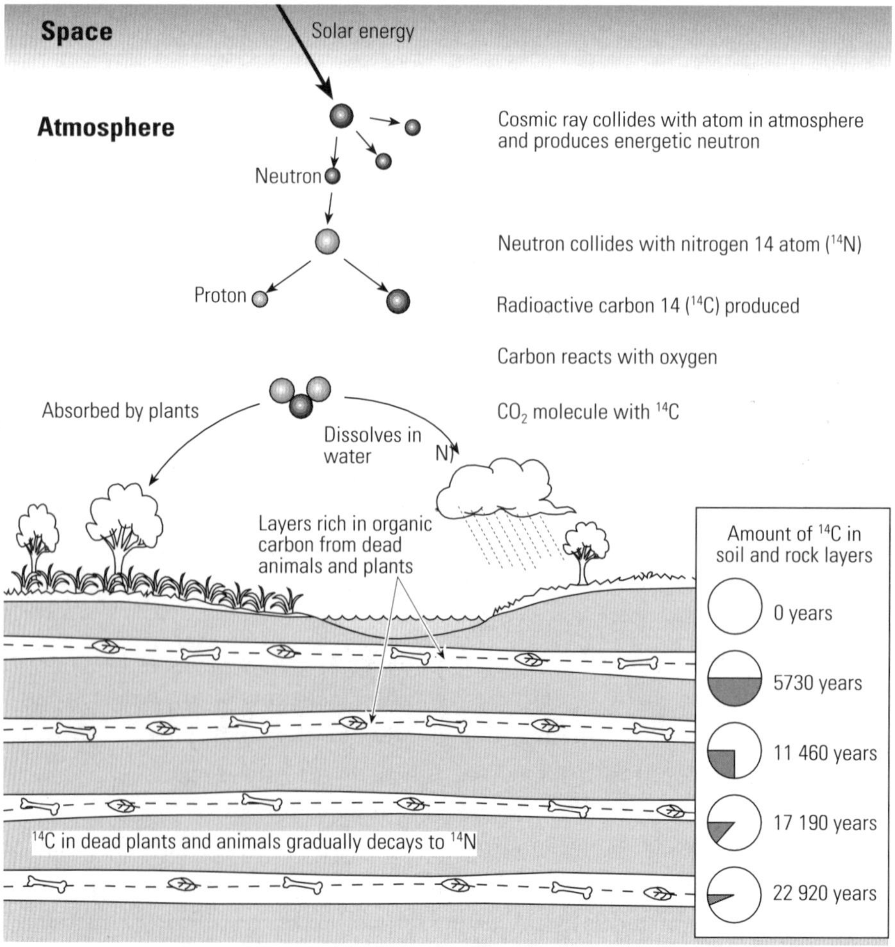

Figure 2.1.2 The carbon 14 radiometric clock.

CHOOSING A RADIO-ISOTOPIC SYSTEM

There are a number of radio-isotopic systems. These include:
- *Carbon 14 dating*—used for measuring organic material.
- *Uranium 238/uranium 235/thorium 232 series*—used for dating rocks.
- *Potassium–argon dating*—used for dating crystals.
- *Rubidium–strontium dating*—used for dating old rocks.

Commonly several radio-isotopic systems are used to date material. Which one(s) should be used is dependent on a number of factors:
- the type of material being tested
- what the material is made of
- the origins of the material
- how much of the radio-isotope was originally present in the sample
- the half-life of the radio-isotope
- whether the radio-isotope is part of a mineral compound.

It is important to choose the correct dating method for the material being tested, as the material's composition and origin may or may not preclude certain radio-isotopes from its composition.

How much of the radio-isotope was originally present in the sample is vital for accurate dating. If the original amount of a radio-isotope is unknown, it is impossible to use that radio-isotope for dating the material. The parent radio-isotope must be sufficiently abundant in the sample in order to be accurately measured. The daughter radio-isotope must be absent in the sample initially or, if it is present, there must be an understanding of why it was there to start with.

The radio-isotope being tested must have an appropriate half-life. For example a radio-isotope with a half-life of days or weeks would be unsuitable for dating old rocks. Also, the half-life must be appropriate for the age of the sample: long enough that some parent atoms are still present, but short enough that accumulation of daughter products has taken place.

The radio-isotope must be part of a mineral compound. An element may have several radio-isotopes present in its natural form. If that element forms a mineral, the ratio of the radio-isotopes will be the same in both the element and the mineral.

Regardless of the radiometric dating method used, every date given to a sample has a range of values in order to allow for any error. For instance when carbon 14 dating has been used to date a shell, its age may be given as 10 000 years, plus or minus 1500 years. This range of values takes into account discrepancies in amounts of parent elements and daughter elements, or accuracy in measuring. If, however, an older rock from the Cambrian period had to be dated, a different radioactive substance would be used; one with a longer half-life. Potassium 40 may be used, for example, because it has a half-life of 1300 million years. The range of possible experimental error may be several million years. In such situations, geologists may refer to the fossil assemblages in the rock sample in order to establish a more precise age for the rock or the material within the rock.

CARBON 14 DATING

One of the most commonly used radio-isotopes to measure the age of organic material is carbon 14. It is found in carbon-based compounds in living material. This radio-isotope is continuously produced in the atmosphere as a result of cosmic radiation from the universe. Carbon 14 has a half-life of 5730 or 5.7×10^3 years. This radio-isotope mixes rapidly and uniformly with other carbon radio-isotopes in the atmosphere. This uniform mixing means that carbon 14 is a constant fraction of all forms of carbon in the atmosphere.

Plants take in carbon from the atmosphere and utilise it in the process of photosynthesis. The carbon 14 radio-isotope ends up in the plant tissues in the form of sugars. Animals feed on plants or other plant-eating animals and the tissues, containing sugars with the carbon 14 in them, then pass into the tissues of the animal. While the organism is alive it will continuously take in carbon 14. At death, no more carbon 14 can be taken in. As radioactive decay takes place, the amount of carbon decreases.

Carbon 14 is very useful in dating animal and plant remains. It is also used by palaeoclimatologists (scientists who study ancient climates) to measure the age of ice in glaciers or the polar ice caps because it is possible to measure the carbon 14 levels in air bubbles trapped in the ice.

FISSION TRACK METHOD

One other way to ascertain the absolute age of particular rocks is by using the **fission track method** on the crystals or mica forming the sample. This method measures the damage to the crystals or mica that are caused by particles released during uranium decay. When these particles hit certain substances they leave tracks, or furrows, in the surface of the crystal or mica. These tracks can be observed by etching a crystal with hydrofluoric acid, which accentuates the tracks, and viewing them with a microscope. The number of fission tracks are counted and related to the decay of uranium. The age of the sample can then be calculated

definition

fission track method
particles from uranium decay leave tracks in glass or mica

REVIEW ACTIVITIES

1
Define the term 'natural radioactivity'.

2
Differentiate between the terms 'primordial' and 'cosmogenic'.

3
Using Table 2.1.3 (page 81), which radio-isotope would be used to date:
a potassium-rich minerals
b volcanic rocks?

4
Summarise how the carbon 14 radiometric clock works.

5
A fragment of zircon crystal is found inside a granite pebble, which is in itself located within a conglomerate. After radiometric dating, the zircon fragment is found to be 2 billion years old. What can be deduced about the age of the granite pebble and the age of the conglomerate?

EXTENSION ACTIVITIES

6
Research the uranium 238/uranium 235/thorium 232 dating process. Describe the steps in the radioactive decay of this series.

7
Explain the main problems with maintaining accuracy in the fission track method.

8
Some ancient placental mammal remains have been discovered in central New South Wales, close to the city of Orange. A lava flow about 3 m above the layer holding the mammal fossils has a radiometric date of 2.8 million years. The fossils appear to be in a stratum formed by a stream bed. A few centimetres above the stratum is a thin layer of volcanic ash. What method or methods would you use in trying to date the fossils more accurately than simply 'older than 2.8 million years'?

DENDROCHRONOLOGY

Dendrochronology, or tree ring dating, is the dating of past events (such as climate change) through the study of tree ring growth. Each year a tree adds a layer of wood to its trunk and branches, creating what are called annual rings. In spring, when moisture is plentiful, the tree devotes its energy to producing new growth. As summer progresses the amount of growth decreases until, in autumn, growth stops and cells die, with no new growth until the following spring. The contrast between the different sized cells is sufficient to form an annual ring. For each year of growth, the tree lays down an annual ring. Therefore a ninety-year-old tree will have ninety annual rings across its trunk. Some of the rings will be thinner than others, depending on the growing conditions in the particular growth season. A cold, dry growing season will produce a thin growth ring, whereas a warm, wet growing season will produce a thicker growth ring.

Certain trees are very slow growing and long lived. They may continue to grow over a period of thousands of years, thus giving dendrochronologists valuable information about an extended period. This method can be used on living tree material and also fossilised tree material. If dendrochronology is used on fossilised material in combination with carbon 14 dating, information can be gathered of paleoenvironments.

REVIEW ACTIVITIES

1 Why is dendrochronology a useful tool for dating?

2 Explain why some trees are more useful to dendrochronologists than other trees.

EXTENSION ACTIVITIES

3 Describe the climatic conditions necessary to produce thick annual rings.

4 Explain why carbon 14 dating should be combined with dendrochronology when gathering information on palaeo-environments.

AGE OF THE EARTH AND MOUNT NARRYER ZIRCONS

Recently, geologists exploring the Pilbara range of mountains in Western Australia recovered samples of zircons in the area around Mount Narryer. These zircons were located in ancient lava tubes, which are tunnel-like spaces beneath the surface of a solidified lava flow. Uranium 238/uranium 235/thorium 232 dating showed that the formation of these zircons occurred between 4 and 4.2 billion years ago. These dates made the zircons some of the oldest rock material to have been discovered in the world.

Recently it has been published in *Nature* magazine that further study of zircons from this area has now resulted in them being dated from 4.4 billion years ago.

REVIEW ACTIVITIES

1 Where is Mount Narryer located?

2 Where were the zircons located?

3 Explain what a lava tube is.

EXTENSION ACTIVITY

4 Explain what a zircon is and how it is formed.

SUMMARY

- Geologists have presented Earth's geological history as a column, divided into eons, periods, eras and, the smallest unit, epochs.

- There are two types of rock dating: relative and absolute (or radiometric).

- Relative dating is based on stratigraphy, which is the study of the layers that form rocks.

- The principle of faunal succession states that each formation contains a unique assemblage of fossils and each assemblage succeeds one another in an orderly and predictable way.

- Absolute dating is based on the decay of radio-isotopes that have very long half-lives, such as carbon 14. The age can be calculated by measuring the proportion of parent and daughter radio-isotopes.

- Other methods of dating rocks and other materials include the fission track method and dendrochronology.

- Atoms are made up of three particles: electrons (negative), protons (positive) and neutrons (neutral).

- Protons and neutrons are located in the centre of the atom in the nucleus.

- Electrons are located in shells, or energy levels, around the outside of the nucleus.

- There are two types of atoms. One type is a stable form and the other type is unstable. The unstable atoms are called radio-isotopes or radionuclides.

- There are three sources of radioactivity: primordial, cosmogenic and human produced.

- Radio-isotopes release energy from their decomposing atomic nuclei. This is called radioactivity.

- Index fossils are fossils that identify and date the strata in which they are located.

PRACTICAL EXERCISE
Dating rocks

Background

A team of geologists from the Australian Antarctic Survey team are sent to Scott Island. This island belongs to a small group of islands and is situated 100 km off the north coast of the Antarctic mainland. The function of the team is to take rock samples and carry out radiometric dating of the samples. In all, they collected twelve samples of rock. These samples are marked on the map of the island. (See Figure 2.1.3.)

ACTIVITIES

Look at the data and complete the following activities.

1
Photocopy Table 2.1.3. Using the graph of radioactive decay (Figure 2.1.4), calculate the age of each sample collected. Complete the last column in your copy of Table 2.1.3.

2
Analyse the data collected. Comment on the result.

3
Sample 12 has produced an interesting age for the rock sample. Explain what the problem is. If the date is true, what are the implications? If the date is incorrect, what could be the possible reasons for it being dated incorrectly?

4
Why were these three radio-isotopes picked and not a radio-isotope such as carbon 14?

5
Two of the radio-isotopes tested were radio-isotopes of uranium. What is the difference between uranium 235 and uranium 238?

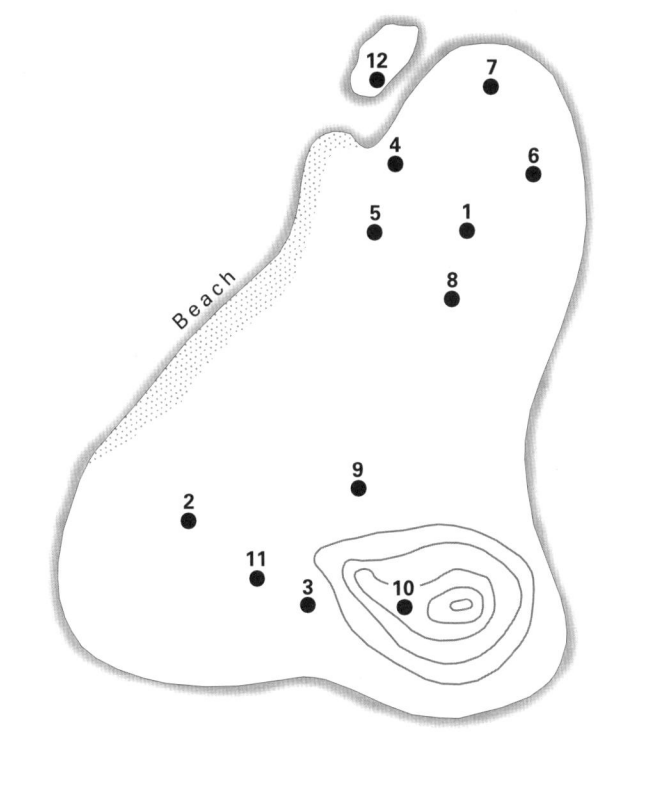

Figure 2.1.3 Sample sites on Scott Island.

Table 2.1.3 Dating rocks using radioactive decay

Sample	Radioactive element	Percentage of original mass remaining	Age of rock (billions of years x 10^9)
1	Uranium 238	79.0	
2	Rubidium 87	98.0	
3	Uranium 235	95.0	
4	Uranium 238	79.0	
5	Rubidium 87	98.0	
6	Thorium 234	92.5	
7	Uranium 235	16.0	
8	Thorium 234	90.0	
9	Uranium 235	80.0	
10	Uranium 235	100.0	
11	Uranium 235	65.0	
12	Rubidium 87	90.0	

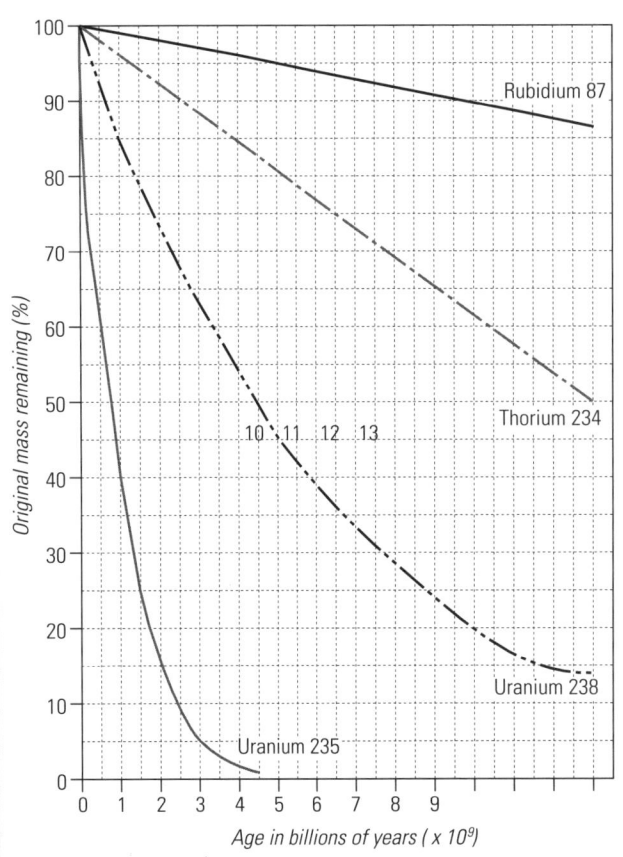

Figure 2.1.4 Graph of radioactive decay.

DYNAMIC EARTH

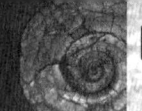

PRACTICAL EXERCISE
Drawing a simulated radioactive decay graph

This is an exercise to simulate the decay of a radio-isotope represented by chocolate M&M's (or cubes with one face painted white). Each time the M&M's (or cubes) are tipped out of the container and the M&M's with the lettering (or cubes with the white surface) facing uppermost are discarded, it represents the radio-isotope's half-life.

Equipment
Each group requires:
- either 100 M&M's (chocolate) or 100 cubes with one side painted white
- a container
- graph paper

Procedure

1 Place the M&M's or cubes in a container. Shake the container and tip them out.

2 Discard those M&M's with the lettering face upwards or those cubes with the white face upwards.

3 Count the remaining M&M's or cubes. Record that number as throw number 1.

4 Shake and tip out the M&M's or cubes. Repeat step 2 and record the number remaining and label that throw number 2.

5 Repeat steps 1 to 3 (recording a different throw number each time) until you have no M&M's or cubes left.

6 Draw a line graph where the X axis is the number of throws and the Y axis is the remaining M&M's or cubes. Plot your results on the graph. The line graph represents the half-life of a substance.

2.2 Plate movements

OUTCOMES

At the end of this chapter you should be able to:

- recall evidence that crustal plates move over time
- recall the relationship between movements of Earth's plates and convection currents in the asthenosphere, and to gravitational forces
- describe the plate tectonic model and use it to explain the distribution and age of continents and oceans.

HISTORY OF PLATE MOVEMENTS

If Captain Cook could return to Australia and disembark exactly where he did 200 years ago, he would get a major surprise. He would be stepping into 2 m of water! Australia has moved approximately 200 cm since his last visit.

Why have the continents moved and how have the main features of the Earth's surface been formed? These questions have led geologists on a search for the last 200 years.

The first steps were made by James Hutton, the founder of modern geology, who proposed in 1795 the principle of uniformitarianism, which can be summed up in the statement 'the present is the key to the past'. The principle put forward the idea that the Earth's present form has developed gradually over a long period of time by processes that still occur today. His ideas were fairly unpopular at the time, as people tended to support the idea of catastrophism. This idea considered that the Earth was young and its present form had been caused by a series of sudden violent events.

A number of other ideas were put forward to explain the Earth's form. They included the contracting earth theory, which was developed between 1859 and 1873 by Dana and Hall. It stated that the Earth started as a molten mass and as it cooled it shrank. Evidence for this could be found in the fold mountains. It did not explain the position and shape of the continents or the presence of rift valleys, for example.

Another theory that was popular during the nineteenth and twentieth centuries was the expanding earth theory. This put forward the idea that the Earth was slowly expanding due to the heating processes within the Earth itself. This heating caused the continents to break into smaller fragments and the cracks formed oceans.

Continental drift hypothesis

The climatologist Alfred Wegener was trying to explain the Carbo-Permian ice age and needed to have Africa and Europe 40° south of their current location to explain the spread of equatorial to polar environments recorded in rocks of the same age. As a starting point he used the jigsaw fit model, which is where the individual continents can be cut out of a world map and they fit together like pieces in a jigsaw. This model allowed him to work out that the continents were at some time joined together and have drifted apart over time. Wegener tested the model with the distribution of terrestrial animal and plant fossils, rock types and fold belts. He discovered that these supported his idea of the continents having been joined together at some stage.

In 1915 Wegener put forward the **continental drift hypothesis**, in which he stated that the continents have not always been in their present locations, but instead have 'drifted' and changed positions. They were originally all united, forming the supercontinent of **Pangea** during the late Paleozoic era. He suggested that at some time later the continent of Pangea started to split, forming smaller continents. The evidence of these splits could be found in places like the rift valleys in Africa, which are long valleys or depressions bounded by steep dipping parallel faults. (Rift valleys are discussed in more detail in chapter 2.4.) These continental fragments then took many millions of years to drift to their present positions.

Geologists rejected his model for a number of reasons. He was an outsider to the field and his ideas challenged their expert views, but most importantly Wegener could not provide a mechanism for the continental drift. The views at the time were that the sea floors had formed at the beginning of time and that continental and oceanic geography were permanent and unchanging.

It was not until Harry Hess's work in the 1940s and 1950s on sea floor spreading that the continental drift hypothesis was given credence. With more evidence supporting the idea, the hypothesis developed into the improved tectonic plate theory. (See Figure 2.2.1.)

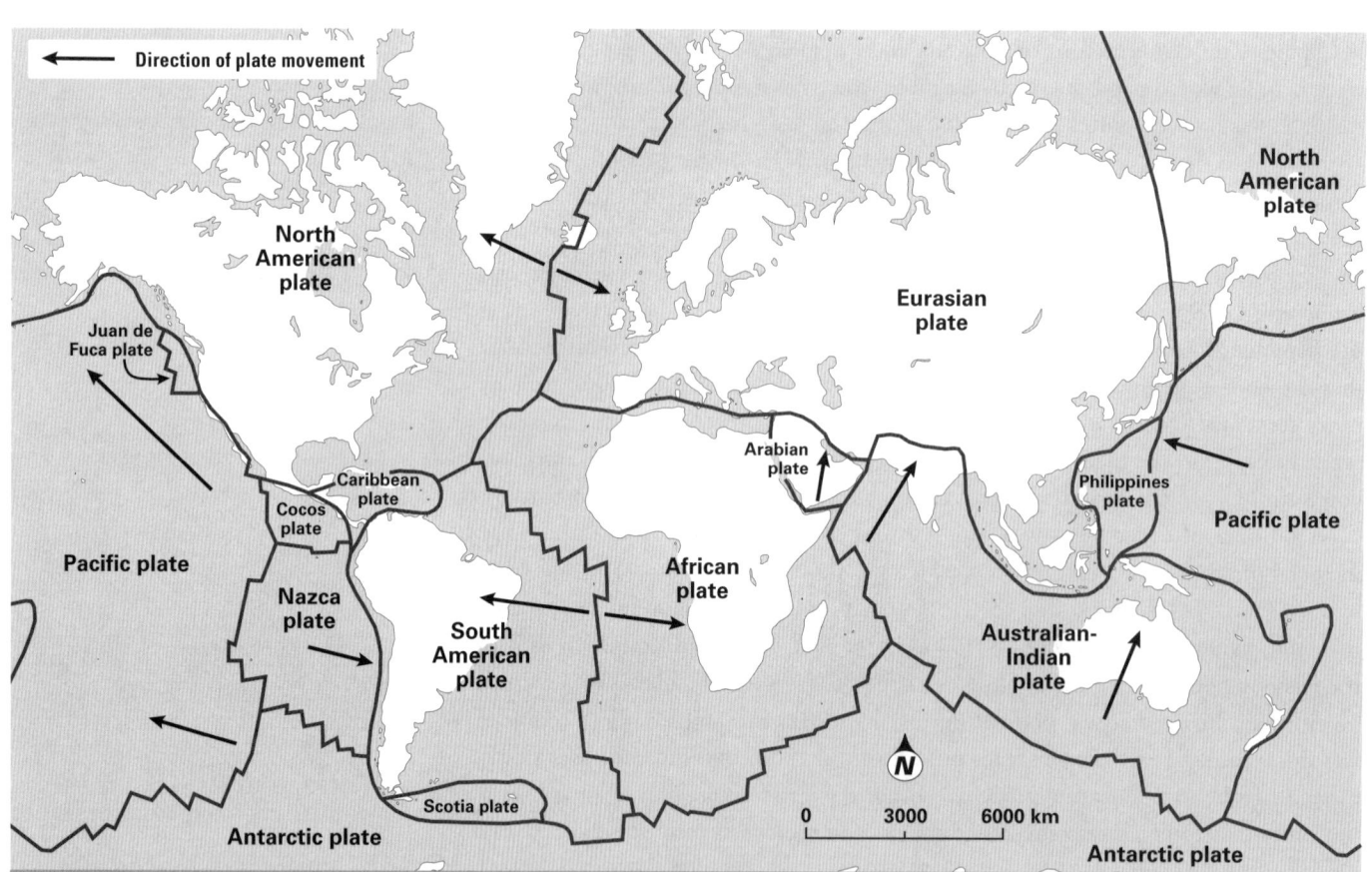

Figure 2.2.1 Earth's tectonic plates.

Tectonic plate theory

The **tectonic plate theory**, or model, states that the lithosphere has fragmented into several large plates, which move. As the continents are part of these plates, they move accordingly. These movements formed mountain ranges, oceans and island chains. Tectonic plate movements are also responsible for volcanoes and earthquakes.

Over the last 200 years geologists have discovered that the planet's surface is not a stable, never changing layer. Instead the crust is a dynamic ever changing layer, that is continuously on the move. It has been found that, despite the dynamism of the

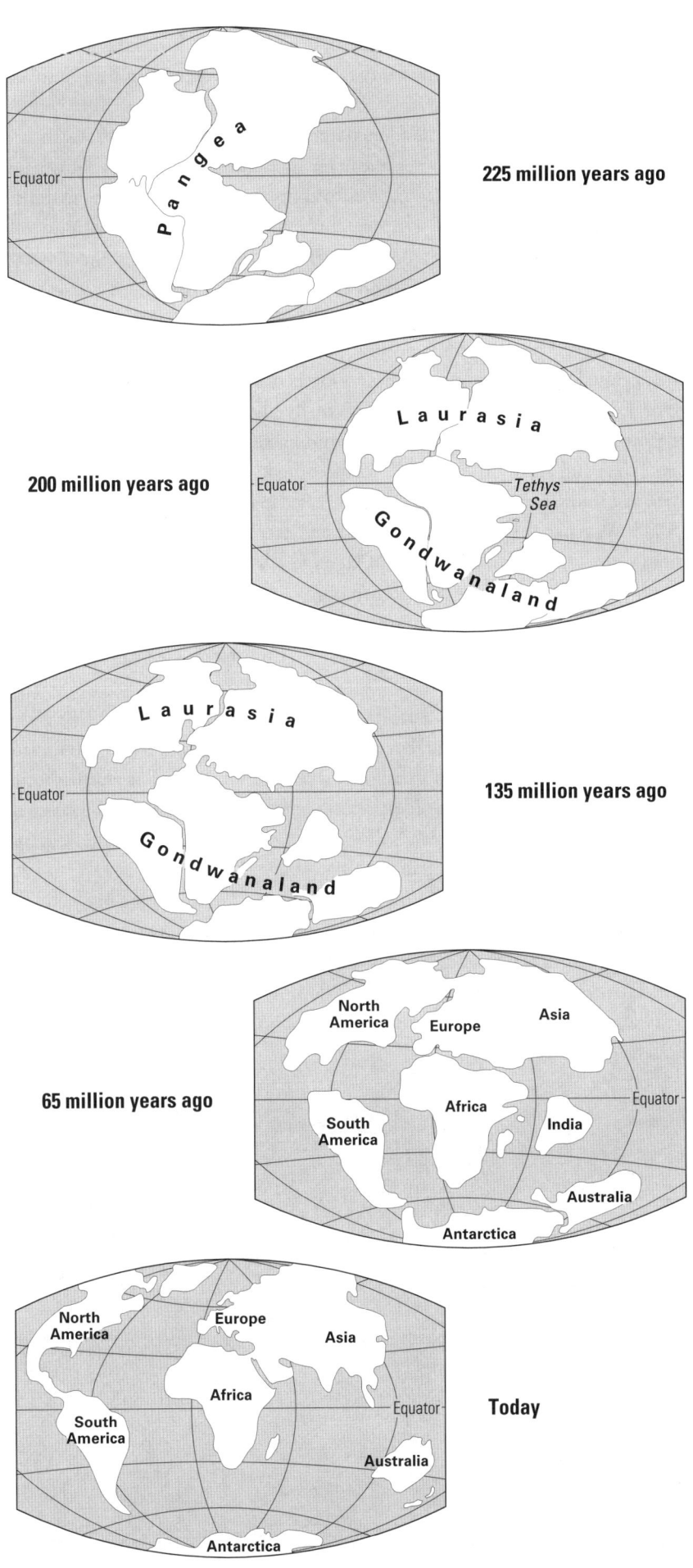

Figure 2.2.2 Continental movements over Earth's history.

definitions

continental drift hypothesis
the theory that the continents have undergone movement

Pangea
a supercontinent that existed about 200–300 million years ago

tectonic plate theory
states that the lithosphere has fragmented into several large plates, which move. As the continents are part of these plates, they move accordingly.

crust, the planet is neither getting bigger or smaller. New lithospheric material appears to be being produced at the same rate as material is moving back into the asthenosphere.

The tectonic plate theory was developed and accepted during the 1960s. It was not a new, novel theory. Instead, rather like many good things, the theory took many years to mature. Alfred Wegener had put forward the initial idea of continental drift in the early part of the twentieth century, but he was not the first to point out evidence supporting the modern tectonic plate theory. In 1620 Francis Bacon noted the similarity in shape of the continents on either side of the Atlantic. The concepts continued to build, with Antonio Sneider in 1858 suggesting that all the continents were at one stage located together on one side of the earth. In 1879 Sir George Darwin suggested that the continents have moved from their original positions. In the 1890s Edward Suess proposed that originally there was a Gondwanaland, a supercontinent made up of the present day landmasses of India, Antarctica, South America, Africa and Australia. As proof, he used the location of mountain ranges as well as the commonality of fossils. (See Figure 2.2.2, page 85.)

In 1924 Wegener further developed the theory of continental drift to explain similarities in the distribution of flora and fauna fossils, rock types and fold belts on both sides of the Atlantic. In 1928 Arthur Holmes suggested that **convection currents** in the mantle were the driving force for the movements of the continents. Harry Hess in the 1960s suggested that the sea floor separated at the **mid-ocean ridge** and the gap was filled with magma.

Two processes can be recognised as shaping tectonic plate theory. The first is the process of sea floor dynamics. It takes approximately 200 million years for plate movements to form an ocean, and then for it to be consumed in a **subduction** zone. Geologists know this because the oldest sea floor they can find is off the coast of Japan; it dates from 175 million years. All older sea floors have been subducted and so the sea floors are relatively young in age.

Secondly, there is the effect of continental dynamics. Continents are buoyant and will not subduct. Continents can grow by accumulation of landmasses on their edges. As continents do not subduct, but grow from accumulation, the rocks that form the continental masses are much older than the sea floor material. As discussed in chapter 2.1, the oldest rock materials date from 4.2–4.4 billion years and are zircons from Mount Narryer in Western Australia.

definitions

convection current
a pattern of mass movement of mantle material in which the outer area is downflowing and the central area is uprising due to heat differences

mid-ocean ridge
occurs when two crustal plates separate and basaltic material wells up through the spreading centre

subduction
the movement of one crustal plate under another so that the descending one is 'consumed' by the mantle

ACTIVITIES

1 Explain why the contracting earth theory lost support.

2 Construct a time line to show how the tectonic plate theory developed.

3 Describe the two theories that try to explain the causes of plate movements.

4 What evidence was used by Edward Suess to justify the existence of the ancient landmass of Gondwanaland?

5 Describe the two processes that have helped shape the tectonic plate theory.

EXTENSION ACTIVITY

6 Research why the expanding earth theory lost support.

CRUSTAL STRUCTURE AND MOVEMENT

Earth scientists think that the lithosphere is divided up into at least eight large plates that change over time. Recent NASA imagery suggests that the Australian-Indian plate is in the process of becoming two plates. During the course of Earth's history, boundaries in the crust (such as mid-ocean ridges and subduction zones) have formed, changed position and disappeared. To understand the processes of the tectonic plate theory it is necessary to understand both the structure and movement of the crust.

Crustal structure

There is a density gradient between the core and crust. The core is made up of material of the highest density (that is iron and nickel), while the crust contains the least dense material.

The rocks that form oceanic crust are darker in colour than the continental crust, because oceanic crust contains rocks rich in magnesium and iron. The oceanic material is denser than the lighter-coloured rocks of the continental crust, which are rich in silicon and aluminium. Oceanic crust rocks belong to a group of rocks called **mafic**. The oceanic crust is formed from two main types of mafic rocks: basalt and gabbro. Basalt forms the upper layers, while gabbro forms the thicker, internal lower layers of the oceanic crust.

Continental crust rocks that are rich in silicon and aluminium are categorised as **felsic**. Continental crust rocks also contain high levels of the mineral **feldspars**. Rocks such as the granites and andesites are major components of continental crust. The continental crust does not have any overall distinct layering. (Crustal rocks are examined in more detail in chapter 2.4.)

Oceanic crust is composed of three distinctive layers and has a fairly consistent overall thickness of 8 km. Whilst the continental crust may have localised layering, it does not have an overall layering.

Continental crust is a lot thicker than oceanic crust—on average 60 km—but thins down to a few kilometres under rifts and up to 80 km under young mountain ranges.

definitions

mafic
a group of rocks containing dark minerals composed of magnesium and iron

felsic
a type of rock containing light-coloured aluminium and silicon minerals. Examples are quartz and feldspars.

feldspars
the most common group of rock-forming minerals that make up 60% of the crust. They are made up of silicates.

Crustal movements

From evidence such as the shape of the continents, the types of animals and plants on different continents and similar rock types on different continents, it is possible to state that a process of crustal movements is taking place. When it comes to explaining why these crustal plates are moving it becomes somewhat more difficult.

The Earth's surface is composed of a number of plates, which carry the terrestrial landmasses. These plates are moving slowly in different directions. Some plates are colliding with each other, others are moving apart and some slide past each other. Exactly what drives plate tectonics is not known. One theory is that convection within the Earth's mantle pushes the plates. (See Figure 2.2.3, page 88.) Another theory is that gravity is pulling the older, colder and thus heavier sea floor with more force than the newer, lighter sea floor.

CONVECTION CURRENTS IN THE ASTHENOSPHERE

The asthenosphere is a hot layer, and the rocks in this layer consist of solid particles, with very small amounts of liquid occupying the spaces in between. It is close to its melting point, with little strength and is able to flow like a liquid. The lithosphere is divided into plates. Some plates carry continents with them as they move, while others carry only oceanic crust.

The lithosphere and asthenosphere are bound together. If the asthenosphere moves, the lithosphere must move too. Conversely, if the lithosphere moves, the asthenosphere has to move also.

It is known that the lithosphere has kinetic energy, and the source of this energy is the Earth's internal energy. It is also known that heat energy passes into the lower surface of the lithosphere from the mesosphere and asthenosphere. All scientists are in agreement that convection keeps the asthenosphere hot and weak.

The basic mechanisms for the movement of the plates are by convection currents in the asthenosphere. The heating of the asthenosphere results in circular convection currents. As the rock heats, it rises to the boundary between the asthenosphere and lithosphere. As it moves further away from the heat source, the rock starts to cool and, when it becomes denser, it starts to sink. Once it sinks, it is heated again and once more starts to rise. This heating-cooling process means a circular convection current is formed. As the current is in motion it then drives the plate sideways away from the hot point: the ridge, which is the edge of the plate. As the plate moves away from the hot point, the material cools and becomes denser. The densest end of the plate sinks back into the asthenosphere at the subduction zone.

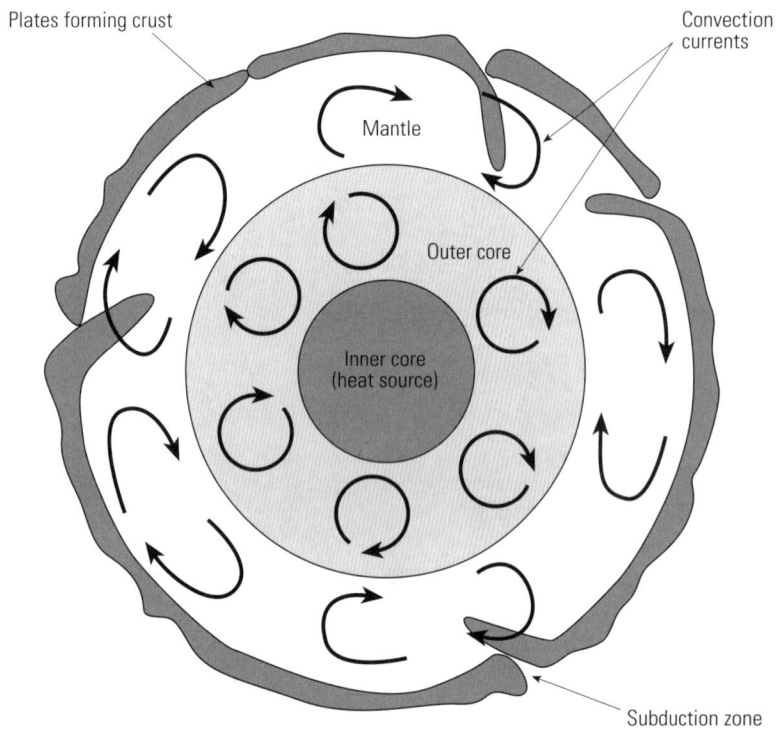

Figure 2.2.3 Convection currents in the asthenosphere.

MECHANISMS BY WHICH PLATES MOVE

Scientists accept that the plates do move, but the processes by which they move are still not completely understood. As yet, scientists are still hypothesising and modelling what causes the convection currents in the asthenosphere and how they make the plates move. However, it is accepted that they are involved in the movement of plates. Convection currents, while plausible, may be only part of the answer.

Gravitational forces are thought to play a role in causing the plate movements. There are two heat sources for the convection currents in the mantle: radioactive decay and residual heat. As radioactive decay occurs (see chapter 2.1), energy is released in the form of heat and this powers the convection currents. A second type of heat is residual heat. This is gravitational energy left over from the formation of the Earth by accretional processes. Gravity also has a role in the sinking of cold, denser oceanic plates in the subduction zones.

Plate motion can be seen at plate boundaries. There are three main types of boundaries:
- divergent boundaries where the plates are moving apart from each other
- convergent boundaries where one plate subducts under another
- transform boundaries where the plates slide past each other.

These boundaries will be discussed in detail in chapter 2.4.

In convergent boundaries one plate subducts under another. This process of subduction has caused geologists to propose a number of different hypotheses to explain plate motions at subduction zones. They include:
- ridge-push mechanism
- gravity-slide mechanism
- slab-pull mechanism. (See Figure 2.2.4.)

The ridge-push mechanism hypothesis states that the plates are moved aside from the diverging boundary and towards a subduction zone by the injection of lava at the mid-ocean ridge. As they move away from the mid-ocean ridge, the plates cool and thicken as they subside. A slope is formed on the side of the mid-ocean ridge, caused by subsidence. Another slope forms on the base of the lithosphere, and new igneous rock is added to the base of the lithosphere by progressive cooling of the lower asthenosphere. The oceanic plate slides down towards the trench by the action of lava being pushed up at the mid-ocean ridge.

The gravity-slide mechanism is based on the difference in elevation between the ridge crest and trench. If the subducting plate falls into the mantle at a steep angle, the trenches and the overlying plate are pulled towards the subducting plate. This pull is due to gravity. The plate slides down the sloping asthenosphere, keeping the central rift valley open and allowing the magma to well up in the gap.

The slab-pull mechanism explains plate movement by suggesting that the hot part of the plate is pulled from the ridge crest by the cold, denser part of the plate that descends into the trench.

The whole mechanism of plate movements is still unknown but involves all three of these processes, together with the effect of convection currents in the asthenosphere. The idea that all three mechanisms are involved in the mechanism of plate movements is open to question, and more study of this problem will have to be made before a final model can be proposed.

a. Ridge-push mechanism

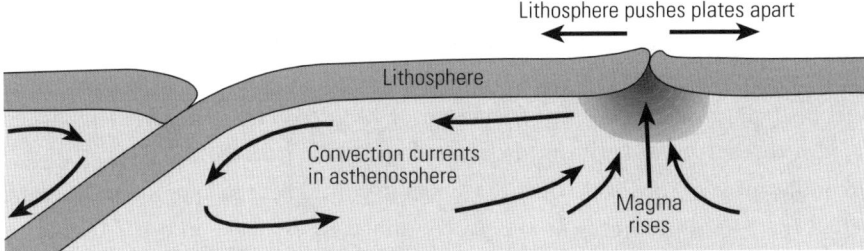

b. Gravity-slide mechanism

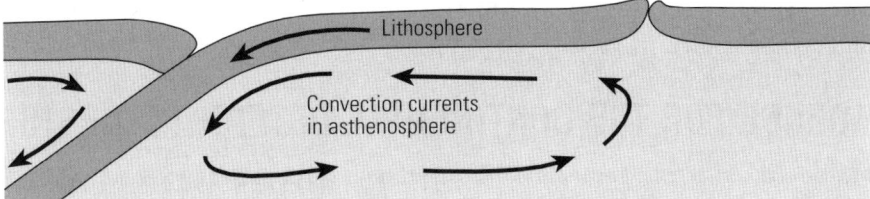

c. Slab-pull mechanism

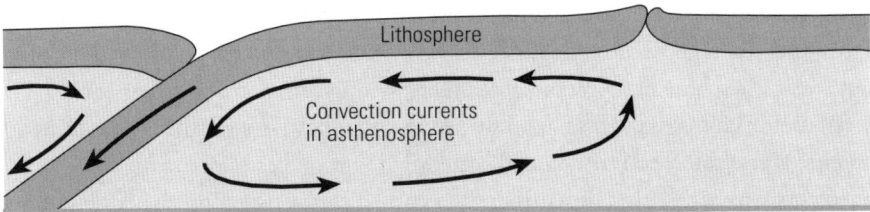

Figure 2.2.4 Three hypothesised mechanisms of plate movement.

REVIEW ACTIVITIES

1 Explain what is meant by the term 'density gradient'.

2 Distinguish between the terms 'mafic' and 'felsic'.

3 Construct a table to contrast the features of the continental and oceanic crusts.

4 Explain the meaning of the terms 'lithosphere' and 'asthenosphere'.

5 Describe the sequence of events in the process of convection within the Earth.

6 Explain the role of gravity in the plate tectonic theory.

7 Distinguish between the three mechanisms of ridge-push, gravity-slide and slab-pull.

EXTENSION ACTIVITIES

8 Explain how scientists know the processes that are taking place in the mantle and causing plate movements.

9 Which tectonic plates are solely made up of oceanic crustal material?

definition

biota
the animals and plants living in a particular ecosystem

REJECTION OF THE CONTINENTAL DRIFT HYPOTHESIS

Wegener's hypothesis of continental drift took over fifty years to be accepted. Originally the hypothesis was rejected for a number of reasons.

Firstly, there was no explanation of how the continents moved. It was generally thought that the continental crust (made up of less dense felsic material) ploughed through the oceanic crust (made of the more dense mafic material).

Secondly, some of the fossil evidence seemed to contradict the hypothesis. Wegener suggested that the supercontinent Pangea continued in existence into the Cainozoic era. Palaeontologists, however, had evidence that the world's **biota** had evolved into distinctive zones since the start of the Cainozoic era, 65 million years ago. Therefore, as there were distinct zones it is more likely that the separation of Pangea occurred. It is now understood that Wegener had made a dating error that misled the paleontologists, as Pangea had started to break up near the start of the Mesozoic era, around 200 million years ago.

EVIDENCE FOR THE CONTINENTAL DRIFT HYPOTHESIS

As mentioned earlier, the continental drift hypothesis is supported by several pieces of evidence.

When Wegener tried to fit the continents together, the fit was not exact enough for many scientists to accept. It was not until the 1970s that Thomas Bullard matched the continent margins using computers. He took into account sea level changes and continental shelves, and thus a better fit was achieved. If the continents are placed together they fit together like a giant jigsaw. In the case of fitting Africa and South America together, the fit is remarkable; there is only an overlap or gap of 90 km between them. This is despite the millions of years that have passed since the continents were joined, and the rock forming and breaking down processes occurring.

Geologists have discovered that there is a correlation between the age, orientation, composition and structure of some rocks on different continents, despite being separated by many thousands of kilometres. Rocks in the north-east of Brazil and West

Africa show strong correlation. The rocks are about 550 million years old, showing that the two continents were joined at that time. There is also a strong correlation between certain mountain ranges in different countries. The Appalachian mountains in the USA are composed of the same material as, and are able to be lined up with, mountain ranges in Canada, Ireland, Britain, Greenland and Scandinavia.

Work into glacial deposits has also yielded evidence to support the hypothesis that the continents were joined. Ice sheets leave scratches and grooves, or striations, in underlying rock layers as well as folds and wrinkles in soft sediments. When South America and West Africa are moved back together using computer simulation, the remains of rock layers affected by ice sheets show scratches, grooves, folds and wrinkles in the same orientation, and radiating out from a central point.

There are a number of living, and fossil, plant and animal groups that show close relationships, despite being on different continents. This suggests a common origin for the organisms in Gondwana before it broke up. (Pangea broke into two continents: the northern continent of Eurasia and the southern continent of Gondwana. Gondwana then broke up into the landmasses of Antarctica, Australia, South America, India and South Africa.)

Members of an ancient plant family called the Proteacae are found in Australia (such as the waratah), while other members are found in South Africa in the form of proteus. Both have distinctive, recognisable features that show they are related, but also have unique features that demonstrate they are different from each other now. At one time the ancestors were the same, but the distance between the two continents and the length of time since the continents separated meant that they have changed over time.

There are visible similarities between the members of the bird family called the rattites (a group of flightless birds), which include the ostriches, rheas, kiwis and emus. However, these birds come from the different continents of South America, Africa and Australia, which are many thousands of kilometres apart. Similar evidence of common features can be found in marsupials that live in Australia and the Americas. Likewise, evidence of commonality is found within members of the camel family: Arabian and Bactrian camels, llamas, vicunas and alpacas. They are all camels, but some live in South America, while others live in Asia. Each of the three families—the rattites, the marsupials and the camels—show evidence of a common ancestry from which they have evolved, and therefore, at some time in the past, the continents they live on must have been linked.

As well as living organisms showing a relationship between species on different continents, scientists also have similar evidence from fossils. Palaeontologists have discovered fossil evidence of the ancient fern *Glossopteris* on the widely separated continents of Antarctica, Australia and South America, as well as South Africa and India. (See Figure 2.2.5, page 92.) Not only does this support the idea that the continents had to be connected in order for this fossil plant to be so widespread, it also supports the theory that the continents have moved to a different latitude. The plant could not have survived in the climatic conditions that Antarctica has today; the climate must have been warmer or Antarctica must have been at a warmer latitude.

Other fossil evidence of continental movements comes from the fossil remains of a freshwater reptile called *Mesosaurus*. (See Figure 2.2.5, page 92.) This reptile lived during the Permian period (245–286 million years ago) and its remains are found in southern Brazil and in South Africa. It would have been impossible for the animal to cross the Atlantic Ocean and so these remains support the idea that the two landmasses were joined. Other fossilised organisms, such as earthworms, are found on different continents and lend support to the theory that these landmasses were once joined.

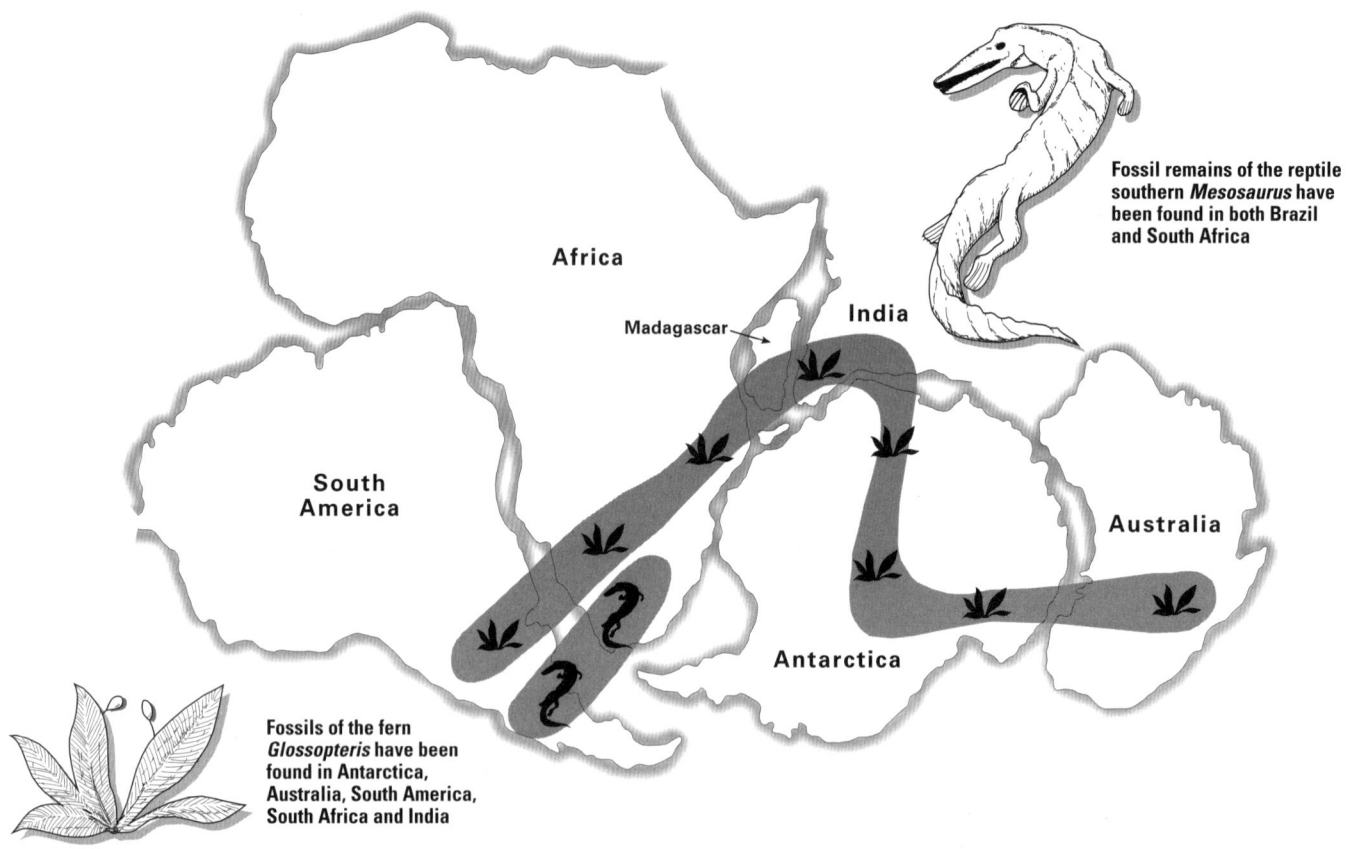

Figure 2.2.5 Location of *Glossopteris* and *Mesosaurus*.

Once all these animals lived on one landmass—Gondwana—which due to tectonic plate action split up into the individual continents and became separated by thousands of kilometres.

Sea floor spreading

Wegener put forward the hypothesis of continental drift at the beginning of the twentieth century. However, it wasn't until the middle of the century that oceanographic work led to the discovery of vital evidence necessary to change the ridiculed 'continental drift' hypothesis into the accepted model of 'plate tectonics'.

Much of the work on the sea floor was as a consequence of the Second World War and the following Cold War with communism. It was essential that the allied forces were able to detect enemy submarines and move friendly submarines through enemy waters without being detected. It was therefore necessary to map the sea floor and also detect the submarines by means of sonar and magnetometers.

During the mid-twentieth century four major scientific developments spurred the formulation of the plate tectonics theory:
- demonstration of the youth of the sea floor
- emergence of the sea floor spreading hypothesis
- precise documentation of the Earth's earthquakes and volcanic activity, which is mainly located along oceanic trenches and submarine mountain ranges (See chapter 2.4.)
- confirmation that there were repeated reversals of the Earth's magnetic field in the Earth's past.

Until the mid-nineteenth century little was known about the sea floor. With the laying of the first trans-Atlantic telegraph cable in 1858 the first evidence of sea mounts in the mid-Atlantic came to light. As geologists gathered more information about the sea floor an anomaly came to light. Geologists thought the age of the sea floor should be around 4 billion years old and, if so, the layer of sediment should be a lot thicker than it actually is.

This problem was solved in the 1960s when geophysicists came up with an explanation. They proposed that the sea floor had split along the mid-ocean ridge. The rocks on either side of the split were moving away from each other. The gaps between the two sets of rocks were being filled by molten material, in the form of pillow lava, forming oceanic crust. This process of sea floor spreading has been in operation over millions of years and has formed a system of mid-ocean ridges 50 000 km long.

This hypothesis of sea floor spreading was supported by several pieces of evidence:
- The rocks are youngest close to the ridge crest and get progressively older the further away from the ridge.
- The youngest rocks at the ridge crest always have the same polarity, as is present worldwide.
- Stripes of rock parallel to the ridgeline show polarity reversal, corresponding with changes in the earth's polarity. (See chapter 2.3.)

Additional evidence came to light during a period of petroleum exploration shortly after the Second World War. In 1968 the *Glomar Challenger*, an ex-CIA spy ship that was designed for submarine rescue, but with the appearance of a geological research ship, was carrying out legitimate research. It was crisscrossing the mid-Atlantic Ridge between South America and Africa and drilling core samples at specific locations. When the ages of the samples were determined by radiometric methods, it provided clinching evidence that proved the sea floor spreading hypothesis.

Further evidence for sea floor spreading came in the 1960s with the Worldwide Standardised Seismograph Network (WWSSN). By the late 1920s seismologists were beginning to identify several prominent earthquake zones parallel to the trenches, which extended several hundred kilometres into the Earth. These zones are known as benioff zones, which are discussed in more detail in chapter 2.4. Evidence of seismicity of these zones came from the WWSSN, which was designed primarily to monitor the above ground testing of nuclear weapons in contravention of the 1963 Test Ban Treaty.

REVIEW ACTIVITIES

1
The discovery of the fossil remains of *Mesosaurus* and *Glossopteris*, each on different continents many thousands of kilometres apart, has given us a number of pieces of information about where they lived. Outline some of the information these fossils have given.

2
Why was the Second World War and the Cold War important in improving our knowledge of the sea floor?

3
What evidence supports the idea of sea floor spreading?

4
Discuss the evidence for and against the plate tectonic theory.

EXTENSION ACTIVITIES

5
Explain what the characteristics of pillow lava are.

6
Describe how a seismograph works.

SUMMARY

- In 1915 Alfred Wegener proposed the continental drift hypothesis, which states that the continents have not always been in their present position but have drifted and changed position.

- Originally the continents were joined together in a super-continent called Pangea.

- In the 1940s and 1950s Harry Hess's work on sea floor spreading supported the continental drift hypothesis and was fundamental evidence in the improved tectonic plate theory.

- The tectonic plate theory states that the lithosphere has fragmented into several large plates, which move as a result of movements in the underlying hot mantle. As the continents are part of these plates, they move accordingly.

- The rocks that form oceanic crust and continental crust are different.
  - Oceanic crust is basaltic in type, darker in colour, mafic (being rich in magnesium and iron) and has a higher density. Overall it is generally composed of three distinct layers and is fairly thin; up to 8 km in thickness.
  - Continental crust is granitic in type, lighter in colour, felsic, rich in silicon and aluminium and has a lower density. It does not have any overall distinct layering and is thicker; an average of 60 km.

- There are a number of proposed mechanisms of crustal movements. The basic mechanism is one of convection currents in the asthenosphere. There are, however, three other mechanisms that may be involved in plate movements:
  - the ridge-push mechanism
  - gravity-slide mechanism
  - slab-pull mechanism.

  Plate movements may be the result of one mechanism or a summation of all of them.

- New plate material is made at the mid-ocean ridge. Old material is returned to the asthenosphere at subduction zones, usually located at deep sea trenches.

- Evidence for the plate tectonic theory includes:
  - the jigsaw fit of the continents
  - correlation between age, orientation, composition and structure of some rocks on different continents
  - correlation of glacial deposits on different continents, evidence being orientation of scratches and grooves in rock layers as well as orientation of folds and wrinkles in soft sediments
  - fossils found on different continents are the same, for example Glossopteris, Mesosaurus and earthworms
  - similar evidence of related living organisms can be found in the marsupials, rattites, camels and the Proteacae plant family.

- Evidence supporting the plate tectonics theory comes from studies of the sea floor:
  - the young age of the sea floor
  - location of volcanic and earthquake activity mainly along oceanic trenches and submarine ridges or mountain ridges
  - evidence of repeated reversals of the Earth's magnetic field in the planet's past.

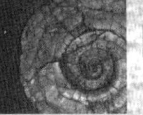

PRACTICAL EXERCISE
Plate boundaries

ACTIVITIES

Using Figure 2.2.1 (page 84), complete the following activities.

1 Name the countries that the Australian-Indian plate passes through.

2 Name three plates that include both continental and oceanic crust.

3 Name two plates that are mainly composed of oceanic crust.

4 Where do most divergent plate boundaries occur?

5 Which is the smallest plate?

6 With which plates is the Australian-Indian plate converging?

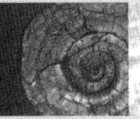

PRACTICAL EXERCISE
Modelling convection currents

This exercise will give you an understanding of the processes that cause convection currents.

Equipment
- 40 ml dark treacle
- 40 ml light corn syrup
- large 1 litre beaker
- Bunsen burner

Procedure

1 Fill the beaker with 750 ml of cold water.

2 Pour in the 40 ml of treacle. This should sink to the bottom of the beaker.

3 Pour in the 40 ml of light corn syrup. This should float on the top of the water.

4 Start to gently warm the water and then complete the following activites.

ACTIVITIES

1 Draw and describe what happens to the treacle and syrup.

2 Propose an explanation as to what happens to the treacle.

3 Is there more than one area of upwelling and downwelling?

4 As heating continues, describe what happens to the convection current.

5 Do the two substances—syrup and treacle—mix or do they remain separate?

DYNAMIC EARTH 95

2.3 Magnetism and the wandering poles

OUTCOMES

At the end of this chapter you should be able to:

- explain how the alignment of magnetic fields of minerals in cooling igneous rocks is an indication of the rock's position relative to the magnetic poles

- assess the significance of apparent polar wandering paths as evidence of continental mobility

- explain the significance of the discovery of magnetic field reversals on the development of a time scale

- analyse the assistance that palaeomagnetism has provided in understanding the process of sea floor spreading and the movement of continents.

THE EARTH'S MAGNETIC FIELD

Before the sea floor spreading hypothesis gained full acceptance, support for the hypothesis came from evidence of magnetic patterns on the sea floor. The Earth's magnetic field is thought to be produced by the moving charges caused when convection moves the liquid outer core around the solid, iron-rich inner core. (See Figure 2.2.3, page 88.) The axis of the magnetic field is not aligned parallel to the Earth's rotational axis, so that the magnetic and geographical poles do not coincide. (See Figure 2.3.1.) The magnetic poles are different from the North and South Poles; the geographical North and South Poles are the axis poles around which the Earth spins.

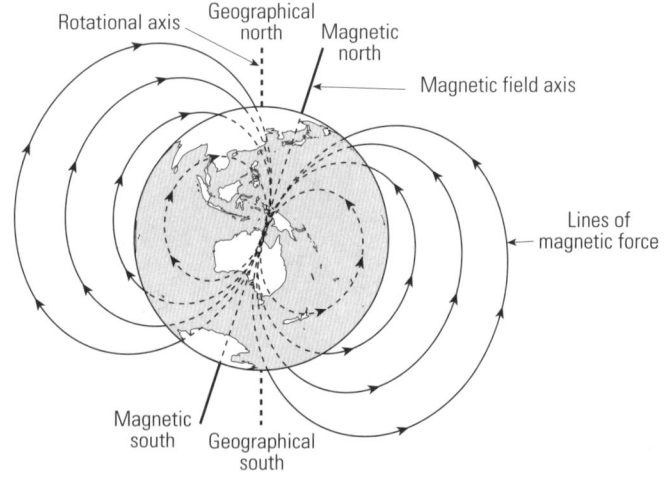

Figure 2.3.1 The Earth's magnetic field.

PALAEOMAGNETISM

Palaeomagnetism is the study of changes in the geomagnetic field during geological time. Some rocks, at the time of their formation, acquire a permanent record of the Earth's magnetic field because of the presence of certain iron-bearing minerals.

The volcanically deposited rocks forming the sea floor possess magnetism. As the igneous rocks cool, magnetic minerals within the rock (such as magnetite and other iron-bearing minerals) align themselves to the Earth's magnetic field. These minerals do so because during the cooling process they become permanent magnets. In order to acquire permanent magnetism, a mineral must cool below a certain temperature, called the **curie point**. The curie point of magnetite is 580°C. Each mineral has a different curie point. At temperatures above the curie point, it is impossible for the mineral to become a permanent magnet.

As lava crystallises and the temperature drops below the curie point, all the magnetite grains become tiny, permanent magnets with the same orientation (or polarity) of the Earth's magnetic field. Once the orientation of the magnetite grains have been fixed, due to cooling, they cannot reorientate. The igneous rock then carries a record of the polarity of the Earth at the time the lava cooled.

Magnetism in sedimentary rocks is a lot weaker than in igneous rocks. The magnetic fields of some very fine-grained sediments become aligned parallel to the Earth's magnetic field when they settle through a still fluid. In oceans, a fine particle settling through a body of water is orientated randomly, according to the local currents. Very near the bottom, where the currents may be still and at great depths, the grains will rotate so that their weak magnetic fields will be aligned to the Earth's magnetic field. As the grains form sediments and become cemented together, they will retain their orientation. The grains record the direction to the magnetic pole and also the inclination of the magnetic field at the time they become cemented together.

MAGNETIC INCLINATION

By examining the magnetism of igneous rocks, it was possible to measure another type of information: magnetic inclination. This inclination varies according to latitude.

If a bar magnet is allowed to swing freely, not only does it point to the North Pole, but it also tilts slightly. This tilt is the magnetic inclination. Near the equator the bar magnet lies relatively flat, but the closer the bar magnet moves towards the poles the steeper the angle of the bar magnet, until at the poles themselves the magnet is nearly vertical.

Using the information of magnetism recorded when a rock forms we can determine the position of the magnetic North Pole at the time of the formation of the rock. Rocks bearing magnetic minerals also give information as to the latitude of the rock at the time of its formation. In other words, studying palaeomagnetism tells us the direction of the magnetic pole and the distance between the rock and the magnetic pole at the time of the rock's formation.

APPARENT POLAR WANDERING

When rocks formed, their magnetic field alignment was parallel with the north–south direction of the Earth's magnetic pole. Ancient rocks retain the magnetism of the Earth at the time of their formation, and the inclination and declination of the magnetite crystals gives the direction and latitude of the magnetic pole at the time the rocks were formed. Given these facts, it is then possible to determine the apparent position of the poles during the formation of the rocks. This led to a number of surprises.

Not all rocks had their magnetic field alignment pointing directly north–south. It was known that the magnetic North Pole moves slightly each year and, hence, the north–south direction of the field moves as well. The movement of the magnetic North Pole was always close to the point of the Earth's rotation. Scientists plotted the pathways of the poles on maps and referred to the phenomenon as **apparent polar wandering**.

The apparent position of the poles is not fixed, but seems to wander. If a number of different rocks on a continent are examined for the orientation of the magnetite crystals they contain, it appears that the poles moved, forming a wander curve or pathway.

Also, measuring the rocks of the same age, from different continents, showed that the pathways for each of the continents were not the same, but were sometimes parallel. This apparent polar wandering showed that it was not the North and South Poles that were moving, but the continents themselves. It was discovered that the path of apparent polar wandering measured in North America was different from

curie point
the temperature below which a mineral can become a permanent magnet

apparent polar wandering
the appearance that the positions of the poles moved

definitions

that measured in Europe. The conclusion that scientists drew was that the two continents had differed in their movement. They had two different pathways between 600 million and 50 million years, after which their pathways coincided.

REVIEW ACTIVITIES

1 Explain how the Earth generates a magnetic field.

2 Describe how both igneous and sedimentary rocks acquire their magnetism.

3 Distinguish between the geographical poles and the magnetic poles.

4 Describe what information may be deduced from the magnetism in the igneous rock.

5 Explain the importance of the phenomenon of apparent polar wandering.

EXTENSION ACTIVITIES

6 Explain what is meant by the term 'permanent magnet'.

7 Research whether the latitude of Australia has changed over its history.

POLARITY REVERSALS

From the study of the magnetism in the rocks, it has been found that, on average, every 300 000 years the Earth's polarity changes. (See Figure 2.3.2.) Instead of the magnetic North Pole being at the geographical North Pole it will suddenly switch to the geographical South Pole. Then, after another 300 000 years, the magnetic North Pole will revert to the geographical North Pole. Scientists as yet do not understand why these magnetic reversals occur, but they do know that these reversals occur rapidly. Records of these polarity changes are locked in the rocks. There is speculation as to whether they are due to the field tilting over, or if it dies down and then builds up again in the opposite direction.

One possible link to polarity reversals is basaltic volcanism. It is possible that material is released from the core-mantle boundary. Hot material rises from the core-mantle boundary, increasing the convection currents. This results in an increase in the strength and stability of the Earth's magnetic field, which becomes strongly positive towards the magnetic North Pole. This increase in positiveness towards the magnetic North Pole is shown in rocks dating from the Cretaceous period between 85 and 125 million years ago. Over the last 2000 years the Earth's magnetic field has been progressively weakening, and possibly in another 2000 years there will be a polarity reversal.

PALAEOMAGNETIC BANDING

Although the evidence of polar wandering was important in giving vital support to the discredited continental drift hypothesis, some scientists still had major reservations with regard to a conclusive demonstration that a supercontinent had actually split apart. The scientists wanted to see a mechanism whereby crust could split open. It was found in the rocks of the Atlantic Ocean floor.

Oceanographers were amazed to find that parts of the sea floor consisted of bands of magnetised rock, displaying alternating bands of normal and reversed polarities.

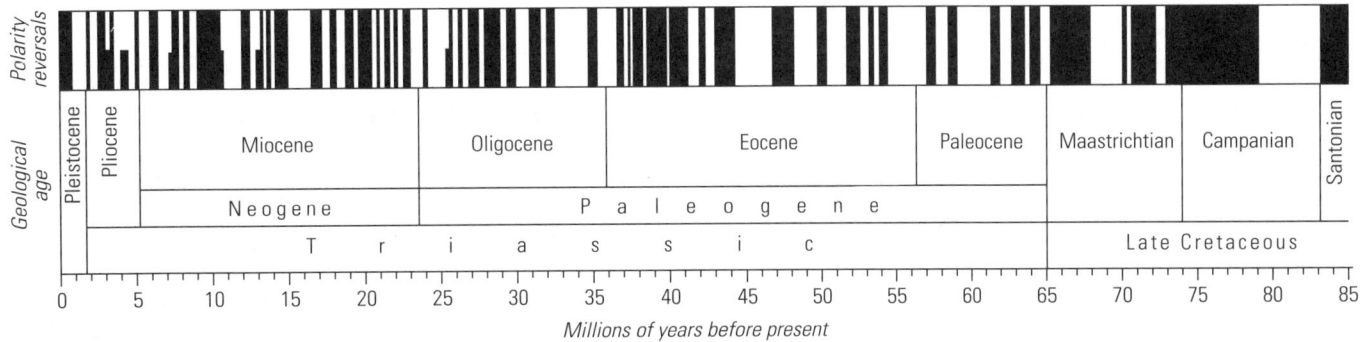

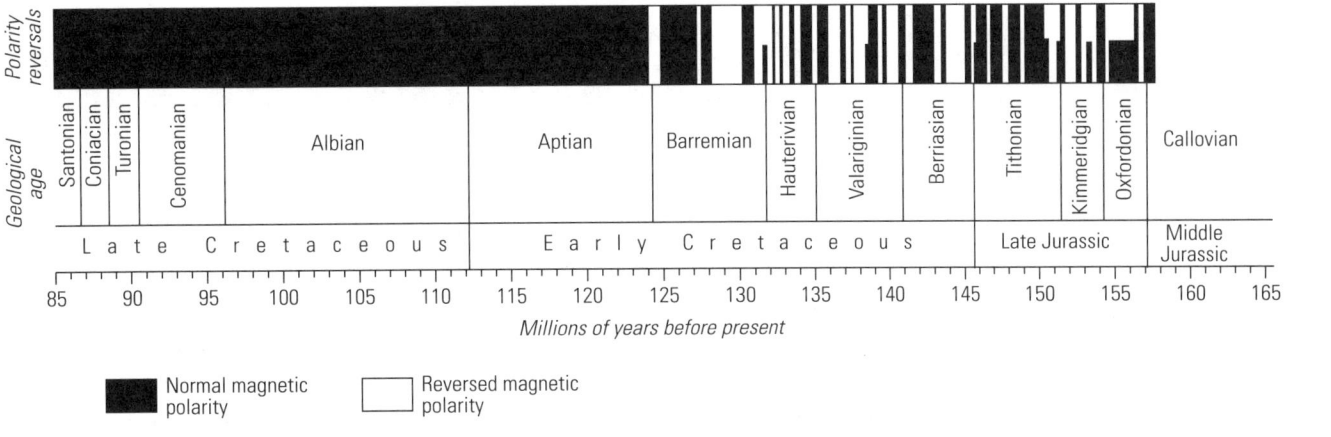

Figure 2.3.2 Magnetic reversal time scale.

(See Figure 2.3.3.) These rocks with differing magnetic intensities are referred to as magnetic anomalies. Scientists found that there were both positive and negative anomalies. Positive anomalies are places where the magnetic field is stronger than normal and the rock's magnetism is aligned with the Earth's magnetic North Pole. When it formed, the rock was aligned with the magnetic North Pole and at that time the magnetic North Pole was in the Northern Hemisphere. Negative anomalies are the opposite; the rocks possess a lower than expected magnetic field and the magnetism in the rock is aligned to the Earth's magnetic North Pole when it is in the Southern Hemisphere.

These bands are up to several hundred kilometres in length and run parallel to the mid-ocean ridge. The bands on one side of the mid-ocean ridge are identical to the corresponding bands on the other side of the ridge. (See Figure 2.3.3.)

This symmetrical pattern at first puzzled the scientists, until they were able to explain the process in terms of sea floor spreading. What was happening was that magma was coming up from the asthenosphere through the mid-ocean ridge and cooling. As it cooled it would be pushed outwards from the ridge in both directions.

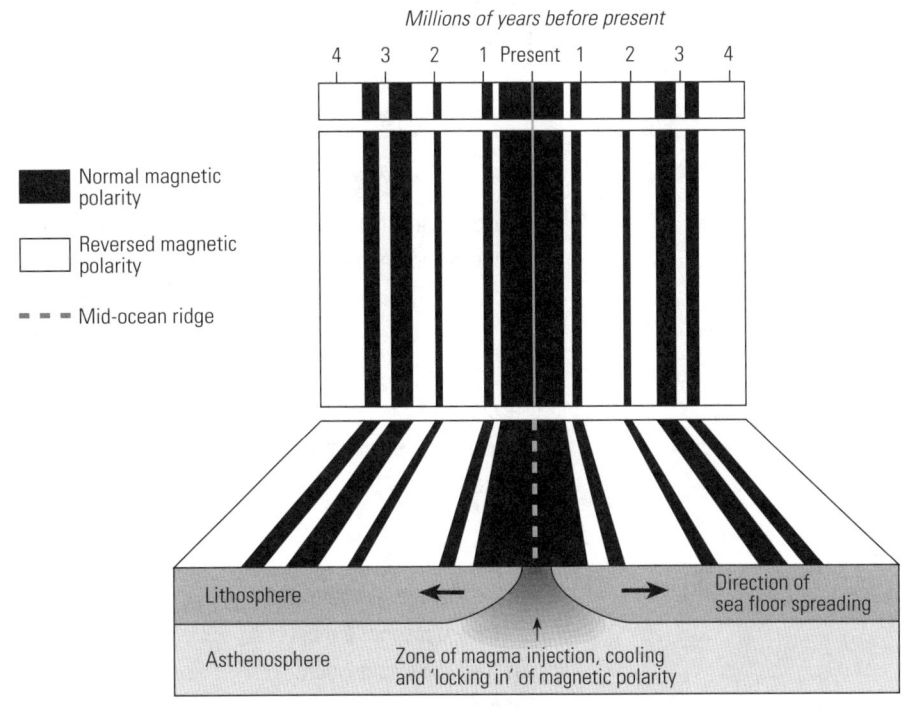

Figure 2.3.3 Magnetic profile of sea floor spreading.

DYNAMIC EARTH 99

REVIEW ACTIVITIES

1 Outline the possible reasons why the Earth's polarity has reversed.

2 Explain how palaeomagnetic banding occurs.

3 Describe why the bands of rocks are the same on both sides of the mid-ocean ridge.

4 Explain why our knowledge of magnetism is important in understanding plate tectonic movements.

EXTENSION ACTIVITY

5 Find out in what ways humans would be affected by a polarity reversal.

SUMMARY

- The Earth's magnetic field is thought to be produced by the moving charges caused as convection moves the liquid outer core around the solid, iron-rich inner core.

- The magnetic poles are different from the geographical, or true, North and South Poles. The geographical North and South Poles are the axis poles around which the Earth spins.

- Palaeomagnetism is the study of changes in the geomagnetic field during geological time.

- Only magnetic minerals, such as magnetite, can acquire magnetism.

- For a mineral to become a permanent magnet it must cool below its curie point. For magnetite this is 580°C.

- As igneous rock cools it fuses the magnetite grains in the same polarity as the Earth's magnetic field at that time.

- Examination of the inclination and declination of the magnetite crystals gives scientists information about the latitude at which the rock was laid down and the direction of the magnetic pole.

- Apparent polar wandering is the apparent movement of the poles, when measuring the direction of the pole from individual continents over millions of years. It is not the pole that is moving, but the continents themselves.

- Polarity reversals occur when the magnetic North Pole switches from the geographical North Pole to the geographical South Pole. After 300 000 years, the magnetic North Pole reverts to the geographical North Pole.

- A positive anomaly occurs when the magnetism is aligned with the Earth's magnetic North Pole when it is in the Northern Hemisphere.

- A negative anomaly occurs when the rock possesses a lower than normal magnetic field and the magnetism in the rock is aligned with the Earth's North Pole when it is in the Southern Hemisphere.

- The sea floor on either side of mid-ocean ridges possesses bands of magnetised rock showing polarity reversals. This gives support to the theory of sea floor spreading.

PRACTICAL EXERCISE
Investigating magnetic time scales

Our understanding of sea floor spreading depends heavily on the information derived from sea floor magnetism. The polarity reversals allow us to determine rates of sea floor spreading and to reconstruct the position of continents in the geological past.

Figure 2.3.4 shows information gained from a core that was drilled in the Pacific Ocean during the early 1960s.

ACTIVITIES

Make a copy of Figure 2.3.4 and then complete the following activities.

1
Refer to Figure 2.3.4.
a At what times does the inclination fall to zero?
b Is the change in magnetic inclination from positive to negative rapid or slow?

c Using the graph, estimate the duration of the change in magnetic inclination.

2
Look at Figure 2.3.5. Which epochs are covered by Figure 2.3.4?

3
Do all the oceans shown in Figure 2.3.6 (page 102) spread at the same rate?

4
Look at Figure 2.3.7 (page 102), which shows one half of the magnetic anomaly profile over the southern mid-Atlantic ridge. Use the scales at the top and bottom of the graph to calculate the average rate of sea floor spreading for the polarity reversals marked A, B, C, D and E.

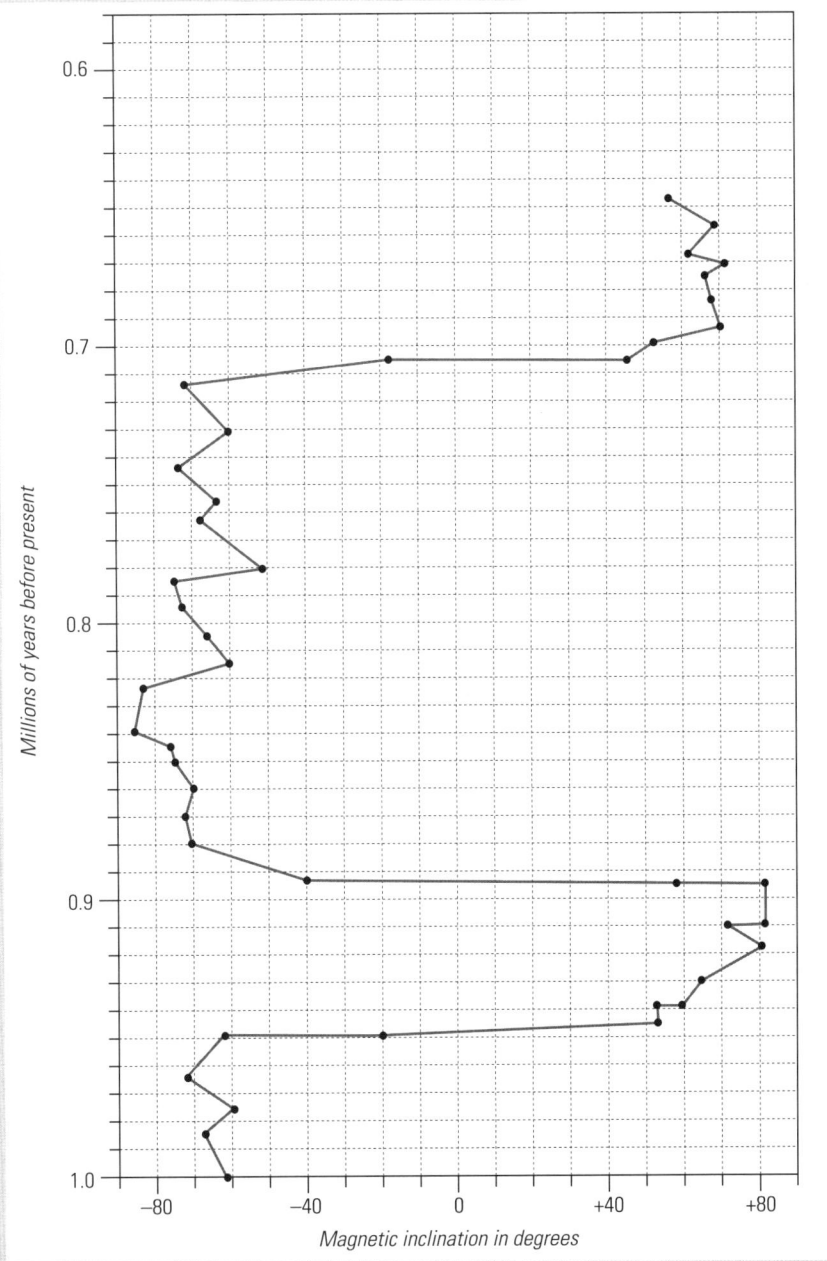

Figure 2.3.4 Magnetic inclination to age of core sample.

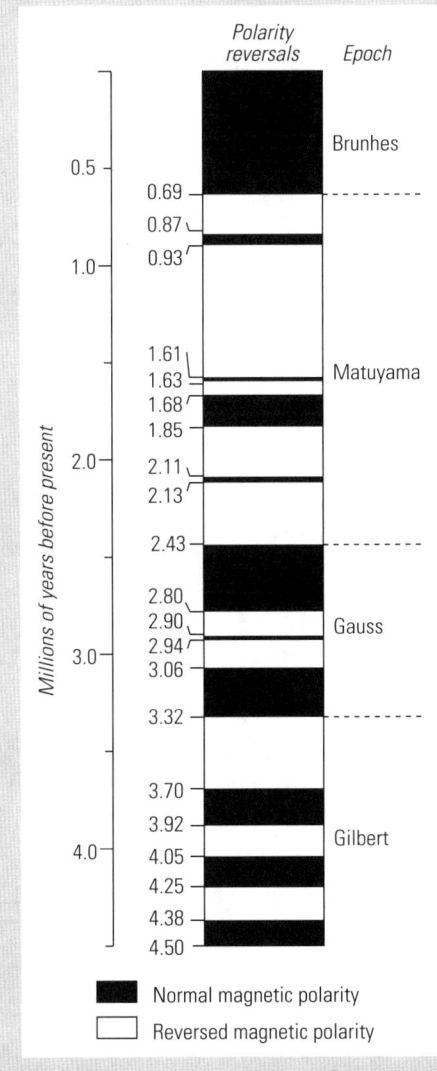

Figure 2.3.5 Magnetic anomaly profile time scale.

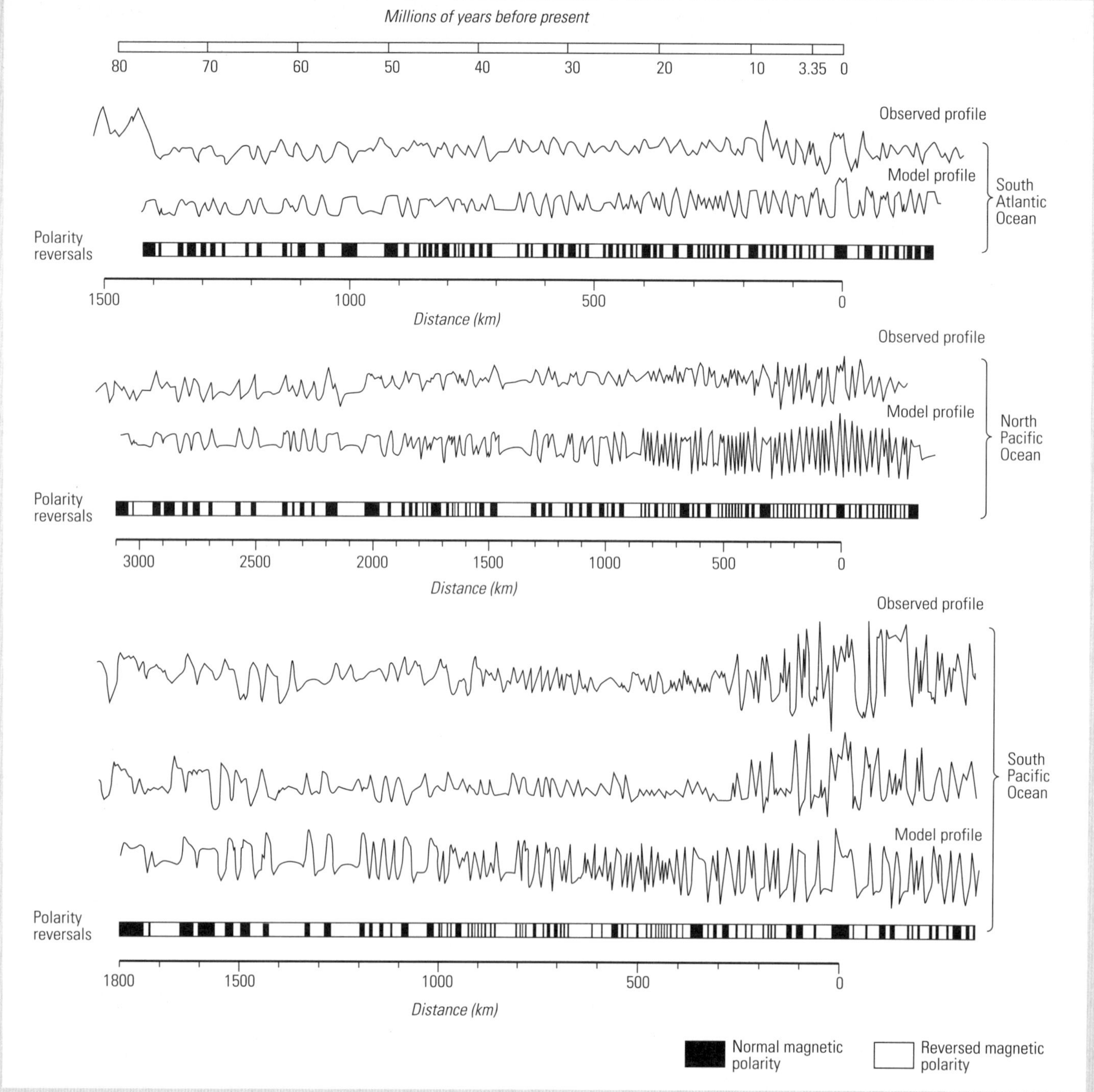

Figure 2.3.6 Polarity data for the South Atlantic, North Atlantic and South Pacific Oceans.

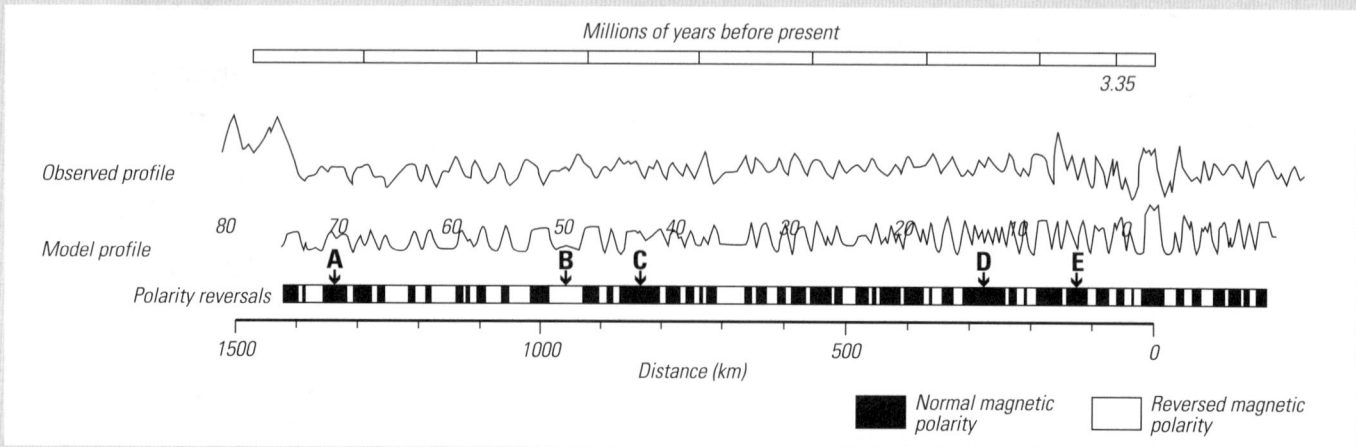

Figure 2.3.7 Magnetic profile of the southern Mid Atlantic ridge.

102 EARTH AND ENVIRONMENTAL SCIENCE: THE PRELIMINARY COURSE

2.4 Tectonic forces

OUTCOMES

At the end of this chapter you should be able to:

- recall how earthquakes, volcanic activity and new landforms result from the interactions at plate boundaries

- define the term 'subduction zone' and identify the geological features that are characteristic of a subduction zone

- describe the characteristics of igneous rocks and volcanic activity associated with subduction zones

- describe the processes that may occur when two plates collide

- explain how granites, basalts and andesites are formed

- identify regions of Australia that provide evidence of past plate movements, and assess the impact of plate movements on past environments of Australia.

PLATE BOUNDARY MOVEMENTS

There are three types of plate boundary movements:
- *Divergent boundaries.* Two plates move apart.
- *Convergent boundaries.* Two plates collide into each other.
- *Transform boundaries.* Two plates move past each other.

As a plate moves it may be undergoing some or all of the above movements, at different points on its boundaries. The plates are usually so large that most of the noticeable effects of the movement only occur at their boundaries. These effects are in the form of earthquakes or volcanoes. Most volcanoes occur along boundaries where plates are moving apart or together, that is diverging or converging. Volcanoes are also abundant along subduction zones where they are the result of partial melting. Plates move apart along zones of spreading. These zones exist where material is coming up from the asthenosphere in the form of relatively low-density magma. It is here that new lithosphere is made. The heat from the asthenosphere helps to swell the new asthenospheric material, which forms a **mid-ocean ridge**.

At other locations, lithospheric plates move back down into the asthenosphere. This generally occurs at **deep sea trenches**, found at the deepest points of the oceans. These regions are called **subduction zones**. Material is descending into the asthenosphere at the same rate as it is being produced at the mid-ocean ridge.

As the plate is subducted it breaks up into slabs. These slabs slowly absorb heat from the asthenosphere around them. As the slabs are colder and denser than the surrounding asthenosphere, they sink further into the layer and some slabs may sink to the boundary between the outer core and mantle. The slabs may take hundreds of millions of years to melt completely.

definitions

mid-ocean ridges
when plates separate in the mid ocean, ridges are formed on either side of the separation zone

deep sea trench
the point where two converging plates meet

subduction zone
the area where one plate goes under another plate

Divergent boundaries

Divergent boundaries of plate margins are called zones of spreading, or rifting. Here the two plates are moving away from each other. Divergent plate boundaries are sites of plate separation and growth. Most of this movement takes place in the oceanic crust, but this was not always so. If a mid-ocean ridge is examined, the boundary between the two plates originally started as a fracture in continental crust. This fracture caused the splitting away from each other of the landmasses of North America, South America, Europe and Africa. The fracture filled with magma, which became oceanic crust and it formed the sea floor of the Atlantic Ocean.

This formation of the Atlantic Basin occurred millions of years ago and is at a mature stage of development. This means that the continental landmasses of the Americas, Europe and Africa are now far away from the seismically active (an area prone to earth movements) mid-ocean ridge. The Atlantic boundaries of these continents are described as **passive continental margins**.

This separation process is being repeated today in the Afar Triangle, where North-east Africa and the Arabian Peninsula meet. Here, at the triangle, Africa is moving away from Arabia, forming an ever increasing Red Sea. This seismically active area is referred to as an **active continental margin**.

Divergent plate boundaries are marked by new plate material being added to existing plates. This new material is caused by upwelling of magma from the mantle. The boundaries can be identified by features such as mid-ocean ridges, rift valleys and chains of volcanoes.

MID-OCEAN RIDGES

The global mid-ocean ridge system is the largest single volcanic feature on the Earth. Here the Earth's crust is spreading, creating new sea floor and renewing the surface of our planet. Older crust is recycled back into the mantle elsewhere on the globe, typically where plates collide. The mid-ocean ridge consists of thousands of individual volcanoes or volcanic ridge segments, which erupt periodically.

Beneath a typical mid-ocean ridge, mantle material partially melts as it rises in response to reduced pressure. This melted rock, or magma, may collect in a reservoir a few kilometres below the sea floor and then upwell to the surface.

There are a number of mid-ocean ridge systems (see Figure 2.4.1):
- the South-west Indian ridge
- the East Pacific rise
- the Mid Atlantic ridge
- the Macquarie ridge.

RIFT VALLEYS

As the continental crust moves over a hot spot, the crust thins and stretches, and the extension causes a central **rift valley** to form. This may occur both under water (forming mid-ocean ridges) and on land (forming rift valleys). This central valley has a high heat flow (as a result of the proximity of the mantle material to the surface), shallow earthquakes, and **basaltic** magma eruptions.

In Iceland, the Atlantic mid-ocean ridge has risen above sea level to form the island of Iceland. In Africa, the East Africa Rift Valley is part of a valley system that intersects another rift valley that formed the Red Sea and the Gulf of Aden. A further example of a rift valley is the Rhine Valley in Europe.

On either side of the central rift valley a ridge is formed from the new material upwelling into the central rift. This rift valley is a feature of divergent plate boundaries. The rift valleys are not only found in mid-ocean ridges, but extend on to dry land at a number of points on the Earth's surface. (See Figure 2.4.2, page 106.)

definitions

passive continental margin
a continental margin that is located far from a seismically active mid-ocean ridge

active continental margin
a continental margin located within the zone of seismic activity

rift valley
occurs when continental crust is splitting apart. The new material forms the valley bottom.

basalt
a dark-coloured, fine-grained basic volcanic rock

continental shelf
a wide, shallow active or passive continental boundary made up from sediments

continental slope
the slope leading to deep water at the edge of a continental shelf

continental rise
a slight slope in the oceanic crust leading up to a continental slope

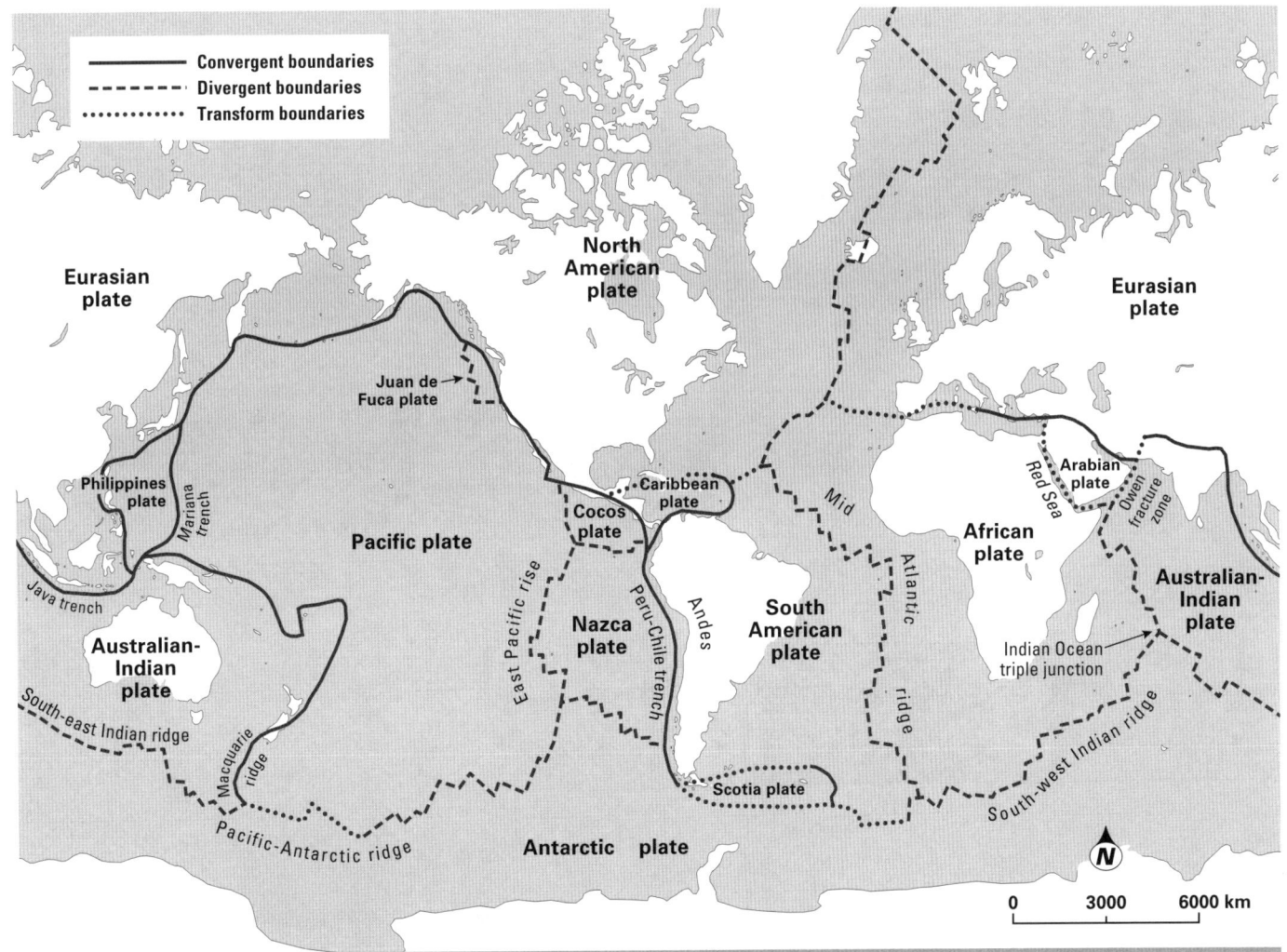

Figure 2.4.1 The major tectonic plates and mid-ocean ridges.

Formation of continental margins

There are two types of margins: passive continental margins and active continental margins. Passive continental margins are far from the seismically active mid-ocean ridge and were formed many millions of years ago. The active continental margins are generally younger than passive continental margins as they are within the zone of seismic activity at the plate boundaries.

In passive continental margins the old oceanic crust has cooled and deepened, and the continental crust close to this area has subsided and is below sea level. This boundary between the oceanic crust and continental crust has accumulated large amounts of sediment, forming a layer many thousands of metres thick. The sediment was washed or blown in from the land.

The underwater layer of sediment forms a wide, shallow continental margin, called the **continental shelf**, which surrounds the continental landmasses. At the edge of the shelf a broad slope leads down into deeper water. This slope is called the **continental slope**. As you travel further from the land, the layer of sediment becomes thinner and forms a very slight slope, called the **continental rise**.

Divergent plate rock types

The major rock type of the volcanic material formed by mid-ocean ridges is basalt. It forms new oceanic crust. Basalt is a fine-textured, dark-brown to black extrusive rock. Gabbro, a closely related rock, is a coarse-textured intrusive rock. An extrusive rock results from the cooling and solidification of igneous materials on the surface of the Earth, while intrusive rocks are those that formed from magma below the Earth's

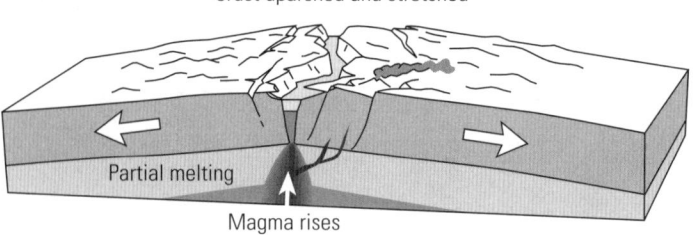
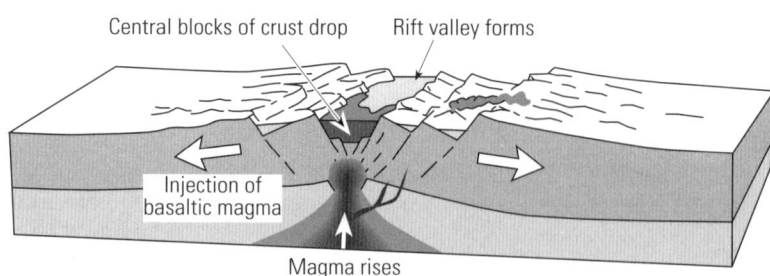
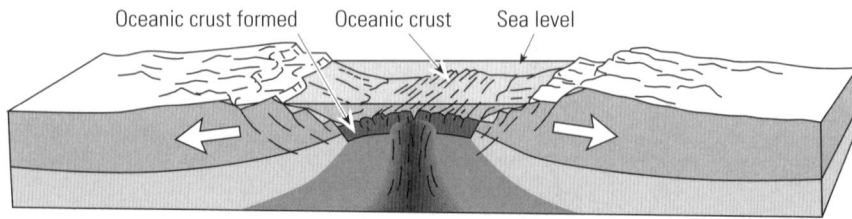
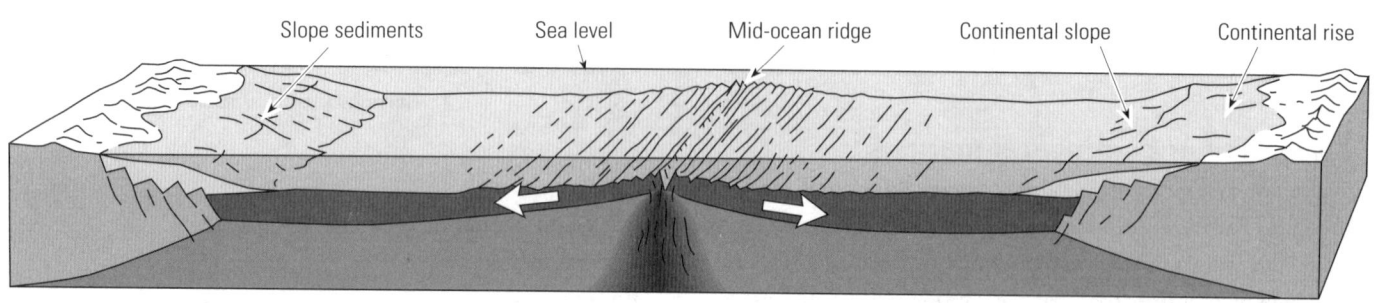

Figure 2.4.2 The formation of central rift valleys and ocean basins.

surface. These mafic rocks compose the entire oceanic crust; basalt forms the upper layers and gabbro forms the thicker internal zone upon which the basalt rests. Basalt forms the islands of Hawaii, the Deccan Plateau of India and the Columbia Plateau of Washington.

Rock types that appear to have been formed on the sea floor and are now emplaced on land are called **ophiolite suites**. These suites are normally composed of deep sea sediments, submarine basaltic lavas and mafic igneous **intrusions**. Often associated with these rocks are massive ore bodies that contain iron and copper, for example.

The normal continental crust rock type is granite, which is a light-coloured, coarse-grained intrusive rock, consisting of quartz and felsic minerals. **Ferromagnesian minerals** are generally absent. Granite and its more mafic variety, granodiorite, are the most common igneous rocks of the continental crust. Andesite, or rhyolite, is an extrusive igneous rock and is also generally confined to continental crust. It is released by violent volcanoes that border subduction zones. Andesite is a grey, fine-grained volcanic rock. Rocks of the andesite family are typical of the volcanic island arcs and continental chains that border the subduction zones, for example Japan.

---

definitions

ophiolite suites
rock types that appear to have been formed on the sea floor and are now emplaced on land

intrusions
the emplacement of magma into cracks in a pre-existing rock

ferromagnesian minerals
minerals containing iron and magnesium

REVIEW ACTIVITIES

1 Distinguish between passive and active continental margins.

2 Give two examples of geological features of divergent boundaries.

3 Draw a labelled diagram of an ocean basin. Include in the diagram a mid-ocean ridge, continental rise, continental slope and continental shelf.

4 Describe the materials that can be found in ophiolite suites.

EXTENSION ACTIVITIES

5 Research the location of two active continental margins and two passive continental margins.

6 A rift valley initially formed the Red Sea. Find out what is likely to happen to the size, depth and shape of the Red Sea in the future.

Convergent boundaries

There are two types of convergent boundaries: subduction zones and continental collision boundaries. The subduction zones can be further separated into ocean–ocean and ocean–continent boundaries.

OCEAN-OCEAN CONVERGENCE

Ocean–ocean convergence results in the production of **island arcs** and deep sea trench systems. Most of these are located within the western Pacific. No two arc, or trench, systems are the same. However, the island arc that forms Japan shows most of the common features found in the arcs. An island arc is a group of islands forming a line or arc, and between the trench and the island arc a fore arc basin is formed. On the other side of the island arc, located between the arc and the continental crust, there is a back arc basin.

In the case of Japan, the Pacific Plate travelled westwards and collided with the Eurasian Plate. This movement is still happening. The Pacific Plate is colder and denser than the Eurasian Plate and so it sinks, or subducts, below the Eurasian Plate at a subduction zone marked by a deep sea trench, to the east of the Japanese island arc. As the Pacific Plate subducts, the sediment that has accumulated on top of the plate is scraped off, forming a ridge called an **accretionary wedge**. As the subducted plate sinks into the asthenosphere, magma is released from the plate when it is about

> **definitions**
>
> **island arc**
> a curved chain of volcanic islands, formed on the landward side of a trench
>
> **accretionary wedge**
> the material scraped off the top of a subducting plate, forming a ridge

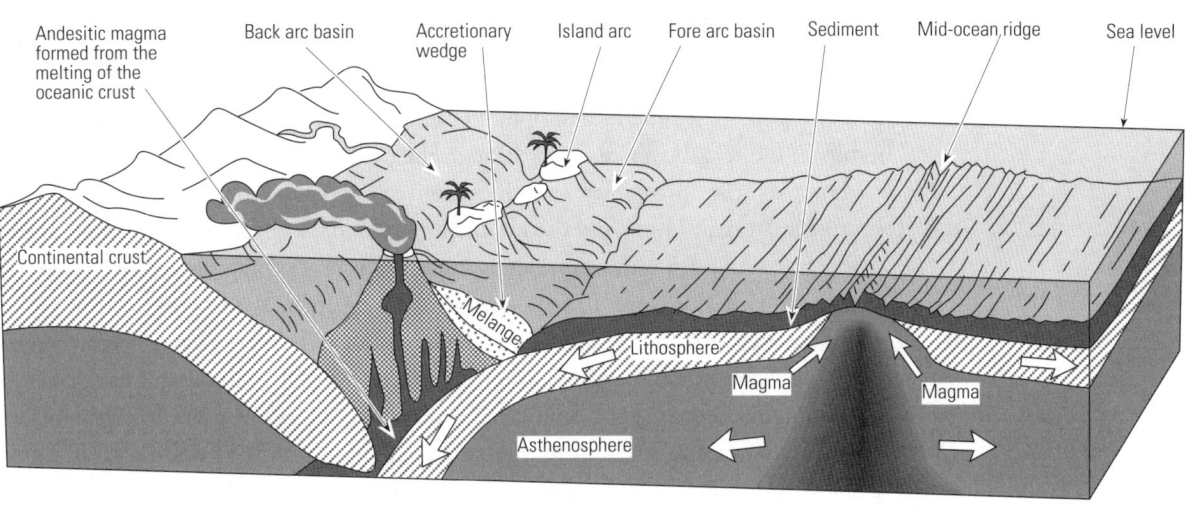

Figure 2.4.3 Island arc volcanic environment.

100 km below the Eurasian Plate. The magma rises to the surface and forms an island arc, made up of volcanic islands. (See Figure 2.4.3, page 107.)

The mechanism for the formation of this magma is as yet unknown. However, geologists have put forward two hypotheses. The first proposes that the asthenosphere starts to melt the subducting plate. The second hypothesis puts forward the idea that the mantle that is trapped between the subducting plate and the lighter plate melts, due to the water being released from the subducting plate. The increasing pressure and temperature causes the water to be squeezed out of the subducting plate. The water rises and, in so doing, lowers the melting point of the hot mantle. This, in turn, generates the magma.

OCEAN–CONTINENT CONVERGENCE

On the opposite side of the Pacific Ocean, the Pacific Plate is a lot younger and thus less dense than the oceanic plate around Japan. By studying the map of convergent plate zones (see Figure 2.4.1, page 105), it can be seen that subduction zones run almost the full length of the western coastline of the Americas. Along this coastline, the Nazca oceanic plate is subducting below the South American continental plate. These plates are undergoing ocean–continent convergence. (See Figure 2.4.4.) The subducting oceanic plate descends into a deep sea trench directly off the coastline. The subduction zones cause a release of magma. Unlike ocean–ocean convergence where island arcs are produced, with ocean–continent convergence the magma erupts onto land. This land eruption forms a chain of volcanic mountains parallel to the coastline and the subducting zone. In South America this volcanic mountain range is called the Andes, and in North America it forms the mountain range called the Cascades.

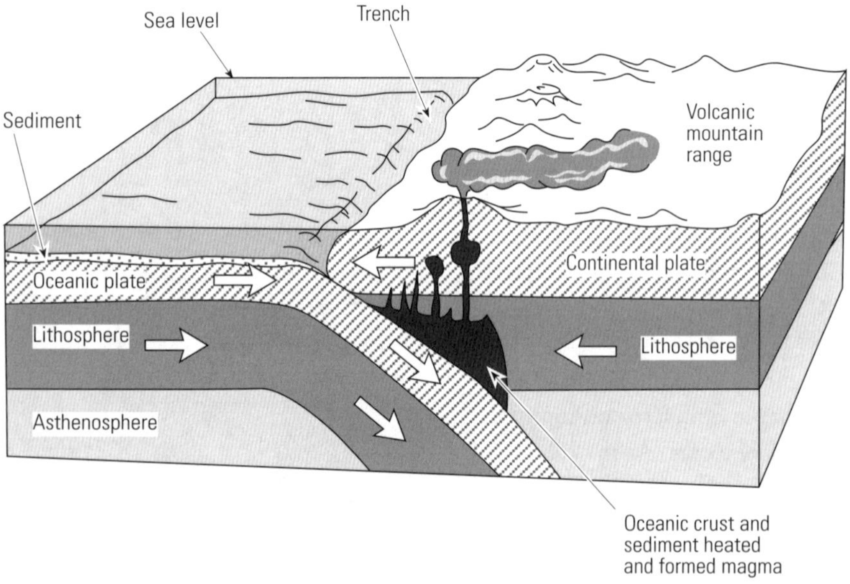

Subduction zones usually occur when a heavier oceanic plate collides with a lighter continental plate. The oceanic plate is forced under the continental plate, with the oceanic plate being pushed down as far as the mantle–core boundary. The rocks formed at this type of boundary tend to be fairly varied and form a melange, which is a chaotic mixture of blocks and fragments of rock of various compositions. They include such rock types as turbidites, which are rocks that have been formed from material deposited after being carried in currents of gas or fluid bodies. Deep sea sediments and ophiolite suites may undergo metamorphism, which results in the rock being changed through the processes of heating and/or pressure.

Figure 2.4.4 Ocean–continent convergence.

Due to the proximity of the deep sea trench to the coastline in a continental–ocean boundary, the continental boundary is very narrow. Most of the sediment coming from the continental landmass is washed rapidly into the deep sea trench and either does not form a continental shelf or the shelf is very narrow.

As the west coast of the Americas is a subduction zone between continental–ocean plates, it means that the coast is an active continental boundary. This makes it susceptible to frequent, powerful earthquakes. As the oceanic plate subducts below the continental plate it grinds past the material around it. The grinding gives rise to a **benioff zone**, which is where a large number of earthquakes are generated.

CONTINENT–CONTINENT CONVERGENCE

Continent–continent convergence, or a continental collision boundary, occurs when two continental plates collide. They are less common than other types of convergent boundaries. With ocean–continent convergence, subduction occurs due to differing plate density. The oceanic plate sinks under the continental plate because the oceanic plate has a greater density than the continental plate. In the case of continent–continent convergence, the plates are of the same density. As the continental plates collide, the rock layers in each undergo folding and uplifting, forming a fold mountain range.

An example of continent–continent convergence is the collision of the Australian-Indian plate with the Eurasian plate, resulting in the formation of the Himalayan mountains. After the break up of Pangea, the plate bearing India started to move northwards towards the Eurasian Plate. Between the two landmasses of India and Asia lay the ancient **Tethys Sea**. Slowly the ocean basin of the Tethys Sea disappeared. The two plates were of the same density and therefore subduction did not occur. The resulting collision pushed the old accretionary wedge from the Tethys Sea into the mountain range of the Himalayas.

Earthquakes occur at convergent boundaries because the plates push against each other and there is a build up of pressure until that pressure is released in the form of an earthquake. The earthquakes felt at continental collision boundaries range from shallow to very deep; they can occur up to 700 km below the surface of the plates.

benioff zone
a zone of grinding between two plates, where one is subducting

Tethys Sea
an ancient sea that was located between the two landmasses of India and Asia

definitions

REVIEW ACTIVITIES

1 Distinguish between the two main forms of convergent boundaries.

2 Define the terms 'melange' and 'turbidites'.

3 Outline the processes that led to the formation of island arcs.

4 Explain the difference in formation between volcanic and fold mountains.

5 Outline the processes involved in the process of subduction.

6 Compare the features of ocean–ocean, ocean–continent and continent–continent convergence.

EXTENSION ACTIVITIES

7 Give three examples of island arcs.

8 What evidence is there that the Tethys Sea once existed?

Transform boundaries

Transform boundaries occur where two plates slide past each other on a horizontal plane. These boundaries may displace continental and oceanic crust. The boundary between the plates is a complex transform **fault** zone.

The most famous example of a transform fault boundary is the San Andreas Fault in California, which has been caused by the transform boundary between the Pacific Plate and the North American Plate. This fault runs in a general north–south direction. On the east of the fault lies the North American Plate and on the west of the fault lies the Pacific Plate. Two major conurbations (cities) have sprung up on either side of the fault: San Francisco on the west on the Pacific Plate; and Los Angeles on the east on the North American Plate. The Pacific Plate is moving

fault
a fracture in a rock along which movement occurs

definition

northwards while the North American plate is moving to the south. When the plates slide past each other, the edges grab and lock together. As this occurs, the rocks bend and flex. When the edges break free, the flexed rock snaps suddenly and an earthquake occurs. Over many millions of years the land on which the two conurbations are located will move until they lie next to each other. Then a sliver of continental crust with Los Angeles on it will split away from the main body of continental crust and form Los Angeles island. Millions of years later this will end up in a subduction zone and will eventually disappear.

THE IMPORTANCE OF PLATE TECTONICS

An understanding of tectonic plates is the key to finding mineral and energy deposits. Over the last one hundred years the need for metals has increased dramatically. Almost all deposits that are located in outcrops on the Earth's surface have been located and exploited. Geoscientists are now looking for deposits located in ancient convergent plate boundaries. Other geoscientists are searching for precipitated minerals on the sea floor, around ancient fumarole sites. These precipitated minerals were released in the hot water coming from fumaroles. As the water cooled, the precipitated minerals sank to the sea floor.

The search for oil has led to an understanding of how it was formed, as well as the conditions necessary for trapping the oil deposits. The most favourable tectonic settings for the accumulation and trapping of oil are in marginal ocean basins. Marginal ocean basins occur between subduction zone trenches and island arcs, both of which occur at the edge of the continental crust. These margins undergo slight folding, which traps the accumulated oil.

It is also very important that humans understand the dynamic nature of the planet so we can predict natural hazards, such as earthquakes, volcanoes, landslides and tsunamis. Such predictions can help to prevent the loss of life and to regulate the building of homes and industry in areas liable to be affected by natural hazards.

EARTHQUAKES AND VOLCANOES

Earthquakes and volcanoes occur primarily along plate margins, and are the most obvious and visible evidence of active plate movements. By studying these generally catastrophic events, geologists have been able to use the information to deduce the shape of the plates. Much work has been done in trying to predict, as well as reduce, the effects of both natural events. The worst earthquake over the last one hundred years occurred in 1976 in the city of T'ang Shan, China. The city of 1 million people was destroyed and 250 000 people were killed. The world's largest explosion occurred when the mountain of Krakatau, Indonesia, erupted in 1883.

Features of earthquakes

Earthquakes are thought to occur because of the slippage of rock, resulting in a sudden movement along a fault line. This idea of slippage along a fault line causing an earthquake is very simplistic, and the cause may well be far more complex.

In 1906 Henry Reid proposed the elastic rebound theory, which put forward the idea that rocks on either side of a fault line or fracture are rough, causing them to lock together rather than slip past each other. The rocks on either side, after locking, will bend. When these rocks bend they store elastic energy. Eventually, when the two sides of the fault slip, the bent rocks spring back to their original shape. The stored energy is released and causes an earthquake. The point of the slippage of the rocks is the earthquake focus. The focus may be just below the surface, or could be hundreds of kilometres below the surface. The point on the surface directly above the focus is called the earthquake epicentre. (See Figure 2.4.5.)

When the epicentres of earthquakes are plotted on a world map it is found that they form narrow zones. Earthquakes are generally formed by the grinding or snapping of one plate as it moves past another at a transform, divergent and convergent boundary.

The earthquake zones can be divided into two types:
- *Shallow earthquakes.* These occur less than 70 km below the Earth's surface, along mid-ocean ridges, transform faults and within plates.
- *Deep earthquakes.* These occur between 300–700 km from the Earth's surface, along trench and island arc systems. The depth increases along the benioff zone at a subduction zone.

At transform boundaries the earthquakes can be either shallow or of intermediate depth (70–300 km). They tend to be very frequent and many are very strong. Convergent plate boundaries produce earthquakes that are sometimes very deep and frequently very strong. Divergent boundaries produce earthquakes that are weak and shallow.

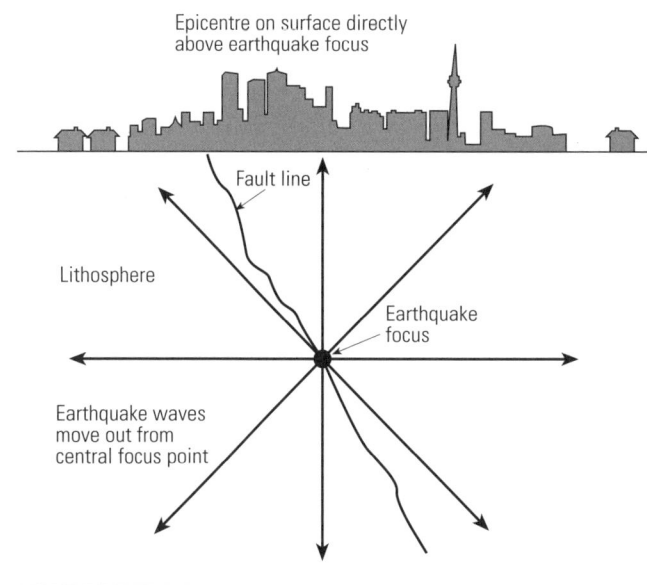

Figure 2.4.5 Earthquake focus and epicentre.

Effects of earthquakes

Earthquakes have six major effects:
- Ground movement, due to shock waves passing through the rock layers and soil near the surface, results in buildings being damaged or destroyed.
- Ground movement breaks pipes or cuts cables, resulting in flooding and fire.
- Where a fault appears on the surface it results in damage to buildings and splitting of roads, for example.
- In regions where there are steep slopes, earthquakes cause landslips and landslides.
- In certain areas the soil may be saturated in water. Because the soil particles are rounded it will result in the soil undergoing **liquefaction**, which turns the soil into quicksand.
- If the sea floor is affected by the earthquake it may result in a **tsunami**.

definitions

liquefaction when soil behaves as if it is liquid

tsunami a tidal wave

REVIEW ACTIVITIES

1 Summarise the types of earthquakes that occur at different plate boundaries.

2 Explain how earthquakes are thought to occur.

3 Distinguish between an earthquake epicentre and its focus.

4 How can earthquakes be used to outline the shape of plates and locate plate boundaries?

5 Outline the potential effects of an earthquake.

EXTENSION ACTIVITIES

6 The Richter and Mercalli scales are methods for measuring the strength of earthquakes. In what ways are the scales similar and how do they differ?

7 What type of earthquake is potentially more destructive: shallow or deep?

Features of volcanoes

The location of volcanoes closely reflects the boundaries of plates. This is especially so around the 'ring of fire', which is an almost continuous chain of volcanoes along the Pacific coastline where the Pacific Plate is bounded by several other plates. (See Figure 2.4.1, page 105.)

Mid-ocean ridges are a chain of shield-type volcanoes under water. (See page 103.) They release basaltic-type lava that helps build the oceanic crust. The lava is extruded from the volcanic vent, and during this process the surface of the lava cools rapidly and solidifies. The pressure builds up behind the solidified lava, from lava within the vent, and it finally forces another blob of lava out into the water. This in turn solidifies, producing another blob. This characteristic type of lava (called pillow lava) can often be seen in rocks forming continents.

Close to convergent ocean–continent plate boundaries there are volcanic island arcs. These volcanoes erupt material that has a composition between that of the oceanic and continental crust. The lava is more acidic than pillow lava that may have come from oceanic crust, and it has an andesitic composition.

COMPOSITION AND TEXTURE OF IGNEOUS ROCKS

Magmas are formed deep under ground and then rise towards the Earth's surface. Occasionally the magma breaks through to the surface. When the extruded volcanic materials cool and solidify they become volcanic, or igneous, rocks. They may also be called **extrusive rocks** because they have been extruded onto the Earth's surface. Extrusive rocks are formed from lava flows and pyroclastic ash or debris.

Some lava that erupts from volcanoes cools very quickly and solidifies without forming recognisable crystals. These rocks have a glassy appearance and are called volcanic glass or obsidian. If the rock has cooled rapidly, it may contain voids where gas has been trapped. Sometimes when cooling occurs rapidly small crystals are formed. These rocks are called aphanitic. Volcanic rocks containing large crystals are classified as porphyritic. There are three main types of volcanic rocks:

- Basalt is the dominant rock in oceanic crust. It is not only found on planet Earth but all around the solar system. These rocks contain such minerals as olivine, pyroxene and feldspar and small amounts of silica.
- Andesite contains an intermediate level of silica, lots of feldspar and some pyroxene. It is named after the Andes Mountains. Andesite rocks contain intermediate levels of feldspar.
- Rhyolite contains such minerals as quartz and high levels of feldspar and silica.

Some magma never reaches the surface. It remains deep under the ground, cooling and solidifying very slowly. Rocks formed in this way are called plutonic rocks or intrusive rocks. These plutonic rocks have large crystals. Plutonic rocks with a coarse appearance and crystals that are visible to the naked eye are described as being phaneritic. Some crystals are unusually large; they are greater than 2 cm, but can be as large as several metres in size. These rocks are called pegmatites. (More on rock types is to be found in section 4.)

There are three main types of plutonic rocks:

- Gabbro is a coarse-grained, dark-coloured rock with the same composition as basalt. It contains low levels of silica.
- Diorite is an intermediate silica-bearing rock and is the plutonic equivalent of andesite.
- Granite is the most common rock type found in continental crust. It has the same composition as rhyolite. Granites generally have medium to coarse sized grains.

Types of eruptions

A volcano is a vent through which magma, rock debris and gases are erupted. (See Plate 13.) Volcanic eruptions are difficult to classify, as most gradually change during the course of the eruption. However, eruptions can be divided into two main types: non-explosive and explosive.

NON-EXPLOSIVE ERUPTIONS

Non-explosive eruptions are those that produce magma with low silica levels, low **viscosity** and, due to its chemical composition, low levels of dissolved gases and liquids. These eruptions tend to be relatively safe and comparatively quiet. Low viscosity lavas tend to spread widely and may create vast, flat lava plains, as can be seen in the Deccan Traps of India. The Deccan Traps are one of the largest volcanic provinces in the world. It consists of more than 2000 m of flat-lying basalt lava flows and covers an area of nearly 500 000 km^2.

There are four main types of non-explosive eruptions. The first is the Hawaiian type, such as found in Hawaii at Mount Kilauea and Mount Mauna Loa, although Hawaiian type volcanoes are not just found in Hawaii. The erupted lava is fluid with a low silica content, and it forms non-explosive fountaining. The volcanoes that release this magma are shield volcanoes, which are volcanoes with gently sloping sides.

The second type of non-explosive eruption are Basaltic flood eruptions, which are similar in character to Hawaiian eruptions but differ in the amount of lava produced. The lava forms extensive sheets of nearly flat flows. Basaltic flood eruptions produce flows with an average thickness of 25 m that can be over 100 km long. Individual flows can cover more than 40 000 km^2. The lava released is very fluid, low silica basaltic magma, which is produced in large amounts. Examples of this include the Columbia River Plateau in the USA and the Deccan Traps in India.

The third type are Icelandic (or Fissure) eruptions, which originate along radial fractures (fractures radiating out from a central point) and lines of small vents. Yellowstone Plateau in the USA and Laki in Iceland are examples of this type of eruption.

The final type of volcanic eruptions are Submarine eruptions. These are often found at the sites of sea floor spreading, such as the mid-Atlantic ridge and the East Pacific rise. The lava produced forms pillow lava and glassy flows. The lava is fluid, possesses low amounts of silica and is basaltic in nature.

EXPLOSIVE ERUPTIONS

Explosive eruptions tend to be more spectacular than non-explosive eruptions. The dissolved gases cause the lava to bubble and form fountains. In many eruptions that produce basaltic rock, bubbling and fountaining will occur at the start of an eruption, but it will then calm down and form a non-explosive eruption. In an explosive eruption the magma has a high silica content and it therefore has a high level of viscosity. This high viscosity prevents the dissolved gases from escaping and instead traps the gases. The magma becomes superheated. As the gases rise to the surface there is a decrease in pressure, allowing the gases to expand. When the gas finally escapes it does so in an explosive manner, throwing out lumps of magma and other material.

Fragments of rock thrown out in an eruption are called pyroclasts or tephra. Pyroclasts come in different sizes. The largest are called volcanic bombs and can be as large as a small car, while the smallest are the size of ash and are therefore called volcanic ash.

Pyroclastic material can erupt in several different ways. A mixture of pyroclastic material and gases, which is denser than air, can form an avalanche of superheated material. The avalanche can travel at over 200 km/h and is called a **pyroclastic flow**.

definitions

extrusive rocks
magma that has been extruded onto the Earth's surface, cooled and formed rock

viscosity
the property of a substance that prevents it from flowing

pyroclastic flow
a hot mixture of gas and particles that moves away from a vent at high speed

Explosive eruptions also come in four main types: Strombolian, Peleean, Phreatic and Plinian. These explosive eruptions are due to a combination of lava and dissolved gases.

Strombolian eruptions are mildly explosive, producing incandescent bombs and ash. The magma type varies. Strombolian eruptions form simple volcanic cones, such as spatter cones or cones made from cinders composed of bombs, scoria and other debris. Strombolian examples include Mount Strobili in Italy and Mount Paricutin in Mexico.

Peleean, or Vulcanian, eruptions usually begin with steam explosions that remove old, solid rock material from the central vent. The main phase of the eruption is characterised by the eruption of viscous, gas-rich magma that forms glassy ash. An eruption cloud, containing ash, develops above the vent. The eruption cloud is shaped like a cauliflower or mushroom and can be grey or black. Lightning in the eruption cloud is common during Peleean eruptions. Airfall and pyroclastic flows can form a cone of ash, surrounded by wide sheets of ash. The magma is rich in silica and viscous in nature. It also contains high levels of dissolved gases. This type of magma results in violent, destructive eruptions. Peleean eruptions are typified by Mount Pelee on the island of Martinique and Mount Vulcan in Italy.

Phreatic eruptions form a steam blast, resulting in steep-sided domes and collapsed **caldera** in the volcano. The rising magma comes into contact with sea water or ground water, causing the water to be converted into steam. The blast is a combination of steam, magma and rock. This results in violent, explosive eruptions with glowing avalanches of lava. The magma is often viscous and rich in silica.

The final type of explosive eruption is the Plinian eruption. Like Vulcanian eruptions, these form stratovolcanoes. Stratovolcanoes comprise the largest percentage (approximately 60%) of the Earth's individual volcanoes. These volcanoes are much larger than simple volcanoes, having been formed over many eruptions. Stratovolcanoes produce large amounts of pyroclastic material, such as lava, as well as magma. The magma is viscous, rich in silica and dissolved gases, and forms andesitic rock. The eruptions produce an exceptionally powerful blast. Examples include Mount Vesuvius in Italy, Mount Krakatau in Indonesia, Mount Pinatubo and Mayon in the Phillippines, as well as Mount St Helens in the USA.

The most violent type of pyroclast is an eruption column or Plinian column, which is a mixture of hot gas and pyroclasts that rises in an explosion into the surrounding cooler air above it. These eruption columns can rise up to 45 km, and it is possible for the pyroclasts to spread thousands of kilometres in the upper atmosphere, where they may remain suspended for several years. Due to the amount of material thrown up into the atmosphere and the length of time it persists there, pyroclasts can have major implications on both local and world climate by blocking out the Sun and reducing the temperature. Sometimes a Plinian column may burst out from the side of the volcano. This is called a lateral blast.

Effects of volcanoes

Volcanoes can be very destructive events, causing damage to buildings, injuries, and deaths. Volcanoes have had a devastating impact on nearby landscapes, blasting away mountains, changing topography and forming surfaces of cooled lava. Volcanoes can lead to massive loss of life and destruction of property. A single volcanic eruption can even result in a temporary change in global climate.

Volcanoes have a number of potential effects, but the number, type and strength of the impacts are dependent on the individual volcano. Some of the common effects are discussed below.

> **caldera**
> a very large bowl-shaped volcanic depression, formed by the combination of the explosion and collapse of the top of a volcanic cone or groups of cones

Pyroclastic flows are so hot and choking that a person caught in one will certainly be killed. Because these flows are very fast, they cannot be outrun. Nuee ardente and ignimbrites are two types of pyroclastic flows. Ignimbrites contain mostly light material whereas a nuee ardente contains denser material. Nuee ardente means 'glowing avalanche' and was named for the pyroclastic flows seen at Mount Pelee in 1902.

Volcanic ash released by volcanoes is very harmful to property, people and the environment. During heavy ash-rains, buildings may collapse and people and animals may die due to a lack of oxygen. When airplanes fly through an eruption cloud a range of damage may occur depending on the concentration of volcanic gas aerosols in the cloud and the ability of the pilot to exit the eruption cloud.

Lahars are mudflows formed by the mixing of volcanic particles and water, and often cause a lot of environmental and financial damage. The direct impact of a lahar's turbulent flow (or of the boulders and logs carried by the lahar) can easily crush, abrade or shear off almost anything at ground level in the path of the lahar. The force of a lahar is so great that buildings and land may become partially or completely buried by one or more cement-like layers of rock debris, if not crushed or carried away. People caught in the path of a lahar have a high risk of death from severe crush injuries, drowning or asphyxiation.

Debris avalanches usually occur on large, steep volcanoes and are one of the most hazardous, but least common, effects of volcanoes. They are mainly caused by instability of the volcano's slope. When a slope of a volcano is not stable it can easily collapse (possibly triggered by volcanic earthquakes), causing debris to be transported away from the slope. Debris avalanches present numerous dangers. When debris from a landslide or avalanche mixes with water a lahar may be produced. Debris avalanches can also dam rivers or cause flooding.

Perhaps one of the most important hazards that can be produced by avalanches or landslides is a tsunami. Tsunamis are large sea waves that have long wave periods. When these waves reach coastal areas, they can go far inland and cause a lot of damage.

A volcanic blast occurs when magma rises asymmetrically into the cone, making one sector of the volcano bulge outwards and become unstable.

Lavaflows burn or bury everything they come across. They can run over houses, roads and any other structures. The speed and power of the lavaflow depends on the type of lava. There are two types of lava: Aa and Pahoehoe. Aa lavas can leave the vent with a speed of about 50 km. The surface on which the lava flows will be changed very quickly, as it places a layer of very new and hot rock onto the pre-existing terrain. The pre-existing terrain is destroyed by the lava flow. It is rare for people to be killed directly by lava. However, lavaflows may start fires.

Hot spots

Not all volcanoes are found at the plate boundaries. Some are found well away from these zones. These non-boundary types of volcanoes are called hot spot volcanoes. Within the asthenosphere there are areas that are hotter than the surrounding magma. This hot spot causes the crust to weaken and melt, allowing a volcano to form. As the plate moves, another area of the crust heats up and another volcano forms. This is repeated a number of times as the hot spot moves across the plate. A well-known example of a series of volcanoes formed by a hot spot is the islands of Hawaii.

A hot spot can be tracked across the continent of Australia. (See Figure 2.4.6, page 116.) The hot spot tracked south as the continent moved northwards. Starting in north Queensland, it moved south, forming the Glasshouse Mountains of southern Queensland, down through the Warrambungles of New South Wales and into Victoria. The hot spot is at present located off the coast of Victoria and the west coast of Tasmania.

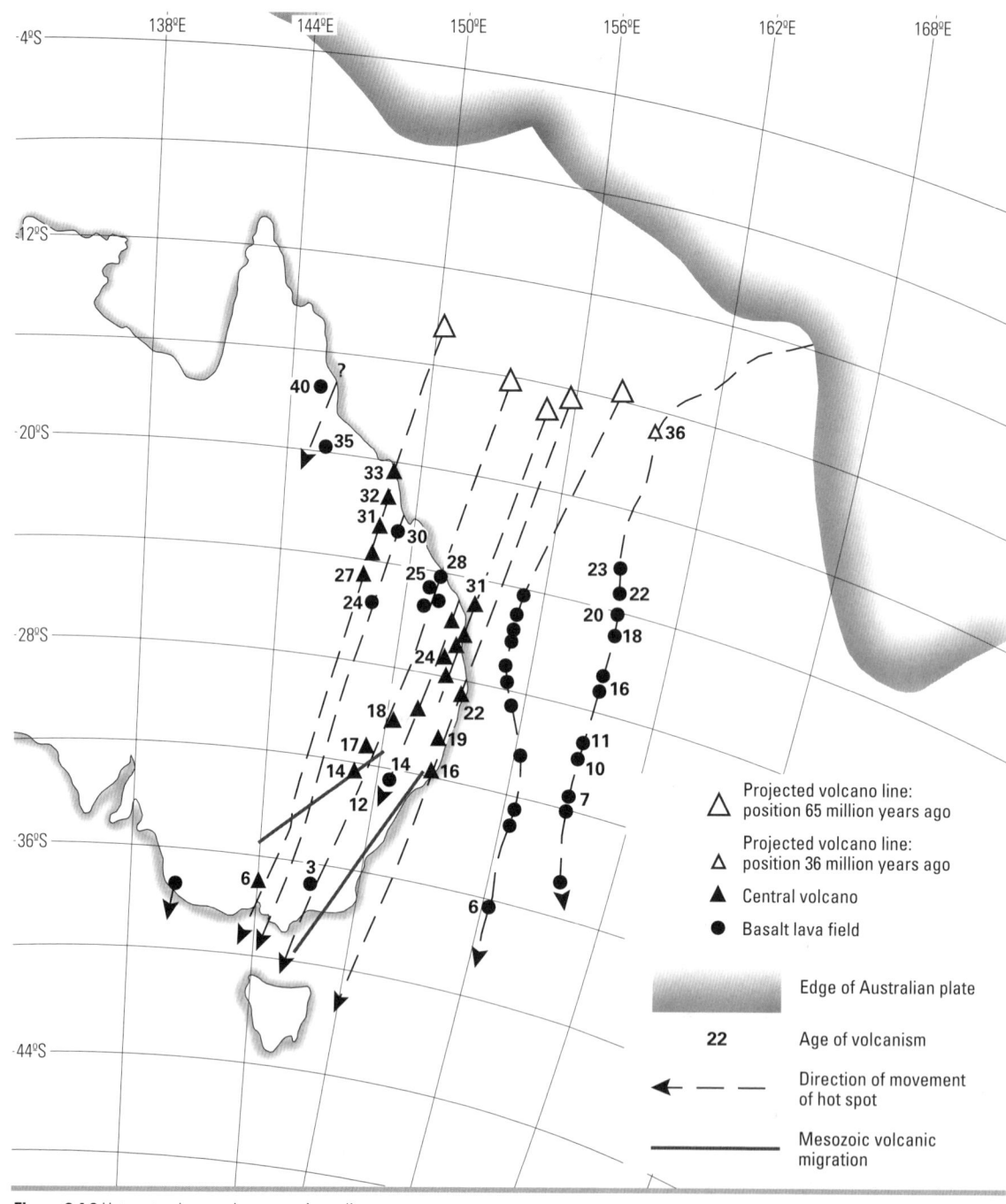

Figure 2.4.6 Hot spot volcanoes in eastern Australia.

REVIEW ACTIVITIES

1 Which kind of volcanoes are andesitic eruptions associated with?

2 Describe the four main types of non-explosive volcanic eruptions.

3 What are pyroclasts?

4 What is a pyroclastic flow?

5 Outline the process forming hot spot volcanoes.

EXTENSION ACTIVITIES

6 Research two recent volcanoes and describe what impacts they had on the local population and landscape.

7 Describe what sort of volcanic eruption occurred at Mount St Helens. What evidence supports your conclusion?

SUMMARY

- There are three types of plate boundary movements:
  - divergent, where the plates move apart
  - convergent, where the plates collide into each other
  - transform, where the plates slide past each other.

- Divergent boundaries are zones of spreading. Here the plates increase in size.

- Divergent boundaries are normally located at mid-ocean ridges.

- At mid-ocean ridges basaltic magma from the asthenosphere wells up and joins the spreading plate, forming the oceanic crust.

- Oceanic crust is composed of gabbro, basalt and sediments.

- Passive continental margins are far enough away from mid-ocean ridges not to be affected by the tectonic activity that results in volcanoes or earthquakes.

- Active continental margins are in close proximity to plate boundaries, and so they are affected by earthquakes and volcanoes.

- A central rift valley forms between two ridges. Magma upwells, forming new material, which forms the ridges.

- The edges of the continental landmasses are surrounded by a layer of sediment, which forms a thick layer called the continental shelf.

- The continental shelf ends in the continental slope, which runs into deep water. At the base of the slope there is a gently sloping and thinning layer of sediment called the continental rise.

- There are two types of convergent boundaries:
  - subduction zones, where the denser oceanic plate sinks back into the asthenosphere under the less dense continental crust
  - continental collision boundaries, where uplift occurs, forming mountains.

- Subduction zones can be divided into ocean–ocean convergence or ocean–continent convergence.

- Continental collisions occur when two plates, each bearing continental landmasses, collide.

- In ocean–ocean convergence the subducting plate slides into a deep sea trench. The sediment that lay on the top of the plate is scraped off, forming an accretionary wedge.

- Magma is released in the subducting zone, which rises to the surface and forms a series of volcanoes in the form of island arcs.

- In ocean–continent convergence the subducting oceanic plate subducts into a deep sea trench close to the continental landmass. The resulting magma upwelling from the subducted plate forms volcanoes on the continental seaboard.

- As subduction occurs, the oceanic plate rubs against the underside of the continental plate, this rubbing zone is called a benioff zone. It is here that a large number of earthquakes originate.

- In continent–continent convergence no subduction occurs. Instead there is uplift of the material at the edge of the plates. By this process, sediments from the now nonexistent Tethys Sea formed the Himalayan mountain range.

- Transform boundaries occur where two plates slide past each other. As the plates move past each other the edges grab and lock against each other. The release triggers an earthquake.

- A fault is a crack in a rock along which movement occurs. The most famous example of a transform boundary is the San Andreas Fault. Here, San Francisco is moving past Los Angeles, and eventually Los Angeles will break away from the mainland.

- An understanding of the dynamic nature of the Earth allows us to:
  - use information about tectonic plates to locate and exploit minerals and energy reserves
  - predict natural disasters and limit their damage to people and communities.

- The earthquake focus is the point from which the earthquake occurred.

- The earthquake epicentre is the point on the surface directly above the earthquake focus.

- In transform faults the earthquakes can be either shallow or of an intermediate depth and are often very strong and very frequent. In convergent plate margins, the earthquakes are very deep and very strong. In divergent boundaries earthquakes are very weak and shallow.

- The effects of earthquakes are ground movement, surface faulting, fire, landslips and landslides, liquefaction of soil and tsunamis.

- Igneous rocks are either extrusive (volcanic) or intrusive (plutonic).

- The most common types of volcanic rocks are basalt, andesite and rhyolite.

- The most common types of plutonic rocks are gabbro, diorite and granite.

- There are two types of volcanic eruption: non-explosive and explosive. Non-explosive eruptions are due to a low level of dissolved gases and magma that is liquid with a low silica content and low viscosity. Explosive eruptions are due to the dissolved gases in the magma and viscous magma with a high silica content.

- The non-explosive types of volcanoes are Hawaiian, Basaltic flood, Icelandic (or Fissure) and Submarine.

- Explosive eruptions may result in the release of pyroclastic material in the form of a pyroclastic flow of magma, gas and rock debris travelling at hundreds of kilometres an hour and with temperatures in the thousands of degrees centigrade. Sometimes the pyroclastic flow comes out from the side of the volcano, forming a lateral blast.

- The explosive types of volcanoes are Strombolian, Peleean, Phreatic and Plinian.

- Plinian columns are a type of explosive eruption resulting in columns of gas, debris and magma.

- The effects of volcanoes are pyroclastic flows, volcanic ash, lahars, avalanches, landslides, tsunamis, blasts and lavaflows.

- Non-boundary volcanoes are hot spot volcanoes. A hot spot is a region of the crust that is hotter than the surrounding crust, where magma has weakened and melted the crust, forming a volcano. This hot spot moves due to the plate moving.

PRACTICAL EXERCISE
Plotting earthquakes

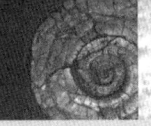

In this exercise you will gather and process data about earthquakes.

ACTIVITIES

1 Make a copy of Figure 2.4.7.

2 Plot the earthquake epicentres shown in Table 2.4.1 on your copy of the world map.

3 Look at the earthquakes you have plotted on the world map and refer to Figure 2.4.1 (page 105). Is there a correlation between the plotted epicentres and the tectonic plate boundaries?

4 Look at your world map and the map of tectonic plate boundaries (Figure 2.4.1).
  a Are the earthquakes more frequent on a specific type of plate boundary?
  b Do any earthquakes occur away from plate boundaries? If so, give an explanation as to why this may be so.

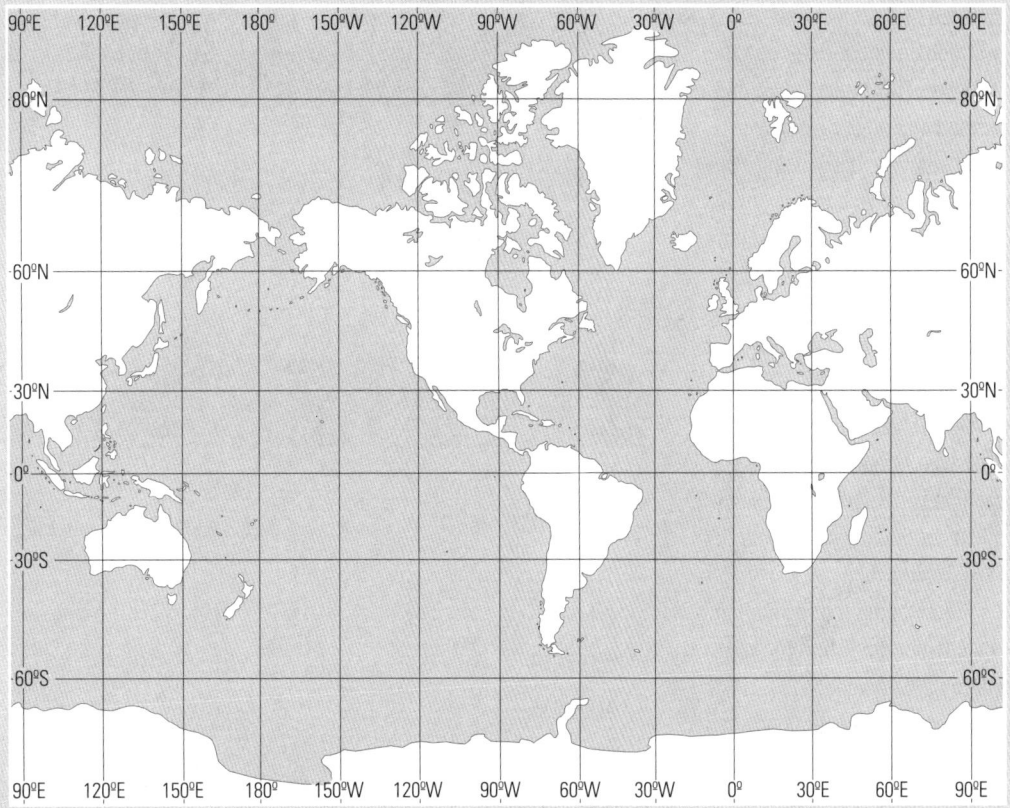

Figure 2.4.7 World map.

Table 2.4.1 Earthquake epicentres

Location	Longitude	Latitude
China	110°E	35°N
India	88°E	22°N
Pakistan	65°E	25°N
Syria	36°E	34°N
Italy	16°E	38°N
Portugal	9°W	38°N
Chile	72°W	33°S
Chile	75°W	50°S
Equador	78°W	0°
Nicaragua	85°W	13°N
Guatemala	91°W	15°N
California	118°W	34°N
California	122°W	37°N
Alaska	150°W	61°N
Japan	139°E	36°N
Japan	143°E	43°N

DYNAMIC EARTH

PRACTICAL EXERCISE
Extrusive rock chemistry

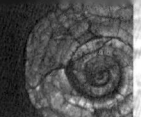

An area of the crust has warmed up and melted due to a hot spot. This melted area has formed a chamber filled with liquid magma. Occasionally the contents exit through a vent on the upper point of the chamber and erupt onto the surface. Over thousands of years the composition of both the magma in the chamber and the erupted material changes. The volcanic rock type formed from the magma is dependent on the silica content of the magma. (See Table 2.4.2.)

The magma that existed in the chamber that first formed contained a number of elements. The most important of these are listed in Table 2.4.3.

Table 2.4.2 Silica content of original magma

Silica content as percentage of magma	Volcanic rock type
45–53	Basalt
54–61	Andesite
62–69	Dacite
70–78	Rhyolite

Table 2.4.3 Key of elements in original magma

Element	Symbol
Silica	A
Magnesium	B
Iron	C
Calcium	D
Potassium	E
Aluminium	F
Silver	G
Gold	H
Copper	I

ACTIVITIES

Refer to Figure 2.4.8 and Table 2.4.4 and then complete the following activities.

1
Photocopy Table 2.4.4. Calculate the number of each element and its percentage of the total amount of the magma within the chamber at each point in time. Fill in the blank columns in your copy of Table 2.4.4.

2
For each of the chamber snapshots (shown in Table 2.4.4), work out the type of rock formed from the magma by using the percentage of the silica in Figure 2.4.8.

Table 2.4.4 Three magma chamber snapshots

Magma chamber when first formed			Magma chamber after 1000 years			Magma chamber after 2000 years		
Element	Quantity of element	Percentage of total	Element	Quantity of element	Percentage of total	Element	Quantity of element	Percentage of total
A			A			A		
B			B			B		
C			C			C		
D			D			D		
E			E			E		
F			F			F		
G			G			G		
H			H			H		
I			I			I		

| Magma chamber at start | Magma chamber at 1000 years | Magma chamber at 2000 years |

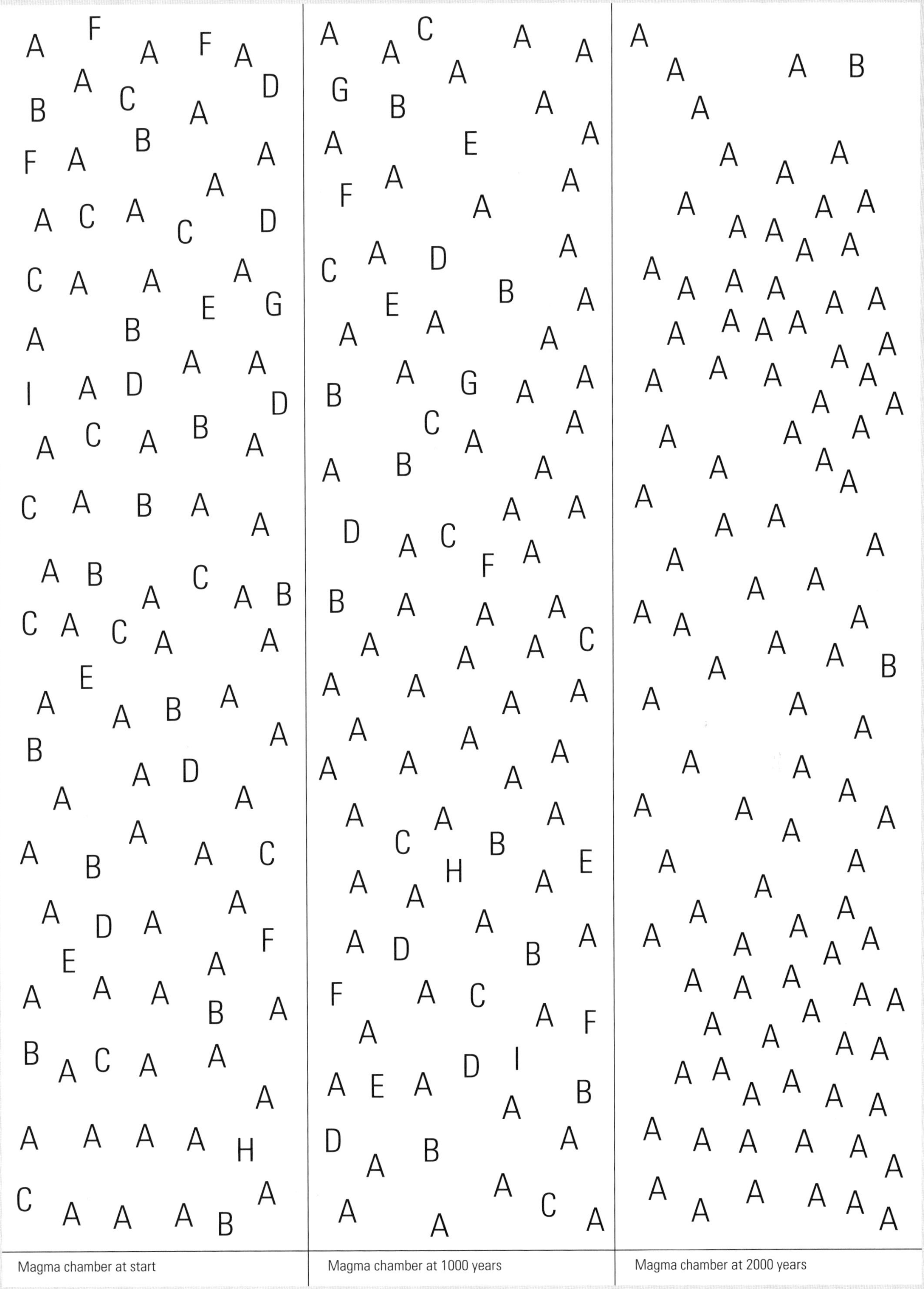

Figure 2.4.8 Contents of the magma chamber.

RESOURCES

Plate tectonics

The Break-up of the Supercontinent Pangea
www.platetectonics.com/book/pangea.

Plate Motions and Crustal Deformation
www.earth.agu.org/revgeophys/demets01/demets01.

Plate Tectonics *www.ucmp.berkeley.edu/geology/tectonics.html*

Plate Tectonics: An Overview
www.jersey.uoregon.edu/~mstrick/geology/Geo_Lectures/Plate_Tectonics.html

Plate Tectonics Self-test
www.ndsu.nodak.edu/instruct/schwert/schwert/geosci/g120/tectonic.htm

Subduction Zones
geo.lsa.umich.edu/~crlb/COURSES/270/Lec13/Lec13.html

Plate tectonics and igneous rocks

Andesite *www.clayir.gly.uga.edu/petrology/redoubt/Andesite*

Igneous Processes and Igneous Rocks
www.uno.edu/~gege/Easley/Physical/Chap03.htm#TOP.

Rhyolite
www.sorrel.humboldt.edu/~jdl1/web.page.images/rhyolite1.

Radioactivity

Atomic Structure and Radioactivity
www.geocities.com/CapeCanaveral/Launchpad/5226/radio

Summary of Radioactivity
www.dbhs.wvusd.k12.ca.us/Radioactivity/Radioactivity.html

Sea floor spreading

Mid-ocean Ridges
www.volcano.und.nodak.edu/vwdocs/vwlessons/volcano_types/spread.

Sea Floor Spreading
visearth.ucsd.edu/VisE_teach/lessons/Sea_floor_act.html

Volcanoes and earthquakes

Kinds of Volcanic Eruptions
volcano.und.nodak.edu/vwdocs/vwlessons/kinds/kinds.html

Plate Tectonics, Volcanoes and Earthquakes
www.soest.hawaii.edu/GG/ASK/plate-tectonics.html

Books

Clark, I. F. & Cook, B. J. *Perspectives of the Earth.* Australian Academy of Science, Canberra, 1986

Dolgoff, A. *Essentials of Physical Geology.* Houghton Mifflen, Boston, 1998

Hamblin, K. & Christiansen, E. *Earth's Dynamic Systems.* 8th edn. Prentice Hall, Provo, Utah, 1998

Kious, J. & Tilling, R. *This Dynamic Earth.* United States Geological Service, Washington DC, 1996

Lumine, J. I. *Earth.* Cambridge University Press, Cambridge, 1999

Murck, B. & Skinner, B. *Geology Today.* John Wiley & Sons, New York, 1999

Skinner, B., Porter, S. & Botkin, D. *The Blue Planet.* John Wiley & Sons, New York, 1999

Plate 1 The space shuttle is used for space exploration.

Plate 2 CSIRO's Parkes telescope is 64 m edge to edge. The dish collects the radio waves and focuses them onto the receiver at the top of the mast.

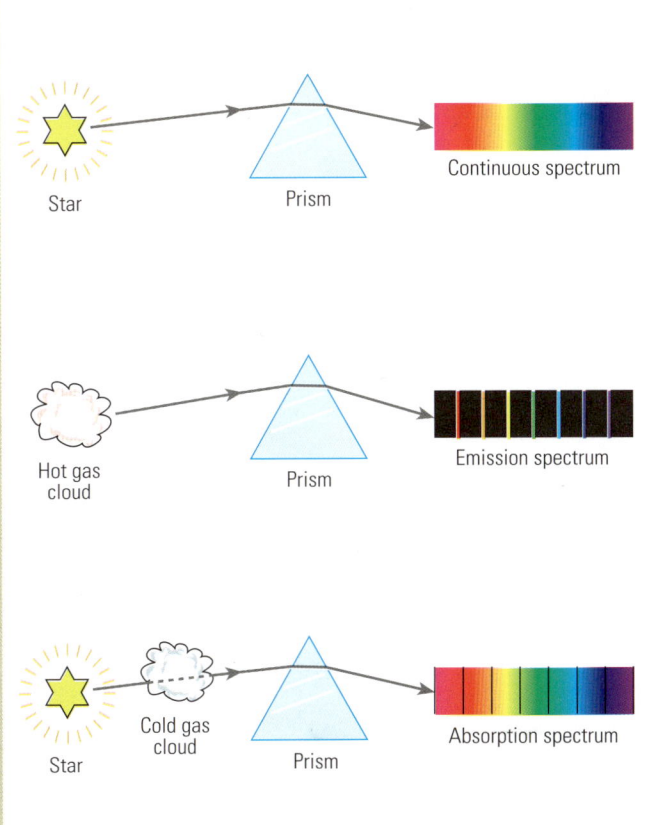

Plate 3 Continuous, emission and absorption spectra.

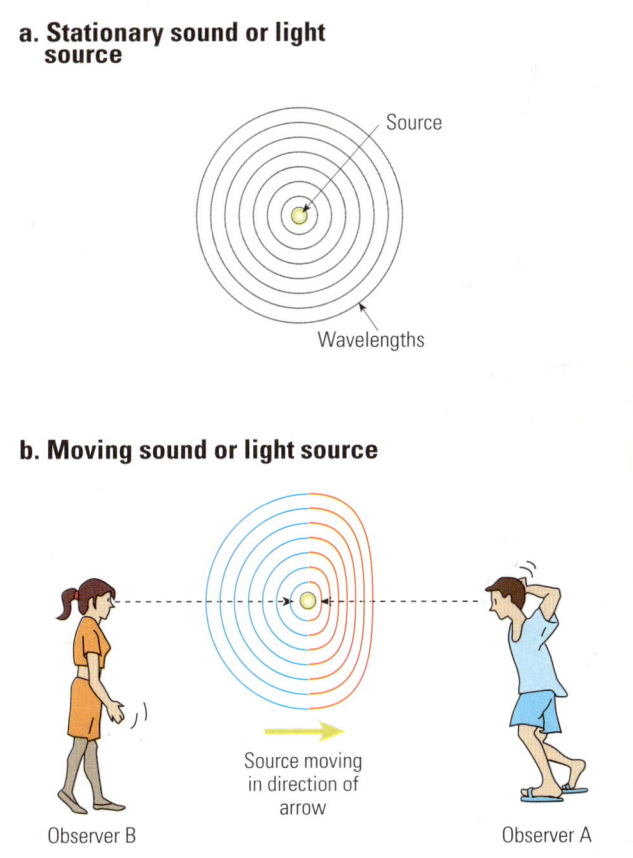

Plate 4 The Doppler effect.

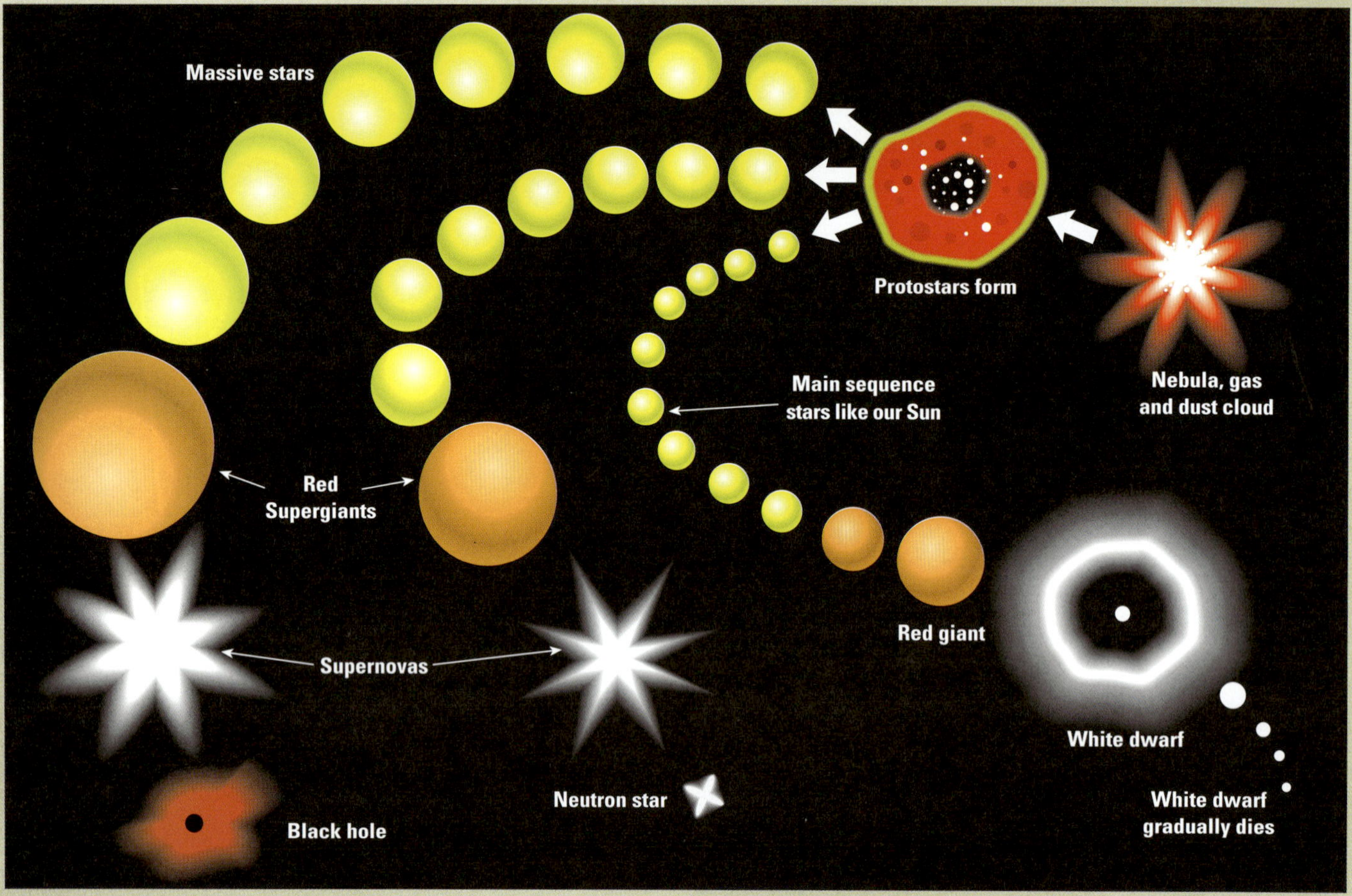
Massive stars
Protostars form
Nebula, gas
and dust cloud
Main sequence
stars like our Sun
Red
Supergiants
Red giant
Supernovas
White dwarf
Neutron star
White dwarf
gradually dies
Black hole

Plate 5 The evolution of stars.

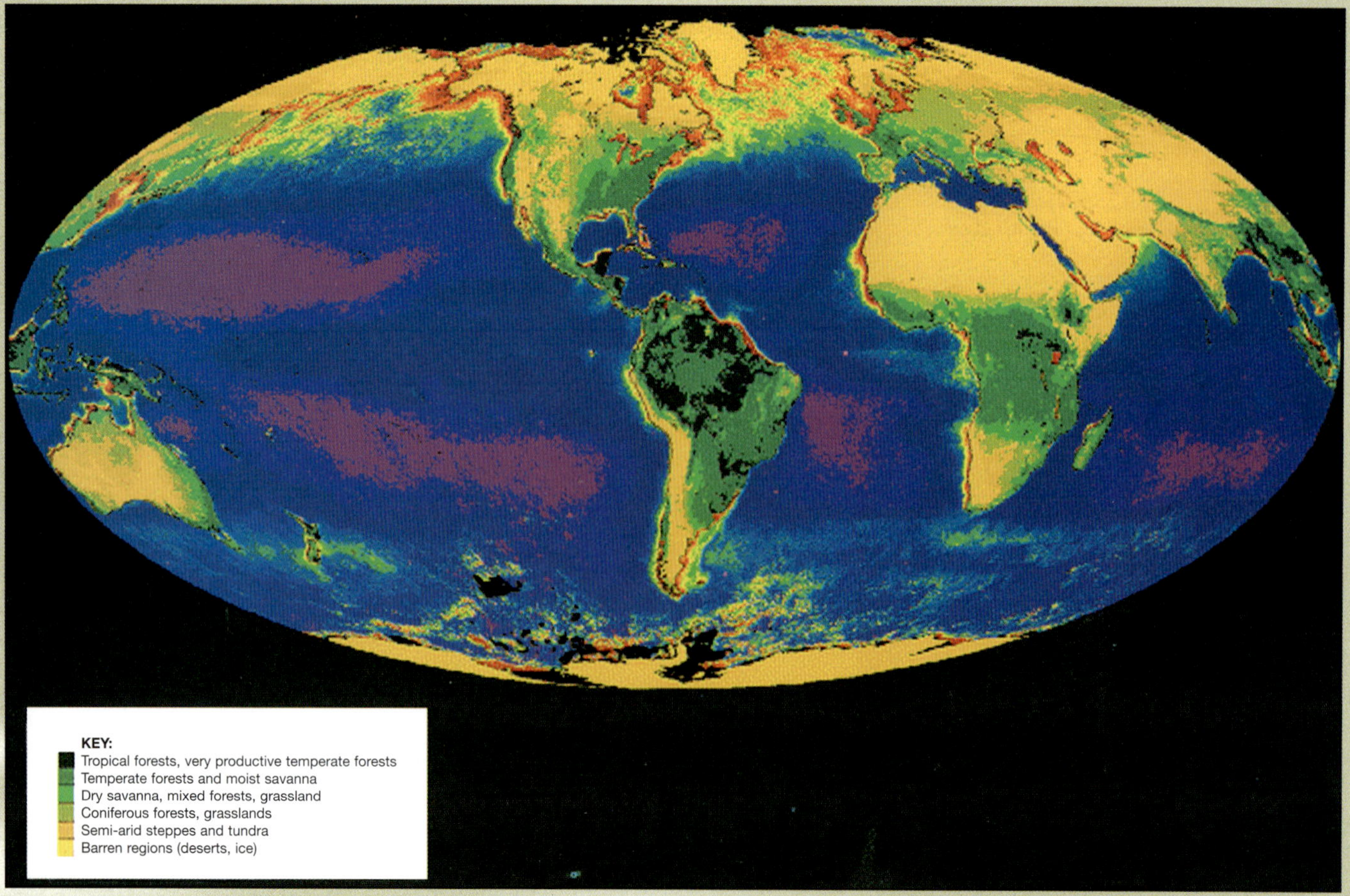
KEY:
Tropical forests, very productive temperate forests
Temperate forests and moist savanna
Dry savanna, mixed forests, grassland
Coniferous forests, grasslands
Semi-arid steppes and tundra
Barren regions (deserts, ice)

Plate 6 A satellite image of the biosphere. The pink–red colour indicates concentrations of phytoplankton.

Plate 7 Wolf Crater in Western Australia is a meteorite impact site.

Plate 9 A view of planet Earth showing a thin layer of the atmosphere.

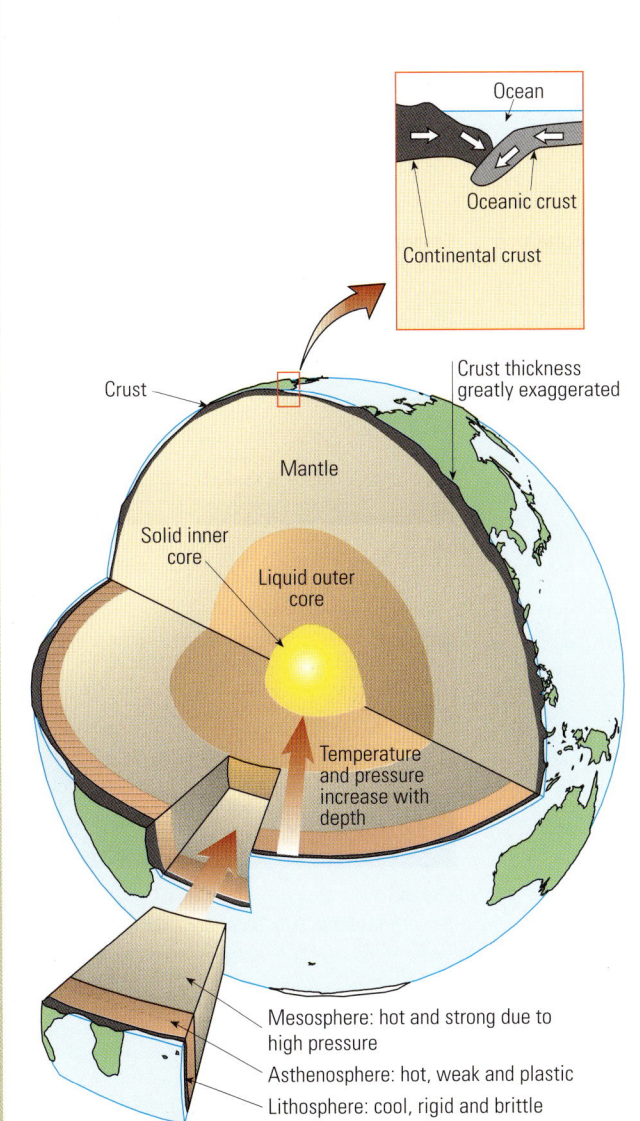

Plate 8 A cross-sectional view of the Earth reveals layers of different composition and zones with differing physical properties.

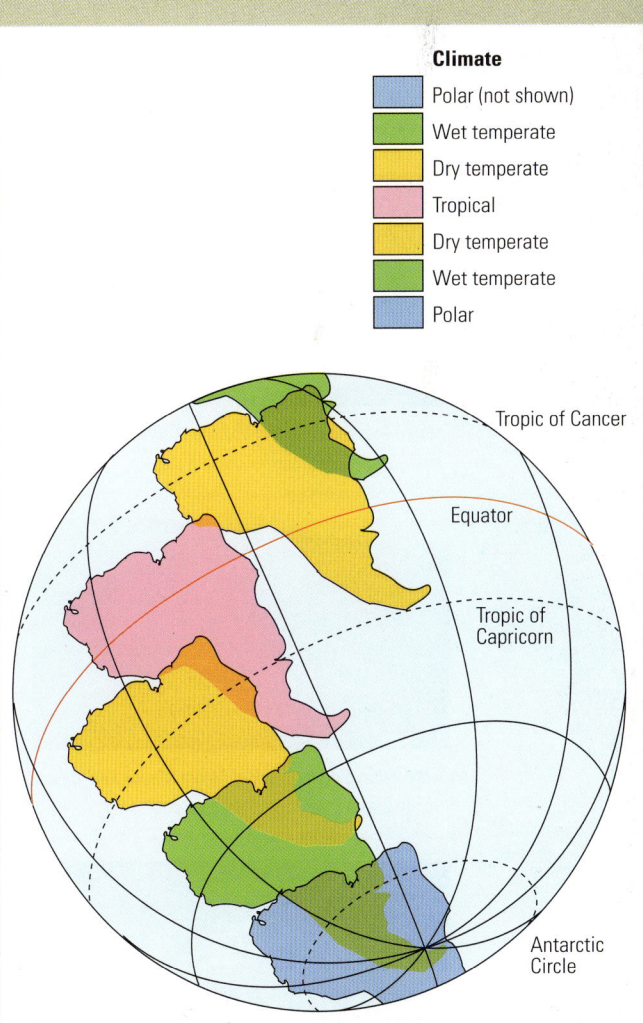

Plate 10 Continental drift.

Plate 11 A deep sea vent (fumarole).

Plate 12 Fossil stromatolites showing layering of different generations.

Volcanic type	Volcanic shape	Eruption composition	Eruption type
Shield volcano	 *Mount Mauna Loa Shield Volcano, Hawaii* Very gentle slopes; convex upward slopes	Basaltic pyroclasts, occasionally andesitic; basaltic lava flows	 Hawaiian *Mount Mauna Loa, March 1984*
Cinder or scoria volcano	 *Cinder volcano on the flank of Mount Mauna Keo, Hawaii* Straight sides; steep slopes; large summit crater	Basaltic pyroclasts, occasionally andesitic	 Strombolian *Mount Strombolini, Italy, December 1969*
Strato volcano	 *Mount Shasta and Shastina, California* Steep sides; alternating layers of lava and pyroclasts	Highly variable basaltic or rhyolitic lava and pyroclastic; has an overall andesitic composition	 Vulcaniun *Tavurvur Volcano in Rabaul Caldera, Papua New Guinea, October 1998*

Plate 13 Shield, cinder and strato volcanoes.

Plate 14 Outcrops, vegetation and soils all affect each other.

Plate 15 Rock type affects topography.

Plate 16 Heathland.

Plate 17 Open woodland.

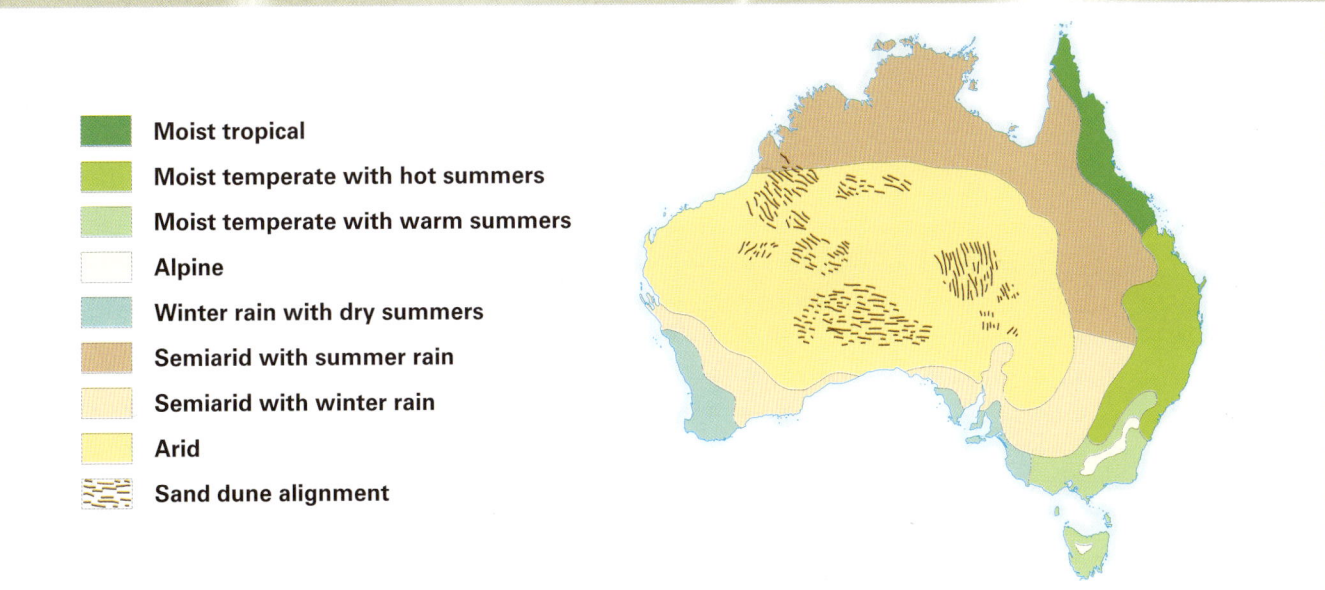
Plate 18 Australia's climate zones. The sand dune arrangement in the centre is related to the dry conditions during the last ice age.

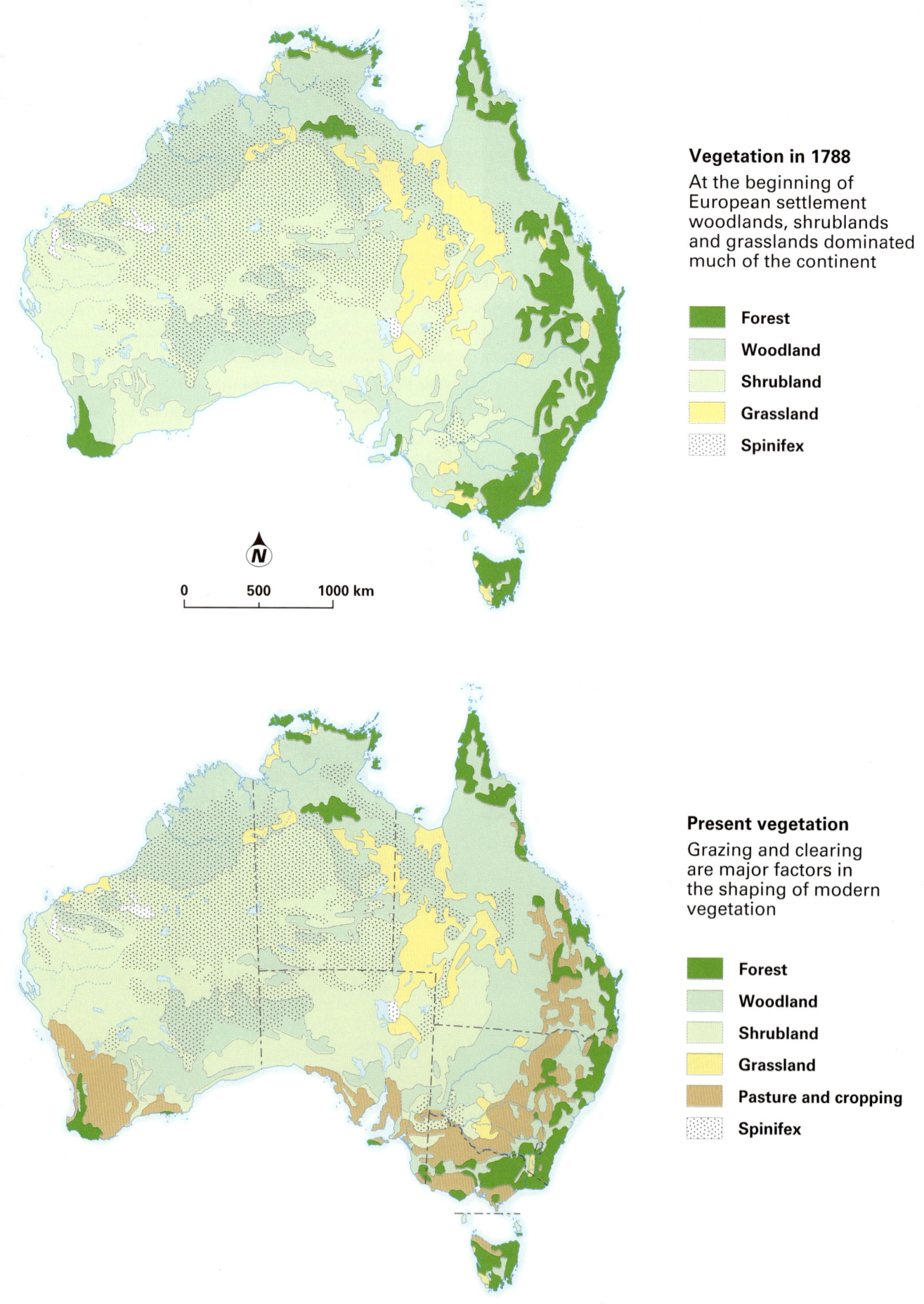

Plate 19 A comparison of vegetation in 1788 and 1988.

Plate 20 More or less resistant rocks lead to structures such as honeycomb weathering.

Plate 21 Water as an agent of erosion. Note the different channel heights.

Plate 22 Cross-bedding reflects the action of running water.

Plate 23 A brachiopod. What does its presence imply about the environment in which it was buried?

Plate 24 Cliff at Redhead Beach looking south. Evidence of a delta is visible.

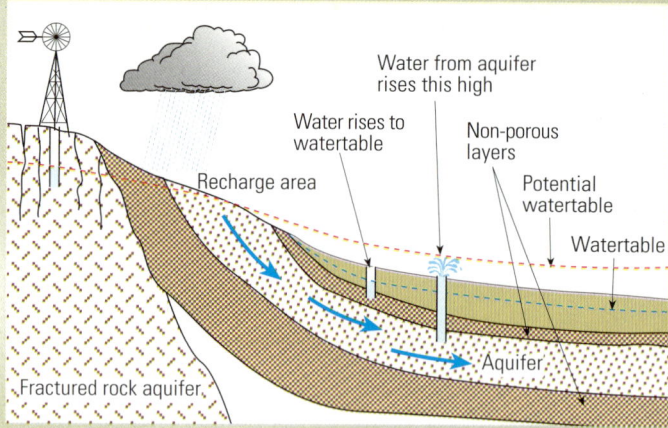

Plate 25 The structure of an aquifer and the movement of water within it.

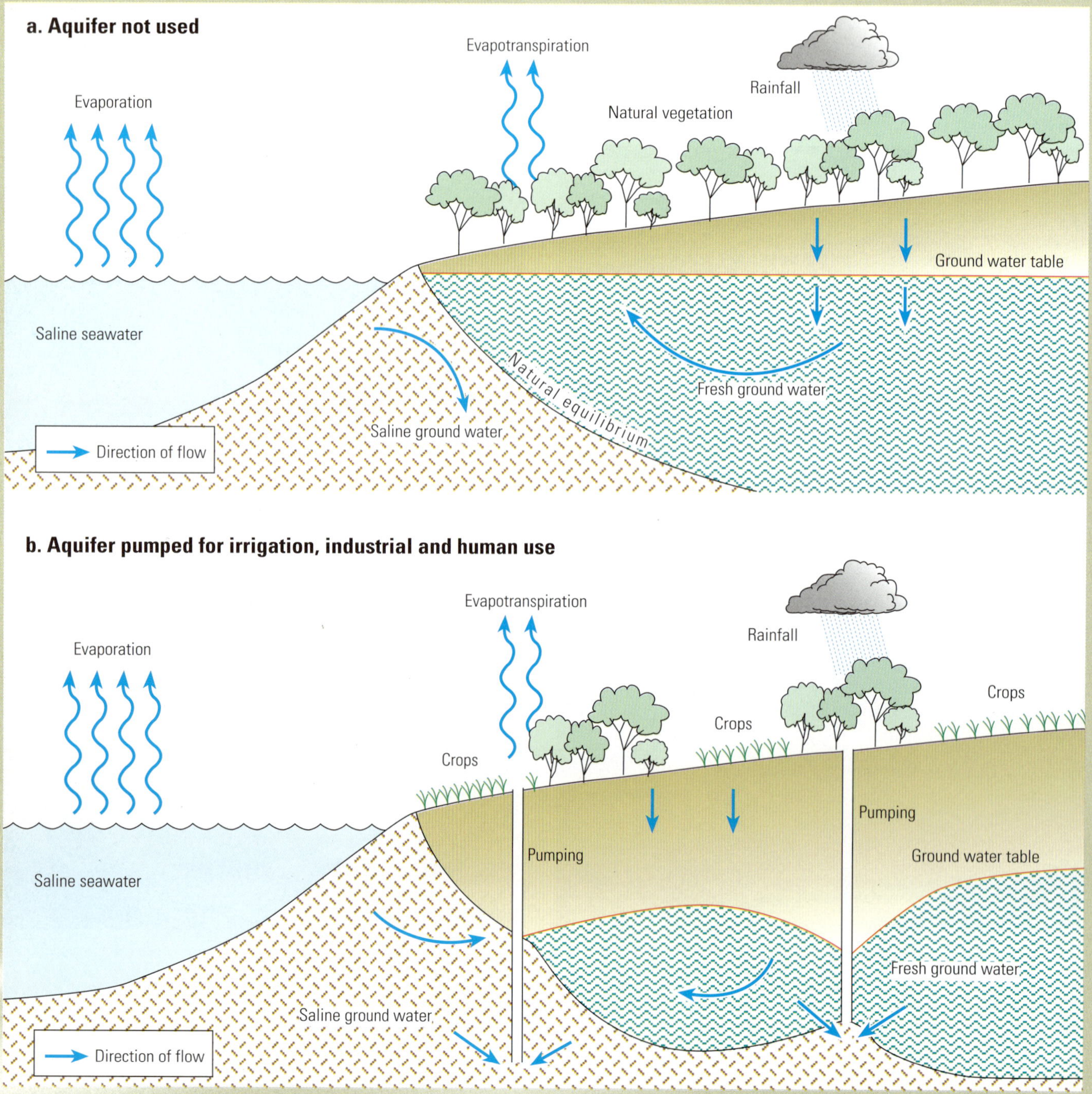

Plate 26 How ground water extraction leads to salt water intrusion. The process is described on page 239.

The local environment

In this section we will use a systems approach to examine some important concepts about the local environment. We will examine, in turn, the rocks that make up our environment, the soils produced from them and the effects of humans on animal and plant diversity. In particular, we will look at ideas that can be applied to the environment in which you live. It is important that you try to apply the knowledge and understanding you acquire to your local environment. By doing this you will gain a better understanding of the ideas presented in this section, learn the concepts and extend the knowledge you already have of the interaction of people and the environment.

CONTENTS

Chapter 3.1	The local environment as a system	124
Chapter 3.2	The elements of lithosphere and landscape	127
Chapter 3.3	Soils: Features and formation	146
Chapter 3.4	Adaptations and survival in disturbed habitats	169
Chapter 3.5	Human impacts on the environment	177
Chapter 3.6	Balancing development and conservation	191
Chapter 3.7	Biodiversity	198
Resources		203

3.1 The local environment as a system

OUTCOMES

At the end of this chapter you should be able to:

- recall that the Earth and the environments that make it up are dynamic systems
- describe the nature of a system
- recall that parts of a system exchange energy and matter
- describe the components of your local environment system
- explain how systems can be influenced by events external to the system.

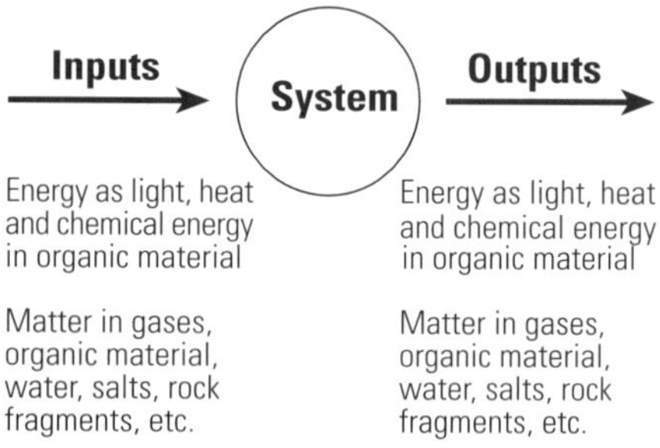

Inputs → **System** → **Outputs**

Energy as light, heat and chemical energy in organic material

Matter in gases, organic material, water, salts, rock fragments, etc.

Energy as light, heat and chemical energy in organic material

Matter in gases, organic material, water, salts, rock fragments, etc.

Figure 3.1.1 A simple system. Matter and energy are inputs and outputs.

definitions

system
something that takes in matter and/or energy, processes it in some way and then gives out matter and/or energy

energy
the quality something has that allows it to cause change

matter
the material that objects are made of. Matter is composed of atoms or the parts of atoms.

faeces
the undigested part of food together with bacteria and chemicals from the gut

Think about the environment in which you live. Is it easy to describe? How well can you explain the way it works? Can you predict with confidence how a change in one part of your environment will affect other parts? When you think about such questions you will probably agree with the suggestion that environments are complicated things which consist of many interacting parts. Now, increase the size of the environment you are thinking about. Consider New South Wales, then Australia, and now the Earth as a whole. Does understanding become easier? Hardly. Environments of all scales are complex and dynamic. They contain many interrelated parts that change over time in many different ways.

Scientists, in studying complex and changing things, often use the idea of a system. (See Figure 3.1.1.) A **system** is something that takes things in (inputs), processes them in some way, and gives things out (outputs). The first step in trying to understand a system is to determine what its inputs and outputs are.

The inputs and outputs of a system consist of both **energy** and **matter**. The environment in which you live receives energy in the form of light from the Sun and in the heat generated by human activities. Photosynthesis converts light energy into chemical energy, on which most ecosystems depend. Matter enters our environment in a number of forms. Rivers, creeks and streams not only bring water to our local environment but also carry sand, mud, plant materials, dissolved salts and micro-organisms into the environment. Water may leave our local system through evaporation and when a river carries water away. When trees are harvested, plant material ends up in landfill and crops are sent to market, we remove matter from our local environment. Such organic material can be regarded as an output.

A system may contain parts that are themselves smaller systems. Figure 3.1.2 shows the components of a local environment and the movement of energy and matter between the parts.

Energy and matter are exchanged between the smaller systems that make up the environment. Animals, for example, depend on the energy and matter in the food they obtain from plants or other animals. Animals, such as mammals, also give out energy in the form of heat and as matter and chemical energy in **faeces**. The soil receives energy from the atmosphere and re-radiates part of it. Water from the atmosphere enters the soil and may leave carrying dissolved nutrients and salts.

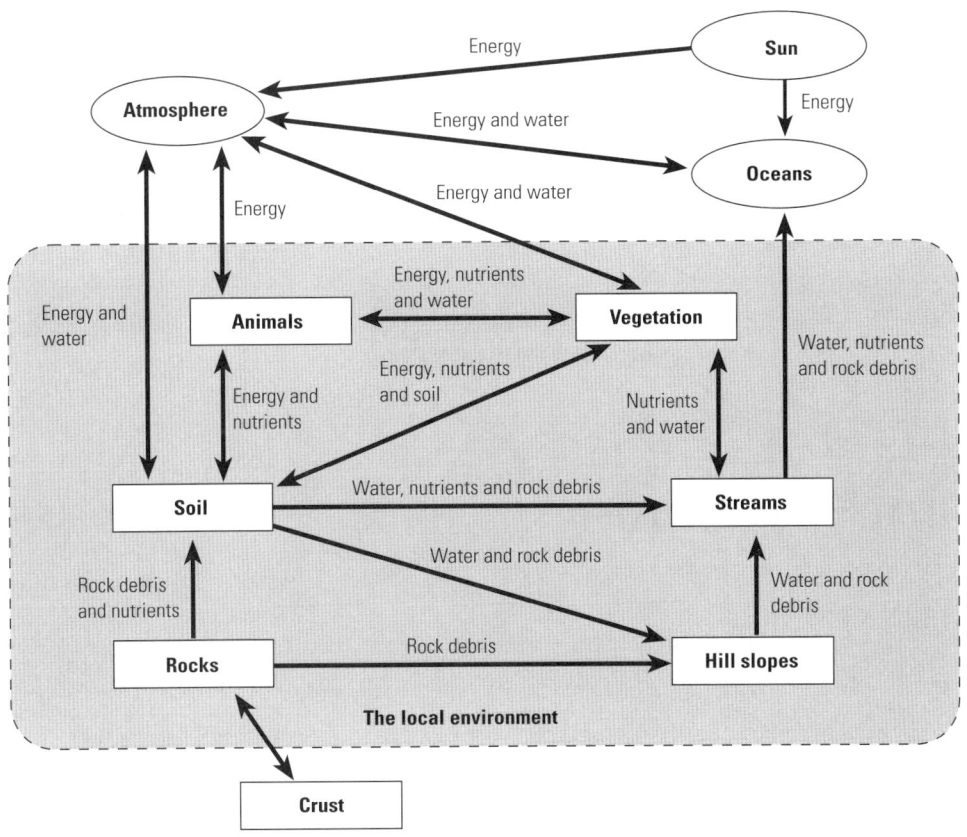

Figure 3.1.2 The flow of energy and matter in a local environment.

If we can understand some of these exchanges we are in a position to understand how changing one part of a system may affect other parts or other systems. As an example, consider your local environment and rain. What effect does a lack of rain have? Rain is something from outside the system—an input. If the rainwater does not reach the soil, vegetation will not be able to obtain the water it needs. Streams will not flow as strongly and less clay, and the nutrients that clays bind, will be carried away from your local environment. Because more water is absorbed, less water run-off will occur and less water leaves the local environment as an output.

REVIEW ACTIVITIES

1 What is a system?

2 Why can we think of our local environment as a system?

3 List, and give examples of, the two things that are inputs and outputs of systems.

4 Make a diagram of your local environment like that shown in Figure 3.1.2. (Your diagram should have more interconnections and boxes than those shown in Figure 3.1.2.)

5 Consider your local environment as a system. Explain how it can:
a be affected by changes to its inputs
b affect other environmental systems by changes to its outputs.

EXTENSION ACTIVITIES

6 Cave environments do not receive light from the Sun. Explain how energy enters a cave environment. Does matter enter and leave such an environment?

7 Explain why a gas stove is a system. Why is an ecosystem harder to understand than a gas stove?

SUMMARY

- Systems have inputs and outputs of energy and matter.
- The Earth can be studied as a dynamic system.
- Systems can be thought of as being composed of other systems.
- A system can be affected by events external to the system.

PRACTICAL EXERCISE
Analysing systems

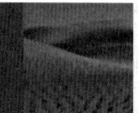

Consider the Earth as a system.

ACTIVITIES

1 A closed system is one that does not take in or give out material or energy but an open system does. Is the Earth a closed or open system?

2 Make a short list of evidence to support your answer to activity 1.

3 The Earth can be thought of as being made up of subsystems: the atmosphere, biosphere, lithosphere and hydrosphere. Make a diagram to show the interaction of the Earth's subsystems. On your diagram show whether energy or matter is being moved from one subsystem to another.

4 Scientists sometimes use comparisons to explore subjects.
  a Consider a hot cup of tea or coffee as a system. What enters the system? What leaves it? What processes are present in the cup?
  b Now consider the atmosphere. What enters this system? What leaves it? What processes are present in the atmosphere?
  c Compare the cup of tea or coffee and the atmosphere. Don't forget to look at similarities and differences.

5 Summarise what you have learnt from this exercise.

3.2 The elements of lithosphere and landscape

OUTCOMES

At the end of this chapter you should be able to:

- distinguish between igneous, sedimentary and metamorphic rock groups in terms of their origins and common mineral composition

- identify and describe the geological features of the local environment that determine its landscape

- classify several common igneous, sedimentary and metamorphic rocks using a key, with particular reference to those rocks in the local environment.

TALKING ABOUT EARTH MATERIALS

How much do you know about the rocks where you live? We often take rocks for granted, but rocks and their structures are very important factors shaping the environment. As we shall see, the breakdown of rocks produces soils. The nature of the soils in an area affects the types of plants and animals that are found there. How rocks weather also influences the landscape. Hard rocks are broken down slower than soft rocks. As a result we find hard rock types forming hills and nearby soft rocks forming lower areas. Rivers and creeks often flow along zones of weakness in rock units and the water carves the landscape. In addition, humans value some rocks above others. Coal, for example, is regarded as of more value than the sandstones and shales found with it. When people mine or quarry an area to obtain rocks for some purpose they frequently change the land surface.

Minerals

Rocks are made up of interlocking or cemented grains of matter. The grains are usually minerals or aggregates of minerals. A mineral is a naturally occurring **inorganic** substance with a definite structure and composition. Recognising minerals is a key part of correctly identifying a rock. While there are more than 2000 minerals described we can successfully identify most rocks using only a few mineral types. Some common minerals and their properties are shown in Table 3.2.1 (page 128).

Recognising minerals not only helps us to identify a rock but it also helps us to identify a rock's overall **composition**. Knowing a rock's composition is important because it affects the materials produced by weathering of the mineral and the elements available to other parts of the environment. In this section minerals will be described in groups that are based on composition. The principle groups are the silicates, carbonates, sulfates and sulfides, and oxides and hydroxides.

SILICATES

Most commonly occurring minerals, about 95%, are silicates. Silicate minerals contain silicon and oxygen arranged in the form of a **tetrahedron**.

Why silicates are so common becomes clear when you examine a list of the elements, and their abundance, in the Earth's crust. (See Table 3.2.2, page 129.) Silicates are composed of the most abundant elements. Understanding silicate

definitions

inorganic
materials not containing carbon compounds other than carbon oxides and similar simple compounds

composition
a description of the things making up an object. A chemical composition describes the elements or compounds in a substance and their relative abundances.

tetrahedron
(plural: tetrahedra) the basic building block of silicate minerals

Table 3.2.1 Common minerals and their properties

Mineral	Chemical composition	Cleavage and fracture	Colour and lustre	Hardness (Moh's scale)	Density (g/cm³)	Comments
Amphibole	$Ca_2(Mg, Fe)_5Si_8O_{22}(OH)_2$	Two at 60° and 120°	Black to green	5–6	3.2	–
Bauxite	$AlO(OH)$	One perfect	White	6.5	3.4	Often forms pea-sized aggregates
Biotite	$K(Mg, Fe)_3AlSi_3O_{10}(OH)_2$	One perfect	Black to dark brown	2.5–3	3.0	Splits into very thin sheets
Calcite	$CaCO_3$	Three perfect; rhombus	Colourless, white or yellow	3	2.7	Bubbles when acid is applied to it
Chlorite	$(Mg, Fe)_5Al_2Si_3O_{10}(OH)_8$	One perfect	Green	2	2.5	Forms foliated masses
Garnet	$CaAl_2Si_3O_{12}$	Conchoidal fracture	Red to brown; vitreous lustre	6.5–7	3.6	Very characteristic of metamorphic rocks
Goethite	$FeO.OH$	One perfect	Yellow-brown to red	5–5.5	4.3	Yellow streak, dissolves in hydrochloric acid
Gypsum	$CaSO_4.2H_2O$	One perfect; two poor	Rusty brown	2	2.3	Yellowish streak
Halite	$NaCl$	Three perfect; cubic	Transparent to white	2.5	2.2	–
Haematite	Fe_2O_3	None	Red to silver-grey; lustre may be metallic or earthy	6	5.3	Red streak
Kaolin	$Al_2Si_2O_5(OH)_4$	One perfect	White to brown	2	2.0–2.5	Common in soils and weathered rocks
K-feldspar (orthoclase)	$KAlSi_3O_8$	Two at right angles	White, grey or pink	6	2.6	White streak
Muscovite	$KAl_3Si_3O_{10}(OH)_2$	One perfect	Colourless to light brown	2–2.5	2.8	White streak
Olivine	$(Mg,Fe)_2SiO_4$	Conchoidal fracture	Green to brown	6.5	3.4	White streak
Plagioclase	$CaAl_2Si_2O_8$ to $NaAlSi_3O_8$	Two at right angles	White to grey	6	2.7	A very common mineral, forming characteristic rectangular crystals
Pyrite	FeS_2	Uneven fracture	Brass like to golden yellow	6.5	5.0	Often forms well-formed cubic crystals
Pyroxene	$(Mg,Fe)SiO_3$	Two at about 90°	Green, dark brown or black	6	3.3	–
Quartz	SiO_2	Conchoidal fracture	Colourless, yellow, grey, purple or other	7	2.7	Light (if it makes one)
Serpentine	$Mg_6Si_4O_{10}(OH)_8$	Splinters when fractured	Green to brown; silky lustre	2.5	2.5	–
Staurolite	$Fe_2Al_9Si_4O_{22}(OH)_2$	One poor	Brown to red	7	3.8	–

structures will help us to understand their properties, why some weather easier than others and why they occur in certain rocks and not others.

In a silicate mineral, tetrahedra may occur individually, in chains, sheets or as complex networks. (See Figures 3.2.1 and 3.2.2, page 130.) The tetrahedra are either joined directly and strongly to other tetrahedra or they are held together, more weakly, by metal ions. When a mineral crystal breaks it breaks along planes of weakness—usually along planes where the metal ions are located. These planes are called cleavage planes and they appear as lines or flat surfaces on mineral grains.

The amount of direct joining between silica tetrahedra in a mineral is reflected in how the mineral weathers and, indirectly, in its colour. As a general rule, the more linkage there is between tetrahedra, the more resistant to weathering the mineral is. Olivine is a green, glassy silicate mineral. Its structure consists of individual tetrahedra held together by magnesium and/or iron ions. (See Figure 3.2.2.) The metal ions help determine the colour of the mineral. Weathering occurs when the metal ions are plucked out of the mineral, leaving the tetrahedra loose and easily pulled apart. Quartz, on the other hand, contains no metal ions in its structure so it is very resistant to weathering and also very hard. The range of colours found in quartz is due to minor amounts of metal ions that are trapped in the mineral.

Whether a mineral is light or dark in colour can be useful in identification. Dark-coloured silicate minerals are usually those in which magnesium and iron are present, acting to hold together individual silica tetrahedra or chains of tetrahedra. The dark silicates that are rich in magnesium and iron are called the **mafic** silicates. Mafic minerals range in colour from black to dark green. Light minerals, such as quartz and feldspars, contain lots of silicon and aluminium. The silicate structures of quartz and feldspars are networks and contain few metal ions that produce dark colours. Such silicate minerals, which are rich in aluminium and silicon, are called felsic minerals.

CARBONATES

While there are some sixty carbonate minerals, the most common carbonate mineral is called calcite. Its chemical name is calcium carbonate and it has the formula $CaCO_3$. Calcite is very common in sedimentary rocks, but it also occurs in igneous and metamorphic rocks. Like quartz, calcite can occur in a range of colours, although white and grey are common. Carbonates give a characteristic reaction when a few drops of acid are placed on the mineral. Gas bubbles are generated as the mineral dissolves. The bubbling is termed **effervescence**. Calcite also has a characteristic set of three cleavage planes. (See Figures 3.2.3 and 3.2.4, page 130 and 132.)

SULFATES AND SULFIDES

Sulfate minerals contain the sulfate group (formula: SO_4^{2-}). Gypsum is the most common sulfate mineral and it occurs in sedimentary deposits. Like calcite it is often white or colourless but it has one plane of cleavage and forms long, flat crystals. Sulfates are important in small amounts for plant growth but in large amounts are harmful to plants and soil organisms.

Sulfide minerals are composed of metal ions and sulfur. The sulfide called pyrite (formula: FeS_2) occurs in all types of rocks. It occurs as a result of metamorphic processes and igneous **hydrothermal** activity and can also form in oxygen-poor environments through the action of bacteria. Sulfides may react with oxygen in water to produce acidic water. This water is an important environmental problem for mines producing metal sulfides because the acidic mine waters can damage plants and harm a wide range of freshwater animals.

Table 3.2.2 The fourteen most common elements in the Earth's crust

Element	Crustal abundance by weight (%)	Crustal composition (parts per million)	Symbol
Oxygen	46.60	456 000	O
Silicon	27.72	273 000	Si
Aluminium	18.13	83 600	Al
Iron	5.00	62 200	Fe
Calcium	3.63	46 600	Ca
Sodium	2.83	22 700	Na
Potassium	2.59	18 400	K
Magnesium	2.09	27 640	Mg
Titanium	0.44	6 320	Ti
Hydrogen	0.14	1 520	H
Phosphorus	0.12	1 120	P
Manganese	0.10	1 060	Mn
Sulfur	0.05	340	S
Carbon	0.03	180	C

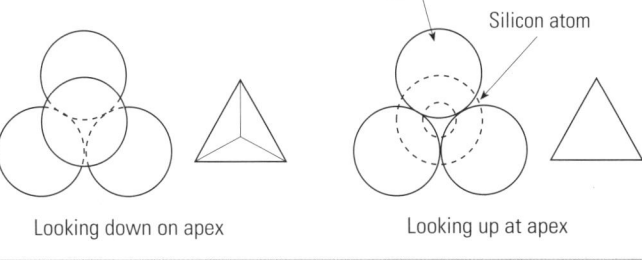

Figure 3.2.1 The silicate tetrahedron.

definitions

mafic
rich in magnesium (Mg) and iron (Fe)

effervescence
bubbles of gas being released from something

hydrothermal
hot fluids created by igneous activity. The fluids carry dissolved minerals and gases.

OXIDES AND HYDROXIDES

Oxides, particularly those of iron and aluminium, are very common. Haematite is an iron oxide found mainly in sedimentary rocks and metasedimentary rocks (sedimentary rocks that have been metamorphosed). Haematite produces the red colour in many sedimentary rocks and soils. Goethite is an iron hydroxide and it is a yellow-brown to red colour. It is the pigment in yellow ochre and gives a characteristic yellow streak. Haematite gives a red streak. Aluminium in soils occurs as an aluminium hydroxide. Diaspore (formula: AlO.OH) is a product of the weathering of silicate minerals and, like other minerals already mentioned, can occur in a variety of colours due to the presence of metal ions, such as iron or manganese.

> **definition**
>
> **oxides**
> compounds containing metal or non-metals combined with oxygen

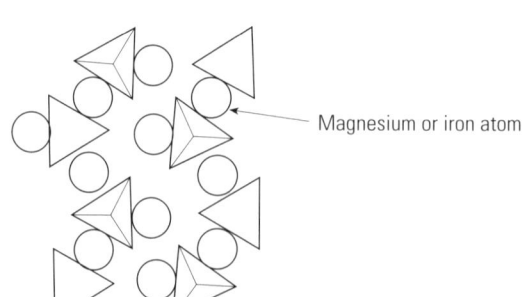

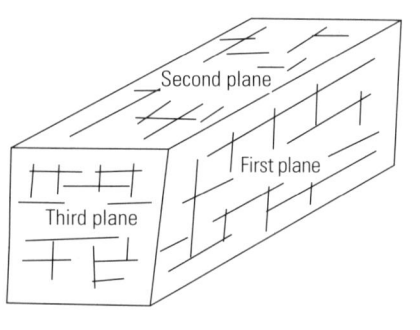

Figure 3.2.3 Calcite and its cleavage planes.

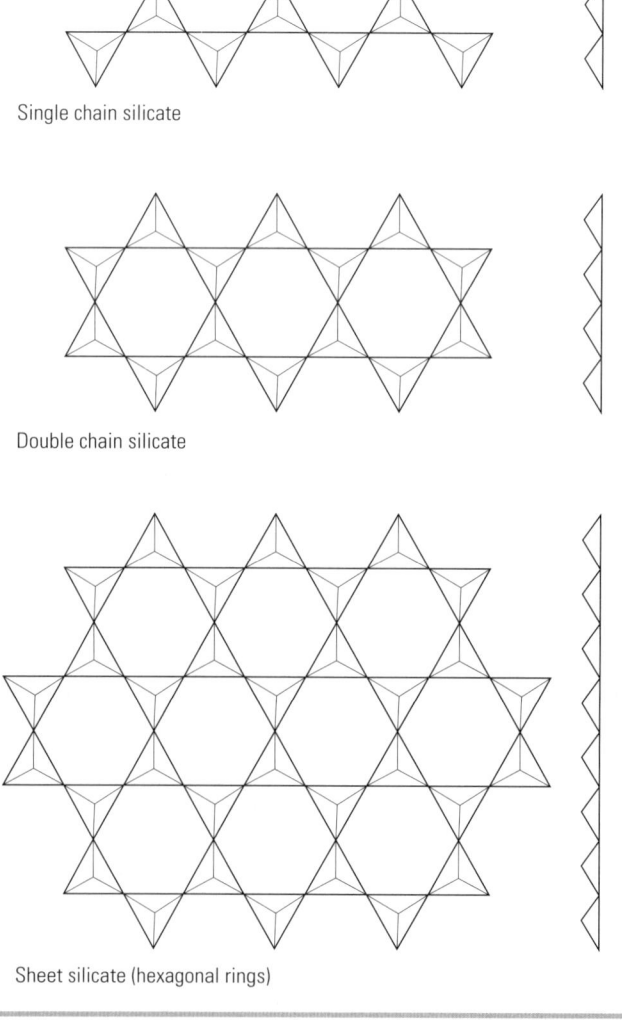

Figure 3.2.2 Four silicate structures.

REVIEW ACTIVITIES

1 Define the term 'mineral'.

2 What are the characteristics of a mineral?

3 Explain the cause of colour in a mineral.

4 Why are silicate minerals common?

5 Use Table 3.2.1 (page 128) to complete the following activities.

  a Compare and contrast the properties of amphibole and biotite.

  b Calculate the ratio of silicon (Si) to oxygen (O) for olivine, biotite and quartz. What trend is shown in the series of minerals?

  c List the different types of silicates, oxides and carbonates shown in Table 3.2.1. Use a table to organise your lists.

EXTENSION ACTIVITIES

6 Zircon and garnet are two minerals with interesting properties. Research the properties and occurrence of each mineral. Compare the two minerals.

7 Compare the oxide and sulfide mineral groups. In what way are these minerals different from silicates?

8 Weathering of silicate minerals can turn a rock like granite into a sandy soil. Work out the minerals that would be first to break down if a granite contains plagioclase and orthoclase feldspar, quartz, biotite and hornblende.

CHARACTERISTIC PROPERTIES OF MINERALS

The characteristic properties by which minerals are identified are colour, lustre, hardness, cleavage and fracture, and density. The form of the mineral—its habit—may also be useful.

Colour

Using a specific colour is not a good guide to identifying minerals. Quartz, for example, exists in many different colours. It may be clear and glassy, have the colour of honey or be grey, white, green, pink or yellow. Opal, with its many colours, is a type of quartz. The colour of a quartz sample is due to the presence of small amounts of metal ions and/or water trapped in the structure of the mineral. While the colour varies, properties due to the crystal structure of the mineral are relatively constant.

Lustre

Lustre describes the way a mineral reflects light. The lustre of a mineral should be judged on fresh surfaces as some minerals react with the air very quickly and lose their lustre. Some terms used to describe lustre are:
- earthy or dull—no shine, like chalk or clay
- metallic—as is characteristic of metals
- non-metallic—shiny but not like a metal.

  Some terms used to describe non-metallic lustres are:
- vitreous—shiny like glass
- pearly—reflects light like a pearl
- silky—reflects light like silk.

definition

lustre
the appearance of a mineral in reflected light

Hardness

Hardness is measured by trying to scratch a clean mineral surface with a substance of known hardness. A set of ten minerals is used as a reference and the range of hardnesses they cover is known as Moh's scale after the mineralogist who invented it. (See Table 3.2.3, page 132.) When testing hardness ensure that any powder produced

Table 3.2.3 Moh's scale of hardness

Hardness	Mineral	Common material
1	Talc	–
2	Gypsum	A fingernail is 2–2.5
3	Calcite	–
4	Fluorite	–
5	Apatite	A knife blade is about 5.5–6
6	Orthoclase	–
7	Quartz	–
8	Topaz	–
9	Corundum	–
10	Diamond	–

definition

cleavage
flat, sheet-like structures along which a mineral tends to break easily

is removed before the scratching is checked. A mineral that is scratched is softer than the mineral that scratches it.

Cleavage and fracture

Many minerals when hit break along flat planes. This is known as **cleavage**. (See Figure 3.2.4.) Cleavage planes are flat and shiny surfaces when fresh. Sometimes you may recognise them as parallel lines on the surface of a crystal. Cleavage is most clearly seen in large crystals. It is very hard to see in small ones because it can be confused with the form of the crystal.

Cleavage is described by the number of cleavage planes a mineral has and their angle to one another, if more than one is present. Be careful not to confuse cleavage with crystal faces. Crystal faces form when a crystal has the space to grow and form flat surfaces.

If a mineral is broken and it does not cleave it is said to fracture. It is possible for a mineral to show both cleavage and fracture. Feldspars are a good example of such minerals. Fractures are usually rough and uneven, but sometimes a mineral may fracture and produce a curved surface. This is called a conchoidal fracture.

Density

Density is the ratio of an object's mass to its volume. Density is a very useful characteristic for classifying some minerals. Details of mineral densities are shown in Table 3.2.1 (page 128). Note the low density of halite and the high densities of some iron minerals. Minerals such as galena have a high density, which is easily noted because small pieces of it are heavy.

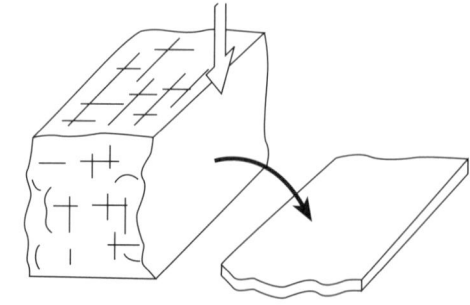

One direction of cleavage
The mineral cleaves into flat sheets
Example: biotite

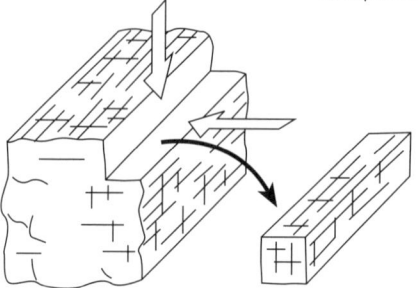

Two directions of cleavage
The mineral cleaves into long fragments
Example: plagioclase

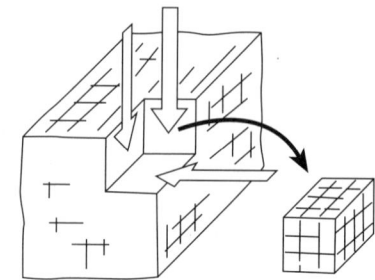

Three directions of cleavage
The mineral cleaves into regular-shaped pieces
Example: halite

Figure 3.2.4 Mineral cleavage.

REVIEW ACTIVITIES

1 Explain why colouring is a poor choice for identifying most minerals.

2 Contrast lustre and colour.

3 Compare a dark mineral and a light mineral from Table 3.2.1 (page 128) in terms of their composition and hardness.

4 A mineral has a hardness of 4. Would quartz scratch this mineral?

5 Refer to Table 3.2.1. List the properties that could be used to distinguish between pyroxene and plagioclase.

6 Summarise the properties of feldspars from the text. Why are feldspars so important?

7 Make a labelled diagram to show fracture and cleavage in a mineral.

EXTENSION ACTIVITIES

8 Select two minerals from Table 3.2.1. Summarise the ways in which the minerals are:
a similar
b different.

9 Select five minerals from Table 3.2.1. Use the properties of the minerals to design a dichotomous key that can be used to classify the minerals. (If you are unsure what a dichotomous key is, ask your teacher or look at Table 3.2.8, pages 144–5.)

DESCRIBING ROCKS

Rocks form under different conditions in the crust of the Earth. Scientists who study rocks—geologists and mineralogists—recognise three major types of rocks: igneous, sedimentary and metamorphic. The three types are defined in terms of their origins. A rock may be formed from molten, or liquid, rock or from the materials produced by changes to a rock at the Earth's surface. It may also be formed from another rock type that has been changed by conditions inside the Earth's crust.

To work out the way in which a rock formed it is necessary to use the clues contained in the rock. It is important to understand that one rock body can give rise to other rock bodies over time. The ways in which this can happen are summarised in what is known as the rock cycle. (See Figure 3.2.5, page 134.)

Rocks are described in terms of both their mineral composition and **texture**. A rock's composition may be given in terms of the types and abundance of minerals or elements present. In this chapter we will use minerals as the basis for describing composition. Texture refers to the size, shape and arrangement of minerals in a rock. Some basic texture terms are defined in Table 3.2.8 (pages 144–5).

Igneous rocks

Igneous rocks are formed when liquid material called **magma** crystallises. Some people think of the lower parts of the crust being entirely made of magma but it is not. Magma occurs only in certain parts of the crust and only about 4% of the crust melts in areas where magmas are formed.

Crystallisation occurs when a material changes from a liquid to a solid. Each mineral in the magma has a different freezing point. Complete crystallisation occurs over a range of temperatures, as different minerals begin to crystallise at different temperatures. At atmospheric pressure, a mixture of the minerals forsterite (a type of olivine) and quartz has solid forsterite **crystals** forming at temperatures as high as 1900°C. However, the first quartz crystals will not form until a temperature of 1700°C is reached.

---

rock
a material made up of minerals bound together

texture
the size, shape and arrangement of particles in a rock

igneous
material formed by cooling and freezing of molten material

magma
molten rock containing some crystals and dissolved gases

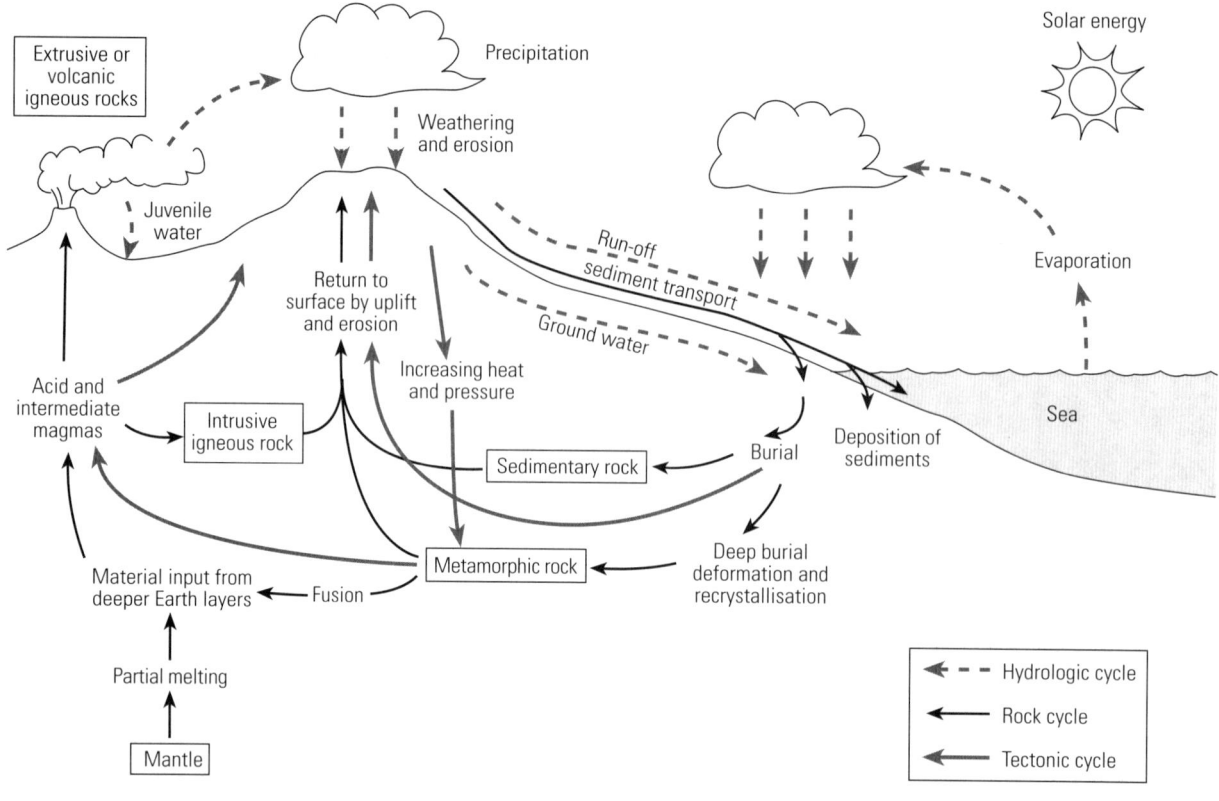

Figure 3.2.5 The rock cycle.

As magma rises through the crust it loses heat and mineral crystals form and grow. As the minerals form they remove elements from the magma. The crystals may settle to the bottom of the chamber or line its walls, but some of the crystals remain suspended in the magma. If the magma contains enough heat to stay liquid until it erupts at the surface of the crust it forms volcanic or **extrusive igneous rocks**. Magmas that cool and freeze in the crust form **intrusive igneous rocks**. (See Figure 3.2.6.)

Igneous intrusions are named according to their shape. A sheet of igneous rock is called a dyke if it cuts across layers of rock and a sill if it is between the layers. A laccolith is similar to a sill in its orientation but it has a flat base and domed top. Batholiths are large irregular bodies of coarse-grained igneous rock that have a large exposed surface—greater than 100 km². A stock is similar to a batholith but smaller. Heat from batholiths and stocks alter the rock around them. The altered zone is called a metamorphic aureole.

Terms that are used to describe the origin of igneous rocks include:
- plutonic—coarse-grained, deep-seated major intrusions
- volcanic—fine-grained extrusive lava flows
- hypabyssal—medium-grained to fine-grained, shallow minor intrusions
- pyroclastic—volcanic rocks formed from air-fall deposits.

The texture of an igneous rock reflects its cooling history. A rock that cools extremely rapidly may not form crystal grains and it will have a glassy texture. The slower the rate of cooling, the larger the mineral grains can grow. A fine-grained (aphanitic) texture is one where most crystals are too small to see with the naked eye. Be aware, however, that many fine-grained rocks still have the occasional mineral grain that is large enough to identify. A coarse-grained (phaneritic) rock contains mineral grains that are large enough to see without a hand lens.

Two other igneous textures to be aware of are porphyritic texture and volcaniclastic texture. A rock with a porphyritic texture contains mineral grains of two distinct sizes, reflecting two rates of cooling. A rock with a volcaniclastic texture is one in which volcanic eruptions have produced igneous fragments, including minerals that have been fractured and broken, ash and larger fragments. These are either welded together when still hot or cemented together by other minerals.

A diagram showing a simple classification of igneous rocks is shown in Figure 3.2.7. On the diagram, line A represents a granodiorite. The five minerals in the rock have areas that are cut by the line. The length of the part of the line represents the proportion of each mineral. In this case, plagioclase is most common and orthoclase rarest. As the composition of the rocks becomes more mafic, the proportion of iron and magnesium rich minerals increases. Mafic rocks do not contain quartz.

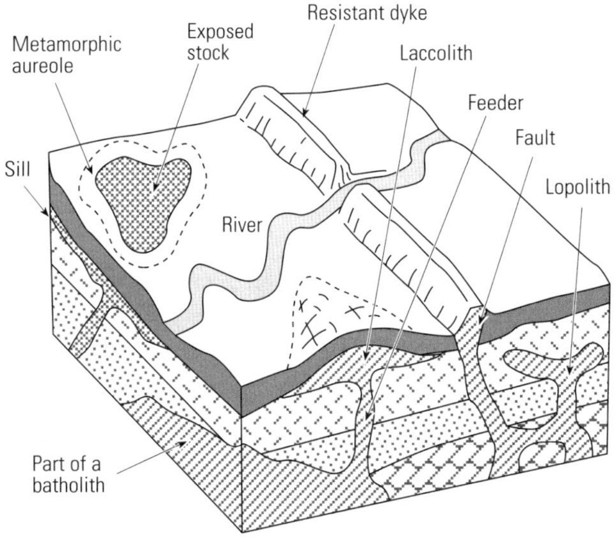

Figure 3.2.6 Forms of igneous intrusion.

Sedimentary rocks

Sedimentary rocks are formed from materials deposited on the Earth's surface by water, wind or ice. This material, called **sediment**, is derived from the mechanical breakdown and chemical change of pre-existing rocks. The resulting sediment is deposited in different environments (see Figure 3.2.8, page 136) and over time it is compacted and cemented together to form a rock. Because sediment is deposited at different times, and gravity levels out the sediment, we find sedimentary rocks forming layers called strata.

In this text, sedimentary rocks are classified into two groups on the basis of their composition and texture. One group is composed of the clastic rocks and the other group contains chemical and organic rocks. Both groups often contain structures formed during and after deposition. These structures are useful for identifying both the origin of the rock and the history of an area. (See Figure 3.2.9, page 136 and Plate 22.)

crystal
a solid formed of atoms arranged in a regular manner

extrusive igneous rock
a rock formed from lava that flowed on the surface

intrusive igneous rock
a rock that solidified from magma within the crust

sedimentary rock
a rock formed from accumulated and consolidated sediment

sediment
material that is transported and deposited by wind, water, ice or gravity

definitions

Texture						
Coarse grained	Syenite	Granite	Granodiorite	Diorite	Gabbro	Peridotite
Fine grained	Trachyte	Rhyolite	Rhyodacite	Andesite	Basalt	

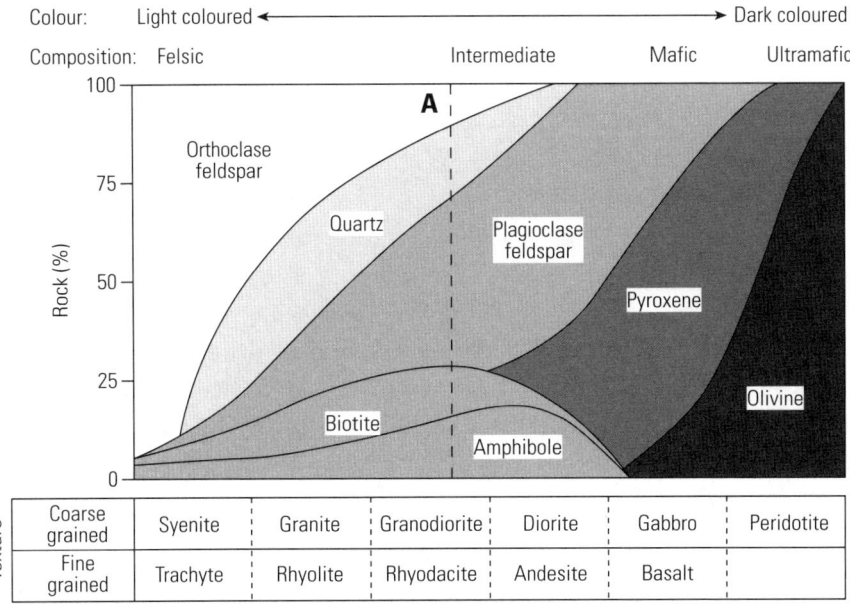

Figure 3.2.7 Classification of igneous rocks using composition and texture. Line A represents the composition of granodiorite.

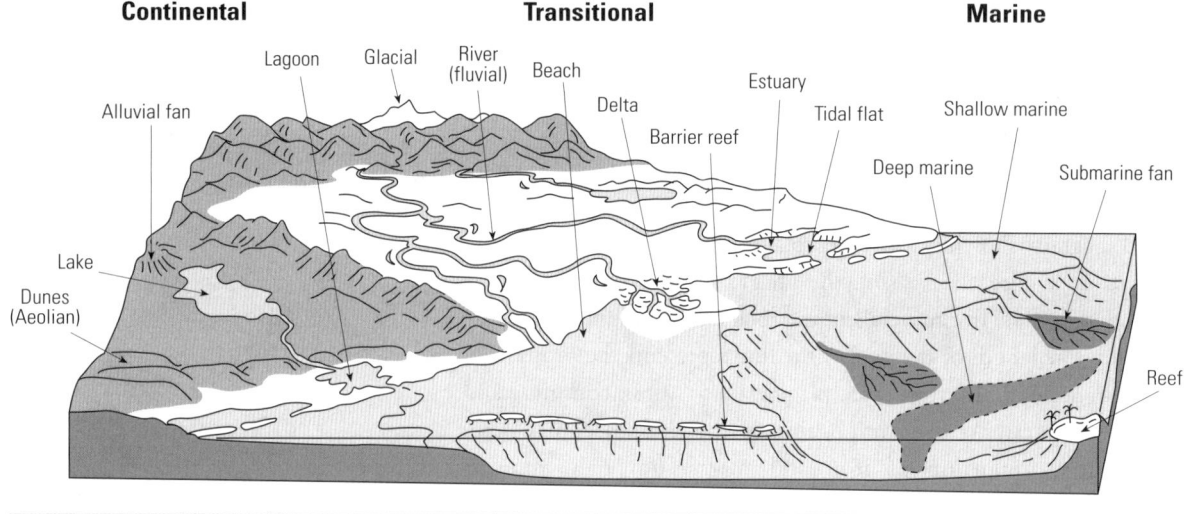

Figure 3.2.8 Environments in which sediments are deposited.

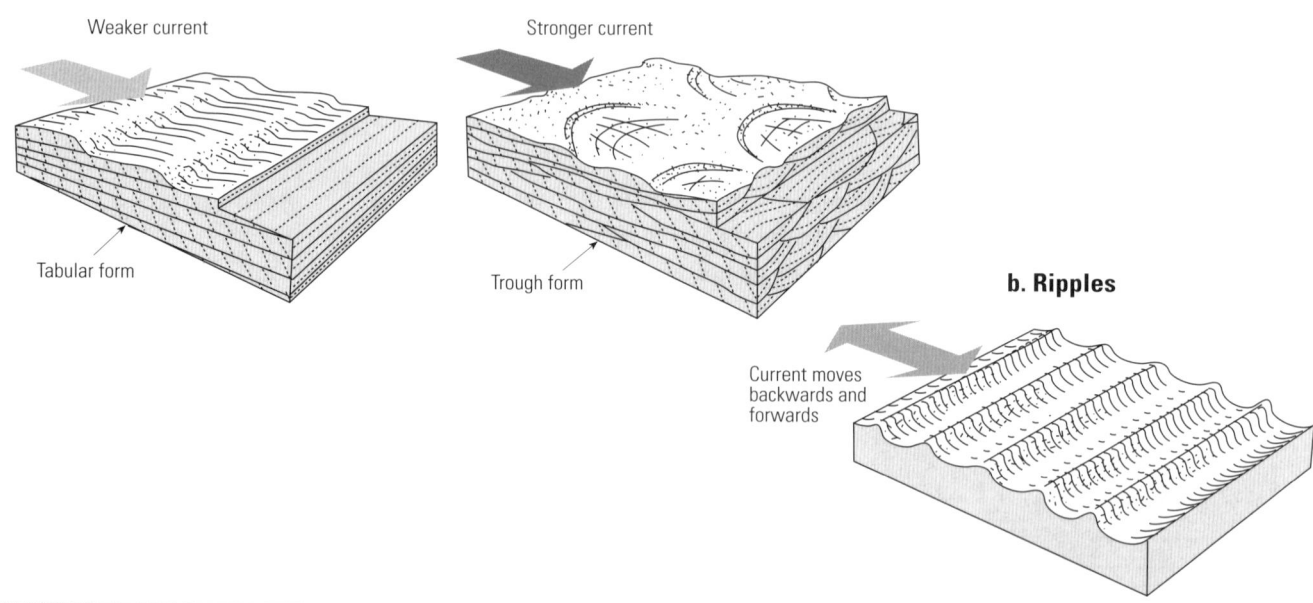

Figure 3.2.9 Sedimentary structures.

CLASTIC SEDIMENTARY ROCKS

Rocks formed from fragments cemented together are called **clastic sedimentary rocks** (see Table 3.2.4) and are said to have a clastic texture. The word 'clastic' comes from the Greek word *klastos*, meaning broken. Clastic sedimentary rocks are very common. The particles that make them up are the remains of rocks that have been weathered, transported and deposited. They are classified on the basis of grain size, which indicates the speed of the water or wind from which they were deposited. Large grains settle out of the water, for example, at speeds where smaller grains are carried along by the current.

CHEMICAL AND ORGANIC SEDIMENTARY ROCKS

The other group comprises the chemical and organic sedimentary rocks. (See Table 3.2.5.) These are rocks formed by chemical **precipitation** or biological activity. They are grouped together because it is sometimes difficult to be sure of the origin of the material they contain.

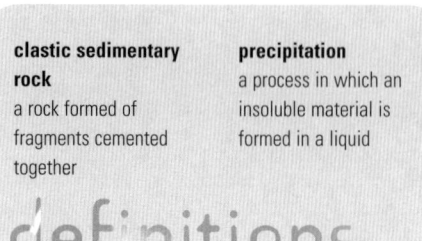

clastic sedimentary rock
a rock formed of fragments cemented together

precipitation
a process in which an insoluble material is formed in a liquid

Table 3.2.4 Classic sedimentary rock types

Rock type	Description	Environment of deposition
Conglomerate	Formed from deposits of gravel, pebbles or boulders with lesser amounts of sand or mud between the larger grains. Fragments (clasts) are usually well rounded and greater than 2 mm in diameter.	Adjacent to mountain ranges and in the headwaters of rivers. Stream channels. Beaches
Sandstone	Composed of sand-sized particles (0.06–2.00 mm) cemented together. The particles may be of any composition but are frequently quartz. The composition, alteration and sorting of grains provide important information about the origin of the sandstone.	Rivers. Fans. Dunes. Shallow marine environments. (Lithic sandstones contain rock fragments. These may suggest the sediment has not travelled far from its source.)
Mudstone	Formed from deposits of fine muds and clays. Particles are less than 0.06 mm in diameter. If the rock contains layers it is referred to as a shale. If it lacks layering we say it is massive. Mudstone is soft and weathers rapidly so exposures are rare.	Quiet water environments. (Black shales with high organic contents may indicate oxygen-poor environments.) Stream channels, floodplains and tidal flats. (Muds exposed to the air for periods of time are coloured by iron oxides, forming red shales.)

Table 3.2.5 Chemical and organic sedimentary rocks

Rock type	Description	Environment of deposition
Limestone	Has a great variety of textures but produces bubbles when treated with acid. Shell and other fossil fragments may be present.	Forms in reefs, precipitates in springs, caves and quiet marine water or forms as pellets in shallow marine areas
Chert	Hard, dense and has various colours. It breaks like glass and has a very fine-grained texture.	May precipitate in cavities or oceans. May also be formed from microscopic organisms that accumulate in marine environments.
Oil shale	Black, fine-grained rock containing oil, clay and fine-grained quartz	Forms in water bodies where plankton grow in the upper surface and are preserved in the oxygen-poor bottom waters
Gypsum	Usually white or grey. May be fine grained or have elongated crystals.	Forms as a chemical sediment in inland lakes or water bodies that are periodically cut off from the sea

Limestone is the most abundant non-clastic sedimentary rock. It may be made up of the skeletal fragments of marine creatures but it may also be formed by chemical precipitation in springs or caves.

Chert is another common rock. It is composed of extremely small crystals of quartz that can form due to biological or chemical processes. Chert may precipitate directly from seawater or it may form from the skeletons of microscopic organisms, such as radiolaria or diatoms.

Oil shale is a clastic shale that contains a great deal of oil derived from altered organic matter.

Gypsum is a chemical sedimentary rock that precipitates in shallow saline lakes.

Metamorphic rocks

A **metamorphic rock** is formed from a pre-existing rock when its minerals change in response to changes in the rock's environment. (See Table 3.2.6, page 138.) The changes may involve temperature, pressure and the composition of fluids within the rock. The minerals change so that they are better suited to the new environmental conditions. A common change involves an increase in crystal size and results in a new rock texture. The increase in a crystal's size means that its surface is smaller relative to its volume and that it is more stable at higher temperatures.

metamorphic rock
a rock formed from other rocks by crystal growth

Table 3.2.6 Common metamorphic rocks

Rock type	Description	Original rock type
Quartzite	Often hard, metamorphosed quartz-rich rock. Some sedimentary structures may be present.	Quartz-rich sandstone
Marble	Large grained and rich in calcium carbonate. Lacks foliation.	Limestone
Hornfels	Very hard and dense, fine-grained, non-foliated rock. Usually dark in appearance.	Often shale or mudstone
Greenstone	Fine grained and non-foliated. Green in appearance due to low-grade metamorphic minerals, such as serpentine and chlorite.	Mafic igneous rocks
Slate	Fine-grained, well-foliated rock. Harder than a shale.	Shale, and some pyroclastics
Schist	Medium-grained to coarse-grained rock with visible mica flakes. Strongly foliated.	Shale, basalt, granite and tuff
Gneiss	Coarse-grained, granular rock. Foliation apparent in alternating layers of light and dark minerals. Layers often strongly folded in outcrop.	Shales, plutonic and volcanic igneous rocks

a. Foliation: layers of platy mica minerals

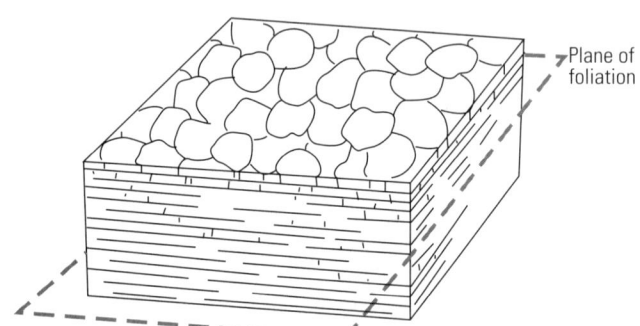

b. Lineation: alignment of prismatic minerals

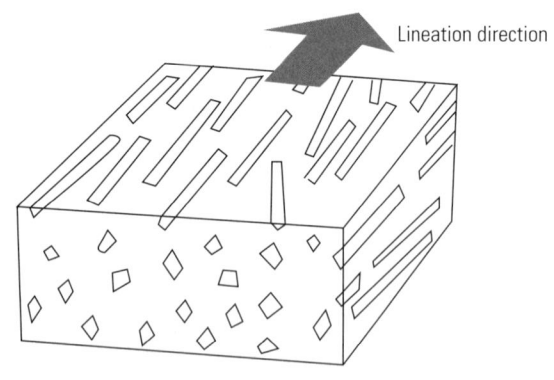

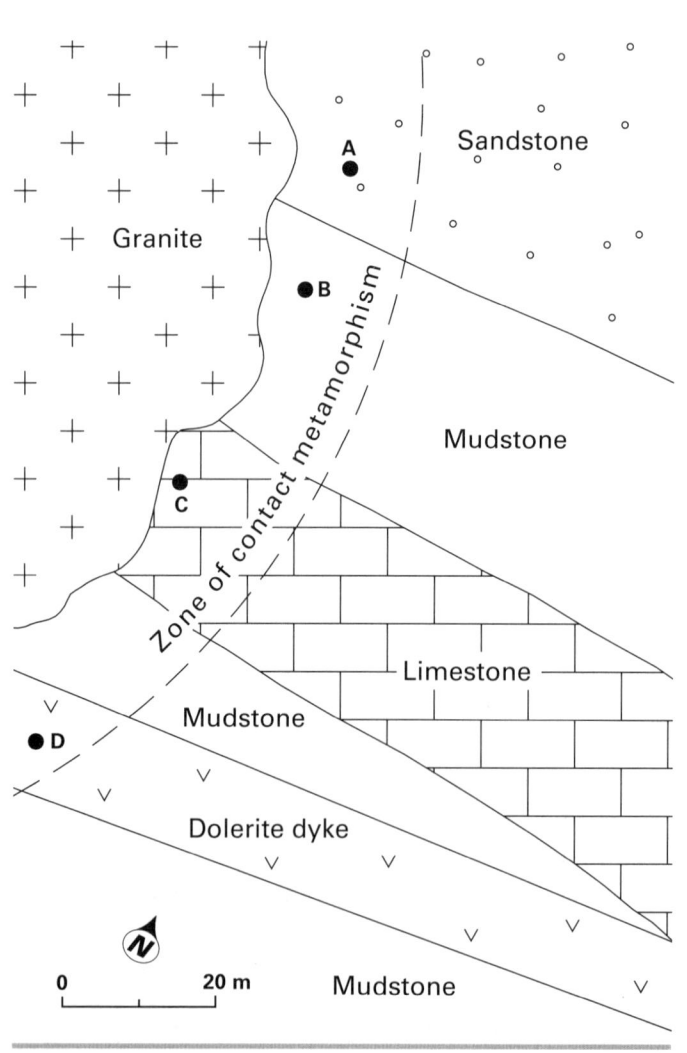

Figure 3.2.10 Metamorphic rock textures.

Figure 3.2.11 A geology map of the area referred to in review activity 8, page 139.

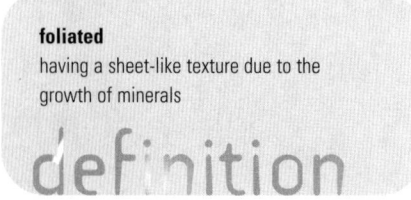

foliated
having a sheet-like texture due to the growth of minerals

Two types of metamorphic rocks are commonly recognised on the basis of their texture. **Foliated** rocks are rocks altered by heat and pressure. (See Figure 3.2.10a.) They contain platy mica minerals forming layers, like the pages in a book. Non-foliated rocks lack this layered structure. Non-foliated metamorphic rocks are those that have been altered mainly by heat. Such rocks are often hard and may contain mineral grains significantly larger than most of the other grains in the rock. The platy minerals in a foliated rock form at right angles to the pressure that is acting on the rock. (See Figure 3.2.12.) Like non-foliated rocks, foliated rocks may also contain large individual mineral grains. When long mineral grains form parallel to one another the rock is said to have a lineation. (See Figure 3.2.10b.)

One of the places where metamorphic rocks form is in mountain belts. The texture of the rocks reflects the conditions within the mountains and, by mapping the distribution of such rocks, earth scientists can work out the nature of events that occurred millions of years ago.

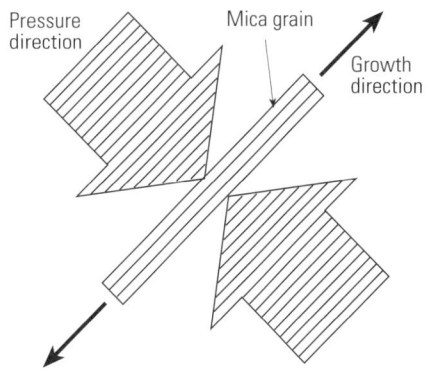

Figure 3.2.12 Pressure and the growth of micas.

REVIEW ACTIVITIES

1 Explain how intrusive and extrusive rocks differ in their texture and origin.

2 What information can the texture of an igneous rock provide?

3 Refer to Figure 3.2.7 (page 135). Compare and contrast the characteristics of a granite and a basalt.

4 Contrast the environments of formation of an igneous rock and a sedimentary rock.

5 Explain how a chemical and organic sedimentary rock differs from a clastic sedimentary rock.

6 List three rock types that contain sediments you would find in a river.

7 Describe how a foliated metamorphic rock is formed.

8 Look at Figure 3.2.11. In the zone of contact metamorphism, heat from the granite alters the surrounding rock. Pressure is not a major cause of alteration. Name the types of rocks you may expect to find at points A, B, C and D. Table 3.2.6 may help you to complete this activity.

EXTENSION ACTIVITIES

9 Draw a diagram to show an igneous intrusion, the metamorphic rocks that surround it and the sedimentary rocks from which the metamorphic rocks have formed. Label the parts of your diagram.

10 Use the diagram you drew in activity 9 to draw a block diagram of the intrusion. Give a name to each rock type shown in your block diagram.

IDENTIFYING ROCKS WITHIN YOUR LOCAL AREA

How likely is it that you will find igneous or metamorphic rocks in your local area? The Earth's land surface consists of igneous and metamorphic rocks covered by a layer of sedimentary rocks and sediments. In places the sedimentary layer is eroded to expose the rocks below. Sedimentary rocks make up about 80% of the rocks exposed at the surface. Mudstones are the most common (60%), followed by carbonates, such as limestones (20%), sandstones (15%) and evaporites (5%). Note that these figures are for the Earth as a whole. In some areas, such as the Sydney

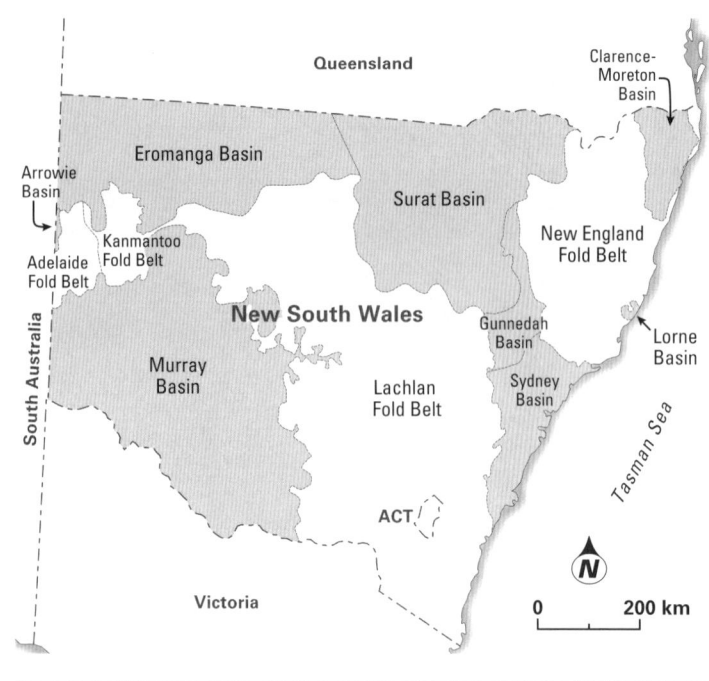

Figure 3.2.13 Structural units of New South Wales.

Basin, sandstones, mudstones and, in some places, volcanic igneous rocks will be common. In other areas, such as Orange or Armidale, wide ranges of igneous, sedimentary and metamorphic rocks may occur. In some western parts of New South Wales, sediments of Tertiary age (66 million to 2 million years ago) cover the surface so that rock outcrops are rare. (See Figure 3.2.13 and Table 3.2.7.)

A structural unit is part of the crust that can be separated from other units by its structures, rock units and history. The basins are composed mainly of sedimentary rocks. The fold belts contain folded sedimentary rocks, igneous intrusive rocks and some metamorphic rocks, particularly around the large igneous intrusions. The Kanmantoo Fold Belt and the Adelaide Fold Belt are the oldest units in New South Wales. They contain high-grade metamorphic rocks. The youngest sedimentary rocks are the Cretaceous rocks of the Eromanga and Surat Basins.

Superimposed on the structural units are a number of areas containing volcanic igneous rocks of Tertiary age.

Table 3.2.7 Characteristics of the structural units of New South Wales

Unit	Age range	Rock types	Tectonic setting
Fold belts			
Adelaide Fold Belt	Late Proterozoic–Early Cambrian period	Originally sedimentary rocks and minor mafic intraplate volcanics. Deformation during Cambrian and Ordovician periods.	Continental rift and foreland fold and thrust belt
Kanmantoo Fold Belt	Late Proterozoic–Early Cambrian period	Sedimentary rocks derived from sediments shed by mountain ranges together with volcanics and intrusive igneous rocks. Deformed during the Cambrian and Ordovician periods.	Active plate margin followed by mountain building during early Ordovician period
Lachlan Fold Belt	Early Cambrian–Early Carboniferous period	A wide range of sedimentary, igneous and metamorphic rocks. Intrusive igneous rocks are relatively common.	Complex series of mountain building, subduction and rifting
New England Fold Belt	Early Cambrian–Late Triassic period	A wide range of sedimentary and igneous rocks with some metamorphic rocks present. Intrusive igneous rocks are relatively common.	Volcanic arc and area between arc and trench (forearc-accretionary prism system). Folded and intruded because of subduction to the east.
Basins (not including very young sediments)			
Arrowie Basin	Cambrian period	Sediments	Deep-water deposition with shallower water to the west and east
Eromanga and Surat Basins	Jurassic–Mid-Cretaceous period	Clastic sedimentary rocks with some coal	Continental and marine basin followed by continental river and lake deposition
Murray Basin	Mid-Oligocene–Late Pliocene epoch	Clastic sedimentary rocks and limestone	Marine to continental deposition in a basin
Sydney and Gunnedah Basins	Late Carboniferous to Mid-Triassic period	Clastic sedimentary rocks and coal measures with some extrusive and intrusive igneous rocks	Continental and marine. A plate margin rift near a Permian period volcanic arc.
Lorne Basin	Early Triassic period	Clastic sedimentary rocks	Continental deposition in alluvial fans and playa (short-lived) lakes
Clarence-Moreton Basin	Late Triassic–Mid-Jurassic period	Clastic sedimentary rocks with some volcanics	Continental setting for rivers and lakes

Some of the rocks are in the form of lava fields and others are the remains of hotspot volcanoes. The volcanic activity began about 80 million years ago, slowed about 35 million years ago and ceased in New South Wales about 12 million years ago. In the south, the Monaro area contains lava fields that erupted about 46 million years ago. Near Orange, Dubbo and Lake Cargelligo are outcrops of igneous lavas that are the youngest volcanic areas in the state.

GEOLOGY AND LANDSCAPE

There are two ways in which geology affects the development of the landscape in an area. Hills, valleys, cliffs, rivers, the shape of the coastline and other features reflect the composition and structure of the rocks making up the environment. (See Plate 14.) It is from rocks that soils form, and soils and rivers are major factors in the establishment of towns and cities. As we will see later in the course, the geology of an area also contains important clues to the area's history.

> **mafic igneous rock** an igneous rock containing minerals rich in iron and magnesium but no quartz
>
> **felsic igneous rock** an igneous rock rich in feldspars and quartz
>
> **resistant** resisting weathering and erosion in the environment where it is found
>
> definitions

An area's geology influences the landscape due to different rocks weathering at different rates. Hard sandstones or non-foliated metamorphic rocks weather more slowly than soft mudstones in the environments they occupy. In warm, wet conditions a **mafic igneous rock** may weather faster than a **felsic** one and in a desert the grain size of a rock may determine the rate of physical weathering it experiences. In cliffs the different rates of weathering (called 'differential weathering') lead to undercutting of cliffs, rock slides and the formation of rock platforms. (See Plate 15.)

Structures influencing landscape development

Geology also influences landscape development due to the structures within the rocks. Types of structures that influence landscape development are bedding orientation, folds, faults, joints and volcanic structures.

BEDDING ORIENTATION AND FOLDS

In horizontal layers, **resistant** rocks form caps above less-resistant layers. The resistant rocks often form vertical cliffs or steep slopes while less-resistant layers produce gradual slopes. When rock layers are tilted due to folding the new orientation of resistant layers may produce scarps, dip slopes and strike ridges. (See Figure 3.2.14.) The shapes of complex folds are easily seen in air and satellite photographs. Hills and valleys are produced by different patterns of weathering in different layers of the folds. Often the weathering and fractures within the folds provide easy paths for water drainage and the patterns of creeks and rivers reflect patterns in the rocks making up the landscape.

FAULTING

Faults control river flow and produce structures, such as horsts and grabens. (See Figure 3.2.15, page 142.) Horst and graben structures can be seen on the freeway trip from Sydney to Canberra. The hills that rise along the edge of Lake George (north-east of Canberra) are due to a fault. As you drive west from Lake George you ascend from a half graben and then descend into a part of a graben in which Canberra is situated. Travelling to the Blue Mountains from Sydney also involves

a. Tilted layers that weather at different rates

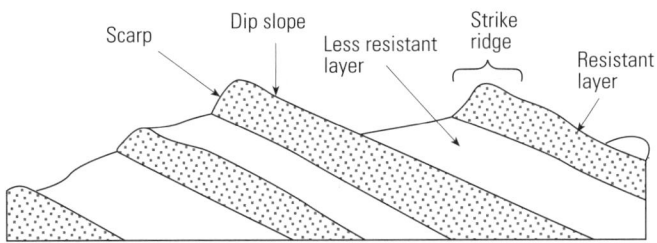

b. Hard layers form raised areas

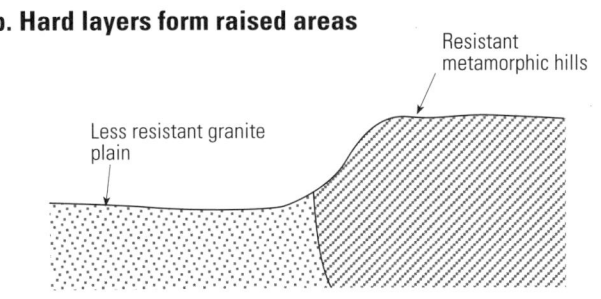

c. Topography formed by folds

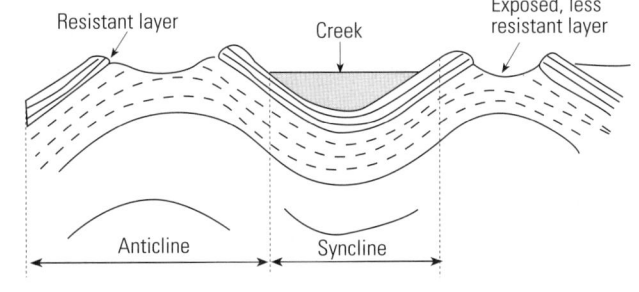

Figure 3.2.14 Examples of how the structure and composition of rocks affect landscape.

negotiating a fault. The Lapstone monocline and Kurrajong fault system mark the boundary between the Cumberland Plain and the elevated plateau that is the Blue Mountains. Where a fault raises an area, rivers may change course. Drainage from elevated areas may be useful in determining the cause of uplift. (See Figure 3.2.16.)

JOINTING

Jointing refers to the cracks that form in large rock bodies near the surface of the Earth. Joint patterns control the development of cliffs and act as lines of weakness along which creeks and rivers sometimes flow. A joint is different from a fault because in a fault the movement occurs along the break in the rock structure. In rock units that are fairly homogeneous a rectangular drainage pattern may form as creeks follow the joint pattern. This is often easy to see where a relatively thin soil layer occurs. Changes in the form of a drainage pattern can sometimes be a useful guide to changes in the underlying geology of an area.

> **definition**
>
> **jointing**
> the formation of cracks in a rock where, unlike a fault, no major movement across the cracks occurs

VOLCANIC STRUCTURES

Volcanic activity may lead to what is called radial drainage as water cuts channels down the sides of a volcano. Volcanic activity can also change a landscape when lava follows and fills pre-existing gullies and creek lines. Examples of this occur in the Southern Highlands near Bowral (south-west of Sydney).

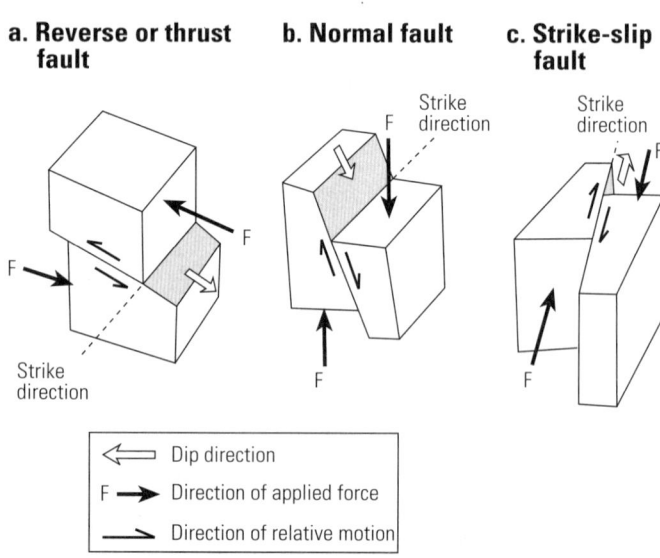

Figure 3.2.15 Types of faults.

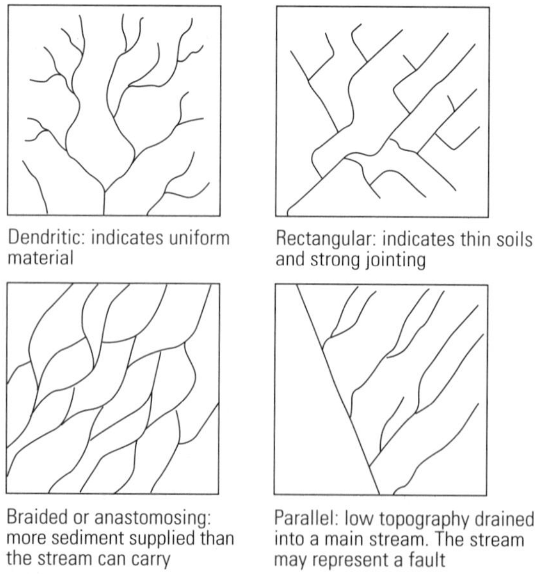

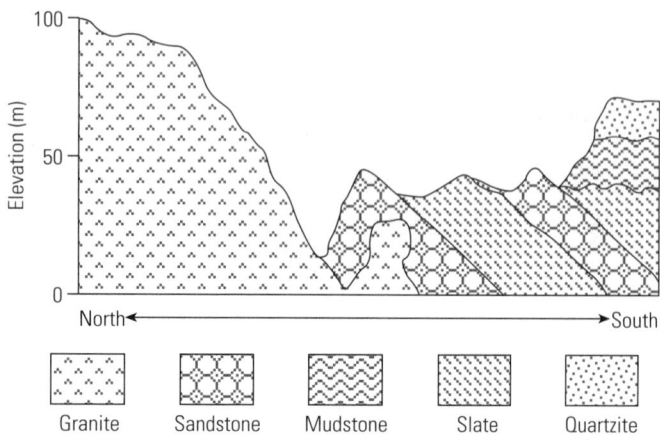

Figure 3.2.17 Cross-section of part of a river valley.

Figure 3.2.16 Four drainage patterns and the geology they reflect.

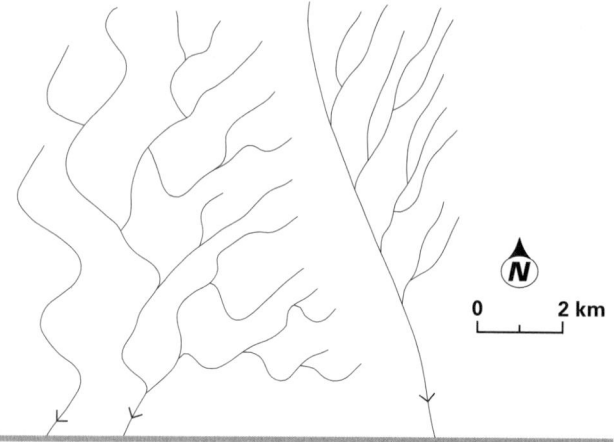

Figure 3.2.18 Map of the drainage patterns over an area.

142 EARTH AND ENVIRONMENTAL SCIENCE: THE PRELIMINARY COURSE

REVIEW ACTIVITIES

1
The nature of the rocks in an area and the structures present determine the shape of the land's surface. Explain how different rock types can produce surfaces of different heights.

2
Describe the sorts of structures that affect the direction of river flow in an area.

3
Figure 3.2.17 shows a cross-section of part of a river valley. Summarise the relationship between the geology of the area and the shape of the land surface.

4
Figure 3.2.18 is a map of the drainage patterns over an area. Copy the map and, on the copy, try to identify the boundary between two rock types and the probable location of a fault.

5
When an igneous intrusion is formed it heats and hardens the rocks around it. Describe how the intrusion and rocks around it will weather if the intrusion is more easily weathered than the igneous rock.

EXTENSION ACTIVITIES

6
Summarise the rock types and structures found in your area.

7
What geological conditions would you need to check before you built a house on a steep, soil-covered hill?

SUMMARY

- Rocks are composed of minerals, which have characteristic properties.

- The most common minerals in rocks are called silicates. Other common groups of minerals include carbonates, sulfates and sulfides, and oxides and hydroxides.

- Minerals are identified on the basis of their properties. The most important mineral properties are colour, lustre, hardness and cleavage.

- Rocks are described in terms of their mineral composition and texture.

- Igneous rocks are formed when magma crystallises.

- Sedimentary rocks are formed from the sediments that are compacted and cemented.

- Metamorphic rocks are formed from pre-existing rocks when they adjust to new conditions.

- Different areas of New South Wales have different types of common rock types.

- In some areas of New South Wales rock outcrops are rare due to sediment cover.

- Geological features influence landscape development.

- Structures that affect landscape include bedding orientation, joints, folds, faults and volcanic structures.

- Some rock layers weather more easily than others and this affects landscapes.

PRACTICAL EXERCISE
Observing and identifying a rock

Wherever you live, the rocks that make up your environment can be identified if you use the sort of procedure given in the activities here. The records you make will help you in determining the overall geology of the area in which you live. Complete the following activities, preferably with a rock specimen you have collected during fieldwork. Be sure to record your observations and answers. Remember that descriptions can involve words and diagrams. If time is available, repeat this procedure for two other rock specimens.

ACTIVITIES

1
When possible, look carefully at the outcrop from which the rock specimen comes.
a Does it have layers or does it cut across other layers? Does it contain any structures?
b Draw the outcrop and label the drawing if you can.

2
a How many minerals does the rock specimen contain? Describe them.
b Use a hand lens and decide whether the rock is a sedimentary, metamorphic or igneous rock.
c Describe the texture and use it to help you decide.
d List the characteristics you use to identify the minerals. Use the mineral table (see Table 3.2.1, page 128) to try to identify them.

3
a Use the key to common rock types (see Table 3.2.8) to identify your rock specimen. Ensure you record the characteristics you use as you work your way through the table.
b Draw your rock specimen and label its characteristic features.

4
Try to predict the environment in which your specimen formed. Explain the reasons for your prediction.

5
Can you identify any changes to the rock caused by being exposed to the weather? Is its surface a different colour? Have some parts disappeared? Describe any alteration that the rock shows.

Table 3.2.8 A key to common rock types

1	a	Rock made up of rock and mineral fragments; fossils may be present	Go to 2	**8** a	Contains almost only quartz	**Quartz sandstone**
	b	Rock not made up of rock and mineral fragments	Go to 12	b	Contains not only quartz but other minerals or rock fragments	Go to 9
2	a	Fragments (clasts) greater than 5 mm on average	Go to 3	**9** a	Contains sand-sized rock fragments	**Lithic sandstone**
	b	Fragments smaller than 5 mm on average	Go to 4	b	Contains feldspars rather than lithic fragments	**Feldspathic sandstone**
3	a	Clasts are rounded	**Conglomerate**	**10** a	Fine layers present	**Shale**
	b	Clasts are angular	**Breccia**	b	No fine layers present	**Mudstone**
4	a	Rock fizzes when tested with acid	Go to 5	**11** a	Composed of microscopic shells	**Chalk**
	b	Rock does not fizz when tested with acid	Go to 6	b	Not composed of microscopic shells but fossils may be present	**Limestone**
5	a	Acid reacts rapidly	Go to 11	**12** a	Texture of rock is glassy	Go to 13
	b	Acid reacts slowly	**Dolomite**	b	Texture is not glassy	Go to 14
6	a	Particles mostly of a similar size	Go to 7	**13** a	Rock contains many pores	**Pumice**
	b	Particles of two sizes (rock fragments in a clay matrix)	**Greywacke**	b	Rock is massive, lacking pores	**Obsidian**
7	a	Medium grained or coarse grained (0.06–2.00 mm diameter)	Go to 8	**14** a	Texture is coarse grained or porphyritic	Go to 15
	b	Not medium grained or coarse grained (fine grained: <0.06 mm)	Go to 10	b	Texture is not coarse grained or porphyritic	Go to 30

15	**a**	Rock reacts with acid	**Marble**	**30**	**a**	Very hard, fine-grained to medium-grained dark metamorphic rock containing quartz	**Hornfels**
	b	Rock does not react with acid	Go to 16		**b**	Not as above	Go to 31
16	**a**	Hard and massive; contains mostly quartz; some evidence of bedding (layers) present	**Quartzite**	**31**	**a**	Grains are lined up in the same direction—foliated	Go to 32
	b	Not as above	Go to 17		**b**	Grains are not lined up in the same direction—non-foliated	Go to 33
17	**a**	Grains are lined up in the same direction—foliated	Go to 18	**32**	**a**	Fine-grained, dark rock with good cleavage	**Slate**
	b	Grains are not lined up in the same direction—non-foliated	Go to 19		**b**	Fine-grained to medium-grained pale and shiny foliated rock	**Schist**
18	**a**	Rock not banded; coarse grains of mica common	**Schist**	**33**	**a**	Texture is pyroclastic	Go to 44
	b	Rock banded; quartz, feldspar, mica and hornblende present	**Gneiss**		**b**	Texture is not pyroclastic	Go to 34
19	**a**	Texture is pyroclastic	**Volcanic breccia**	**34**	**a**	More than 20% of the rock is quartz	Go to 35
	b	Texture is not pyroclastic	Go to 20		**b**	Less than 20% of the rock is quartz	Go to 38
20	**a**	More than 20% of the rock is quartz	Go to 21	**35**	**a**	Composed of microcrystalline quartz	**Chert**
	b	Less than 20% of the rock is quartz	Go to 26		**b**	Contains feldspars	Go to 36
21	**a**	Orthoclase (see Table 3.2.1, page 128) greater than or equal to plagioclase	Go to 22	**36**	**a**	Orthoclase greater than or equal to plagioclase	Go to 37
	b	Orthoclase less than plagioclase	Go to 23		**b**	Orthoclase less than plagioclase	**Dacite**
22	**a**	Orthoclase equal to plagioclase	Go to 24	**37**	**a**	Orthoclase equal to plagioclase	**Rhyo-dacite**
	b	Orthoclase greater than plagioclase	Go to 25		**b**	Orthoclase greater than plagioclase. Folded layering may be present	**Rhyolite**
23	**a**	Texture is porphyritic	**Dacite porphyry**	**38**	**a**	Some quartz or biotite present	Go to 39
	b	Texture is not porphyritic	**Granodiorite**		**b**	No quartz present	Go to 41
24	**a**	Texture is porphyritic	**Rhyo-dacite porphyry**	**39**	**a**	Orthoclase greater than or equal to plagioclase	Go to 40
	b	Texture is not porphyritic	**Granite**		**b**	Orthoclase less than plagioclase	**Andesite**
25	**a**	Texture is porphyritic	**Rhyolite porphyry**	**40**	**a**	Orthoclase equal to plagioclase	**Trachy-andesite**
	b	Texture is not porphyritic	**Adamellite**		**b**	Orthoclase greater than plagioclase	**Trachyte**
26	**a**	Some quartz or biotite present	Go to 27	**41**	**a**	Fine-grained dark rock	Go to 42
	b	No quartz present; dark mineral mainly pyroxene	**Gabbro**		**b**	Medium-grained dark rock	Go to 43
27	**a**	Orthoclase greater than or equal to plagioclase	Go to 28	**42**	**a**	No orthoclase or chlorite present	**Basalt**
	b	Orthoclase less than plagioclase	Go to 29		**b**	Chlorite present	**Greenstone**
28	**a**	Orthoclase greater than plagioclase	**Syenite**	**43**	**a**	No orthoclase or chlorite present	**Dolerite (Dibase)**
	b	Orthoclase equal to plagioclase	**Monzonite**		**b**	Chlorite present	**Amphibolite**
29	**a**	Texture is porphyritic	**Porphyrite**	**44**	**a**	Light colour; quartz present; biotite may be present	**Rhyolite tuff**
	b	Texture is not porphyritic; dark mineral mainly amphibole	**Diorite**		**b**	Darker colour; little or no quartz; amphibole present	**Andesitic tuff**

3.3 Soils: Features and formation

OUTCOMES

At the end of this chapter you should be able to:

- recall the difference between biotic and abiotic features of the local environment

- outline the characteristics of a local soil

- summarise the processes that produce soil

- examine a soil and describe it in terms of:
  – the horizons present
  – the characteristics of each horizon

- analyse the ways in which the vegetation of an area can be influenced by the soil composition and climate or microclimate of a region

- relate the presence of particular animals in the local environment to their requirements within the local environment

- identify, gather and process first-hand or secondary data to identify the dominant types of plants and animals in the area studied.

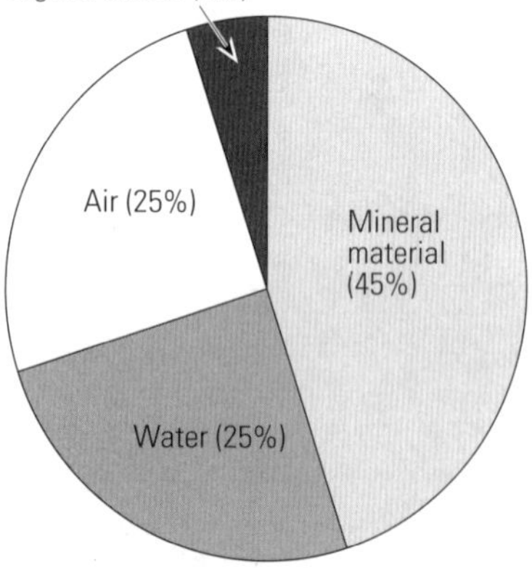

CHARACTERISTICS OF SOILS

Soils are the places where the atmosphere, water and living things interact with the rocks of the lithosphere. Soils are complex systems, containing both abiotic (non-living) and biotic (living) parts. (See Figure 3.3.2.) They cover large areas of the Earth's surface and contain a record of the environments in which they form. Sometimes soils can be preserved in sedimentary rocks as structures called palaeosols.

We rely on soils for food and resource production, but other living things in our environments also rely on soils. Many living things (such as bacteria, fungi and invertebrate animals) live in soils. Plants grow in soil and obtain water and mineral nutrients from it. The diversity and number of plants growing in a soil affects the diversity of animals there. So, the diversity of living things in an area is closely related to the types of soils present.

A soil is a system consisting of solid, liquid and gaseous parts. (See Figure 3.3.1.) The solid part of a soil consists of mineral materials and organic matter. The minerals are derived from the rocks on which the soil forms and the organic matter is derived from the living things that grow on, or in, the soil. A soil may contain 45% minerals and 5% organic matter, with air and water each making up 25% of the soil's volume. These are all averages. Soils in different places differ in the amount of each component present, and in a single soil the amount of these things may vary over time. For example the amount of water present in the soil will be affected by the addition of water through rainfall and the removal of water by plants, evaporation and seepage. Similarly, the air present will depend on the amount of organic activity in the soil and the types of mineral grains present because these things determine the cavities within which the air is found.

The minerals and organic material in a soil determine many of its physical properties. The minerals present vary in their weathering rate, size and composition. These varying properties affect both the structure of the soil and the speed of soil-

Figure 3.3.1 Volume composition of a soil.

soil
a material formed by the weathering of surface materials

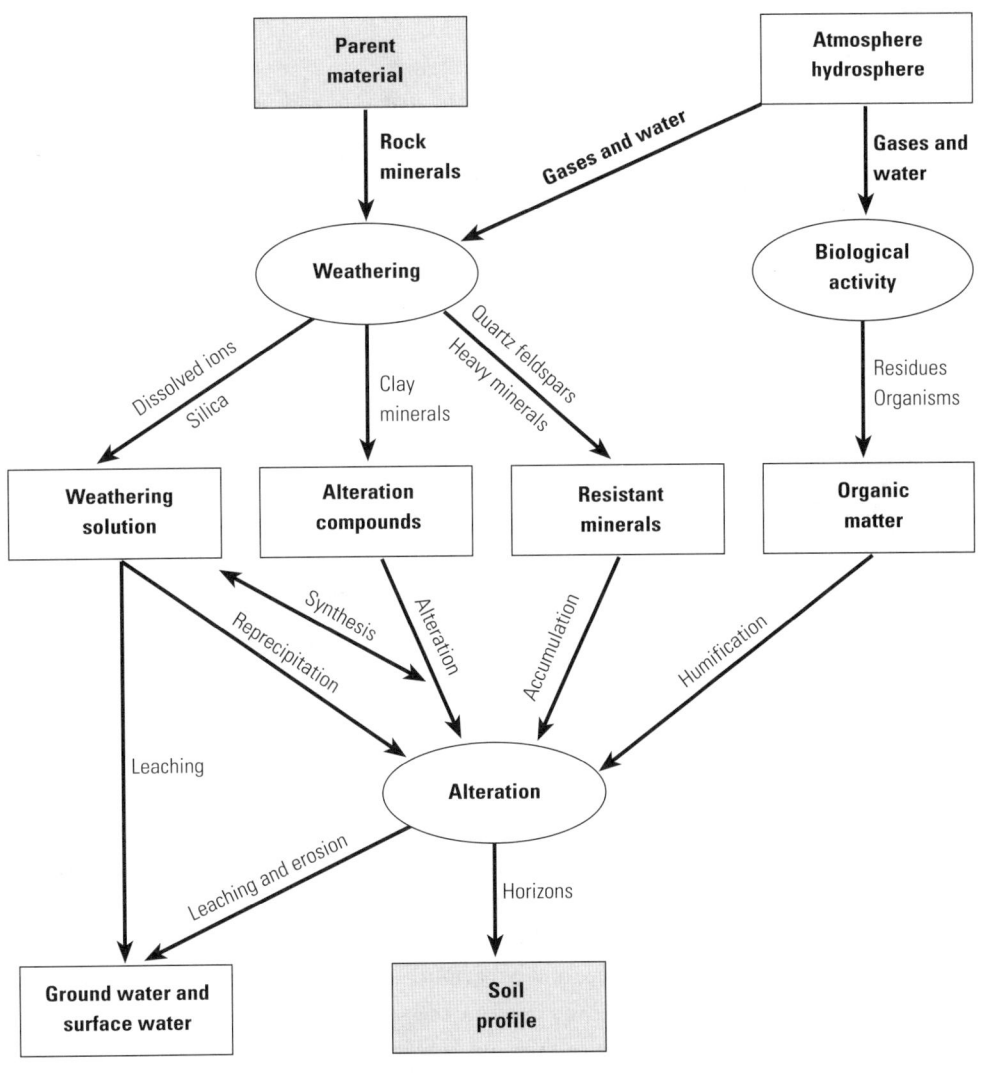

Figure 3.3.2 The interaction of systems that form soils.

making processes. Clay minerals, for example, determine the structures that develop in different parts of the soil. The organic matter present also affects the structure of the soil and its fertility.

REVIEW ACTIVITIES

1 Identify the abiotic and biotic components of a soil.

2 Explain two reasons why we rely on soils for our wellbeing.

3 Explain why the soil and the diversity of living things in an area are related to each other.

4 Construct a bar graph to show the average make up of a soil.

5 Describe how the composition of a soil affects its physical properties.

EXTENSION ACTIVITIES

6 Analyse the origin of the components making up a soil.

7 Define what is meant by the pH of a soil. Explain how the addition of sulfide minerals and carbonate minerals would each affect the pH of a soil.

8 Compare the nitrogen and carbon cycles. Assess the effect of climate on how rapidly nitrogen and carbon material moves in an environment.

DESCRIBING SOILS

Soils vary enormously in their composition, structure and responses to change. We can describe a soil in terms of its physical and chemical properties. The physical properties of a soil include what is called its soil profile. Soil profiles will be discussed on page 156. In this section we will examine the other important properties of a soil and how they are assessed.

Physical properties

The important physical properties of soils include texture, structure, porosity, permeability, field capacity, infiltration rate and aeration.

Table 3.3.1 Particle sizes of soil separates based on the FAO-UNESCO scheme

Soil separate	Particle diameter (mm)
Sand: Very coarse	2.0–1.0
Coarse	1.0–0.5
Medium	0.5–0.25
Fine	0.25–0.10
Very fine	0.10–0.05
Silt	0.05–0.002
Clay	Less than 0.002 mm

TEXTURE

The texture of a soil is a description of the amounts and sizes of mineral particles making up the soil. Leaving out large stones and rocks, the mineral part of a soil consists of three groups of particles: sand, silt and clay. The relative amounts of these groups, or soil separates, determine the name given to the texture of the soil. While you have probably used the terms sand, silt and clay before it is important to note that they are terms relating to size not composition. Not all sands, for example, are composed of quartz. The size of various particles is shown in Table 3.3.1. You may note that this is similar to, but slightly different from, the size scale used to determine particle sizes in sedimentary rocks.

A soil's texture can be assessed in the field by examining its appearance and behaviour. To measure the texture, first break up any clods and examine the sample closely with a hand lens. Note what you can see and how it feels between your fingers. Is it, for example, smooth or gritty? Now, moisten the sample with a little water. Next, try to work the material into a ball and then, using your thumb and forefinger, try to make a ribbon. Do this by pushing forward with your thumb over your forefinger. See how long a ribbon you can make and assess the feel of the sample. Is it, for example, smooth, gritty, soapy or silky? A simple guide to assessing a soil's texture from your results is outlined in Table 3.3.2.

The texture can also be determined in a laboratory. This involves drying the sample well and using a set of sieves to separate fractions of different sizes. Fine fractions may require the addition of a dispersal agent, such as weak sodium hydroxide, to stop fine particles sticking together. The weight of each size fraction is measured and then the amount of each soil separate is expressed as a percentage of the total mass. A number of textural classifications have been proposed and one is shown in the triangular diagram in Figure 3.3.3.

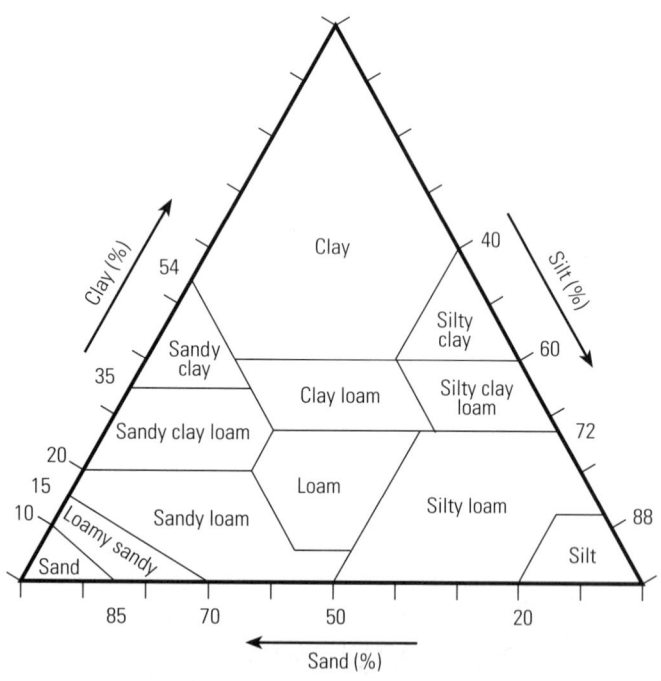

Figure 3.3.3 Textural classification of sediment.

A soil's texture is important for two reasons. First, a soil texture is a guide to other physical soil properties and, second, the texture is an indicator to how plants and other life interact with the soil.

The physical properties that a soil texture is related to include:
- *Drainage.* The larger the particles, the larger the amount of space between them. Coarse-textured soils drain more freely than fine-textured ones.

Table 3.3.2 How to determine the texture of a soil

Texture	Appearance under a hand lens	Feel between fingers		When rolled between fingers
		Damp or dry	Wet	
Sand or loamy sand	Mostly sand	Very gritty	Forms a flowing mass	Does not form a ball. Loamy sand leaves a film on fingers when rubbed but sand does not.
Sandy loam	Sand grains obvious	Gritty	Not plastic (able to be shaped with fingers). Only slight sticking together of mass.	Forms a ball, but only just. Using a thumb and forefinger you can form a ribbon 1–2 cm long.
Loam	Similar amounts of sand, silt and clay	Gritty organic matter may make it feel greasy	Slightly plastic	Ball forms; feels a little spongy. Can form a ribbon about 2.5 cm long.
Silty loam or silt	Sand grains present but mostly silt	Smooth, soapy or silky feel	Smooth, silky feel	Forms a good ball; feels a little spongy. Forms a ribbon about 2.5 cm long.
Silty clay loam, clay loam or sandy clay loam	Some sand grains present	Slightly gritty, soapy feel and sticky when moist	Moderately plastic	Forms a good ball; sandy to smooth as amount of clay increases. Forms a ribbon 3–5 cm long.
Clay, silty clay or sandy clay	Large grains rare or absent	Smooth and non-gritty or slightly gritty	Very sticky and easily shaped	Forms a smooth plastic ball. If mainly clay, it can be moulded like plasticine. Forms a ribbon from 5 cm to longer than 8 cm.

- *Aeration.* Again, the size of particles determines the spaces between them. Coarse-textured soils, such as sands, allow more air to move through the soil than clays and clay loams.
- *Water-holding capacity.* Fine-textured soils, such as clay loams and clays, have fine spaces between the particles. This stops water moving easily through them. Soils with high clay contents hold water better than sandy soils.
- *Shrinking and swelling behaviour.* Soils with certain types of clay swell and shrink as they respectively take up and lose water. This may affect the sorts of structures present. (See below.)
- *Ease of cultivation.* For farming or land regeneration, sandy soils are often easier to cultivate than clay soils. Texture is also, however, an indicator of potential problems. When they are wet, clay-rich soils are prone to compaction and smearing of soil structures. This leads to problems for plant growth and germination. Sandy soils stand a higher risk of structural damage when the soil is too dry. Understanding how soils react to our treatment of them is vitally important for good land management.
- *Ability to hold nutrients.* Clay minerals and organic compounds in the soil have the ability to bind metal ions, such as calcium, magnesium and potassium. This prevents these mineral nutrients from being carried away by water moving through the soil. This leaves the nutrients available for use by plants growing in the soil.

STRUCTURE
Structure refers to the way the mineral particles and organic matter in a soil are held together to form clods or peds. You may think of the peds as the building blocks of the soil. The cracks, pores and spaces between the peds allow water and air to enter the soil. These spaces are also important for the soil microfauna—the animals that live in the soil—because they provide them with habitats and a suitable amount of air.

Structure can be observed in the topsoil—the upper part of the soil—and in the deeper layer known as the subsoil. These layers are described more fully in the later

definition

structure
the way in which soil particles form units, called peds

section on soil profiles. (See page 156.) When determining soil structure, it is important to observe undisturbed soil material. Sometimes a spade sliding through the soil will smear the soil and obscure the soil structure. You should also avoid looking at structure in an area that might have been compacted.

First, you should establish whether peds are present and, if so, how well developed the peds are. Terms you may use for this include weak pedality, moderate pedality or strong pedality. If you see peds, describe their size and type.

There are five or six basic structure types you may find in a soil. They are illustrated in Figure 3.3.4. Two types are effectively structureless. Sands, where individual grains are not bound together, form a structureless layer referred to as single grained. In some soils with a great deal of clay another structureless layer may form. This form has no visible cracks or peds and is referred to as massive. Some peds form long vertical columns. The structure where the columns have flat upper surfaces is referred to as prismatic. If the columns have rounded tops the structure is referred to as columnar. Sometimes the peds form horizontal, flat sheets. This structure is called platy. If the structure resembles an aggregate, or collection, of fine clusters of particles it is called granular. Such small clusters are usually less than 2 mm in diameter. If the structure is an aggregate of larger angular or subangular shapes it is called a blocky structure.

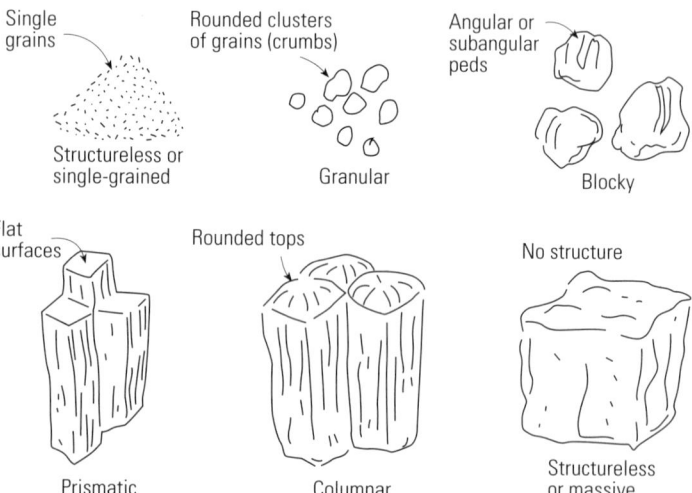

Figure 3.3.4 Structures found in soils.

Good soil structure both benefits and depends on living things. Organic materials, which are breakdown products of soil bacteria and fungi, bind the mineral particles together and help to form the peds. The peds then provide the benefits to the soil microfauna described earlier. When soils lack the organic materials that bind soil particles together, water will cause the particles to break down. This is called slaking. If the clays break down further to form a milky solution the process is referred to as dispersion. As a result of these processes, clays are more easily eroded and hard-setting crusts may form at the surface. This is a problem because the crust can prevent water penetration and the emergence of seedlings from the soil.

POROSITY AND PERMEABILITY

Porosity refers to the amount of pore space in a soil. (See Figure 3.3.5.) It can range from 10% to 70% of a soil. Not all pore spaces are the same size. Water moves easily through soils with many large pore spaces, cracks and animal burrows. Such soils are said to have a high **permeability**. Permeability describes the ability of water to pass through a soil. It is closely linked to the structure and texture of a soil. If water cannot drain quickly away the water may displace air and stop plant roots carrying out respiration. Waterlogged soils can stunt or kill plants that are not adapted to them. An idea of the permeability of a soil may be determined by carefully observing the structure, cracks and pores in the soil in the field.

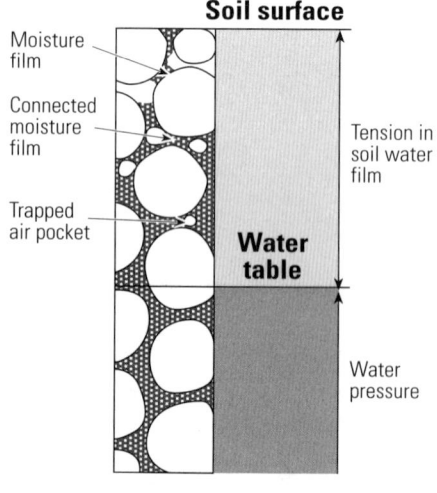

Figure 3.3.5 Pores and water-holding ability.

FIELD CAPACITY AND INFILTRATION RATE

When a soil has been well watered, the water that remains after some water has drained away is referred to as the field capacity of the soil. If a pore is less than about 0.05 mm in diameter, water will be held to the sides of the pore by capillary forces. It is this water that plants use and it is referred to as available water. When a plant cannot take up the available water it will wilt and possibly die. Available soil moisture is a major factor limiting plant growth, and being able to measure the water-holding capacity of a soil gives useful information about the plants that live on the soil.

The field capacity of a soil can be measured by placing a known weight of soil in a beaker, which has also been weighed. Fill the beaker until the soil is well saturated. Drain off the excess water and weigh the sample in the beaker. Dry the sample in an oven or warm place for two days and weigh it again. The loss of weight represents the moisture in the soil at field capacity. Note that the more disturbed the soil sample is the more likely it is that results may understate the field capacity.

The infiltration rate describes how easily water can enter the soil. Initially a dry soil will drain more quickly than if it is wet. Over time, however, the infiltration rate is fairly constant. While infiltration rate gives an indication of porosity it is prone to a great deal of variation depending on cracks in the soil and the vegetation present. A simple way to measure the infiltration rate of a soil is as follows. Carefully clear vegetation and leaf litter away from a small area of soil. Measure the diameter of a metal can or plastic tube. Push the can or tube gently into the ground. Pour a known amount of water into the can or tube—about 500 mL (500 cm^3) is appropriate. Time how long it takes all the water to enter the soil. Repeat adding water to the tube until the results are the same or very similar. Calculate the infiltration rate using the following formula.

$$\text{Infiltration rate cm/hour} = \frac{\text{Volume of water used (cm}^3\text{)} \times 3600}{3.14 \times \text{diameter of tube (cm}^2\text{)} \times \text{time taken (s)}}$$

AERATION

Soil air is very important for plant roots and the microfauna that live in the soil. The amount of air present, or **aeration**, is affected by the amount of water present in the soil and the composition of soil air may vary over time. Soil air is different from air in the atmosphere because it is nearly saturated with water vapour and is relatively rich in carbon dioxide gas. The carbon dioxide is due to bacterial decay of organic matter. Gas diffuses into the soil and the process is slow. If the soil becomes waterlogged and the bacteria use up oxygen, the oxygen level falls and the soil may become **anaerobic**. Plants will not grow well in such conditions because root cells require oxygen for respiration. Anaerobic bacteria find the conditions fine and they may alter iron compounds so that the soil takes on a grey appearance. The process is known as gleying.

Chemical properties

The chemical properties of a soil depend in part on the rock from which the soil formed and the action of living things, water and the atmosphere. Features that may be measured and described include colour, organic content, pH and the presence of particular elements, such as calcium, nitrogen, phosphorus and sodium.

COLOUR

The colour of a soil reflects both the parent material from which it is derived and the processes that have created the soil. It is a very important characteristic used in describing and classifying soils. In your fieldwork you should record the colour of the soils you study, particularly each of the layers of the soil profile. (See page 156.) In a laboratory a soil's colour is determined using a soil colour chart and is described in terms of the dominant spectral colour (hue), the relative intensity of the colour (value) and the relative purity of the colour (chroma).

ORGANIC CONTENT

The organic content of a soil is the material derived from the living things added to the soil and decomposed by soil organisms. (See Figure 3.3.6, page 152.) Soil organisms include large organisms (such as earthworms and insects), small invertebrates, protozoans, bacteria and fungi. (See Figure 3.3.7, page 152.) The soil

---

porosity
the amount of pore space in a material

permeability
the ability of a fluid to pass through a material

aeration
the amount of air present in the soil

anaerobic
the absence of free oxygen

definitions

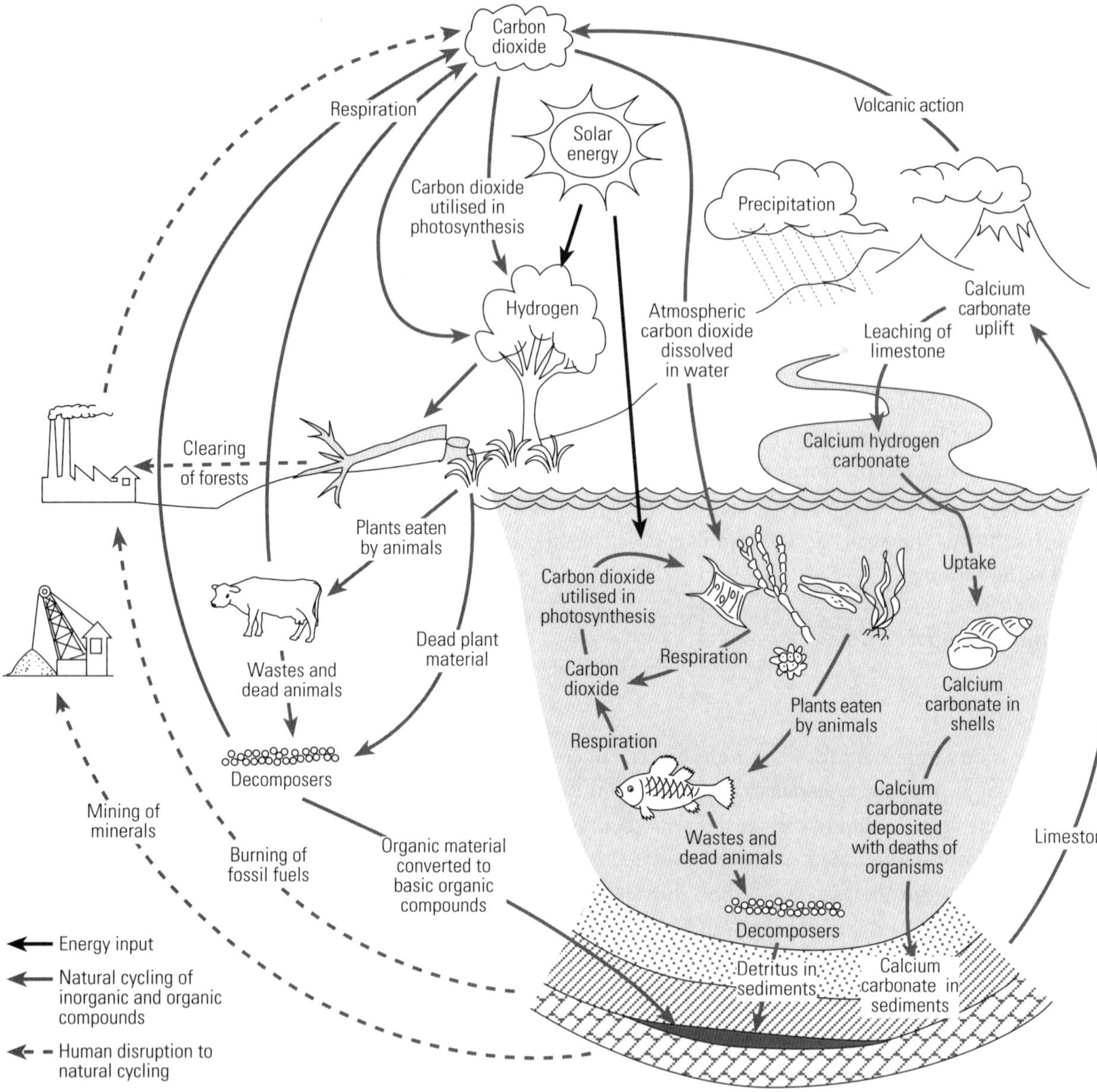

Figure 3.3.6 The calcium and carbon cycles.

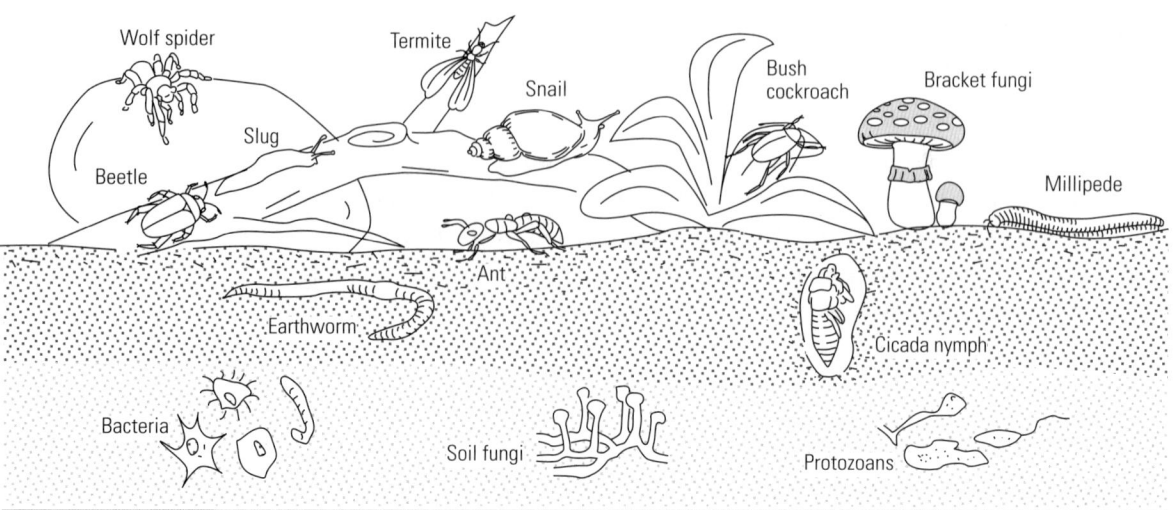

Figure 3.3.7 Soil organisms.

ecosystem can be diverse and the number of individual organisms enormous. A single gram of fertile soil, for example, may contain more than 2000 million bacteria. Soils typically contain 1–10% organic matter.

Organic matter is important for a number of reasons. The role of organic matter in influencing a soil's structure and its water-holding capacity has already been mentioned. A second important role is the storage and provision of nutrients for plants and micro-organisms. The chemistry of organic compounds in soils is quite complex. The compounds trap metal ions and hold them so they are not as easily lost by leaching. Clay minerals do this too, but not as efficiently as organic compounds.

Organic compounds in the soil provide important nutrients, such as nitrogen, sulfur and phosphorus. Nitrogen and sulfur are derived from bacterial proteins and used by plants to form proteins. Over 90% of the nitrogen utilised by plants comes from the organic matter in the soil. A third role of organic matter is in influencing the heating and cooling of the soil. Dark soil heats and cools more rapidly than a light-coloured soil.

The decomposition of organic matter produces humus, which gives the topsoil its dark colour. Humus is composed of broken-down material as well as substances produced by the bacteria and fungi that live in the soil. In soils where biological activity is high, the organic material is well dispersed and occurs in relatively large amounts. The stable, well broken-down material is called mull. In soils where the biological activity is low, decomposition is slow and the humus layer is thin. It may have a layer of leaf litter on the surface with a zone of fermentation beneath it. The material produced by this slow decomposition in the absence of oxygen is called **mor**. It is found in acid soils where the bacterial population is reduced and fungi carry out much of the breakdown. An even more restricted environment for micro-organisms is in waterlogged situations. The absence of oxygen leads to the very slow breakdown of organic material and the formation of dark organic matter called peat.

A simple way to measure organic material in soil involves measuring the amount of carbon present. This is done by strongly heating a known weight of dry soil and measuring the change in weight. This will represent the organic material that has been converted to carbon dioxide and lost from the soil.

The life in soils can be assessed in a number of ways. Agar plates can be used to culture micro-organisms from soil to give an idea of micro-organism numbers, but this needs to be done with care as some organisms in soil can cause diseases in humans. Larger organisms can be extracted from leaf litter using mechanical methods or heat.

pH

The **pH** of a soil indicates the amount of hydrogen ions it contains. (See Figure 3.3.8.) A soil's pH is easy to measure and is important because it influences both plant and micro-organism growth in a soil. A neutral soil has a pH of 7. If a soil has a pH lower than 7 it is acid. If a soil's pH is above 7 it is said to be alkaline. Many plants need a soil pH of between 5.5 and 7.0 to grow well.

The pH of a soil is influenced by such things as **leaching**, the organic content of the soil and the presence of materials, such as inorganic fertilisers. The pH also affects the availability of essential elements for plants. (See Figure 3.3.9.) Some plants are, however, adapted to a high or low pH. A very low pH (acid) soil may contain free elements that are toxic to plants. For example aluminium toxicity in some plants is related to acidity in soils.

definitions

mor
a layer of acidic organic matter with leaf litter on top and fermenting plant remains below

pH
a measure of the amount of hydrogen ions present. A measure of how acid or alkaline something is.

leaching
a process in which soluble material is carried from a material by water flowing through it

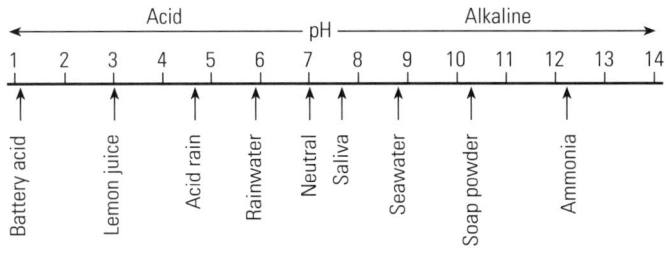

Figure 3.3.8 The pH scale.

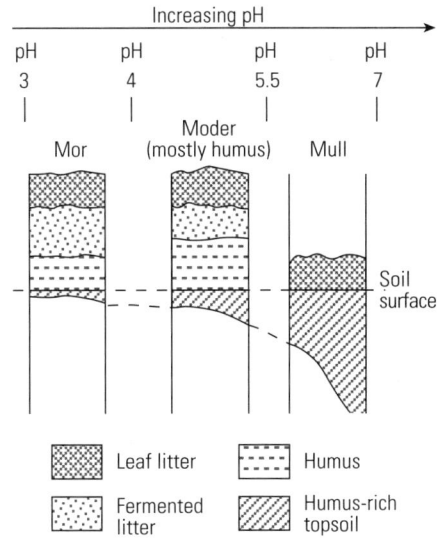

Figure 3.3.9 Surface organic matter and pH.

Measuring pH is fairly straightforward. It is done by mixing a few drops of universal indicator into a small soil sample using a clean stick or stirrer. It is best to mix the soil and indicator on a tile or the surface of a Petri dish. Care needs to be taken when handling the indicator because it will stain clothes and skin permanently.

Next, a fine coating of barium sulfate powder is placed on the surface of the soil/indicator mixture and after a short time a colour develops on the white powder. This may take a couple of minutes. The colour chart that comes with the indicator is then used to determine the pH.

PRESENCE OF ELEMENTS

The elements calcium, nitrogen and phosphorus are important for plant growth and so their presence influences the communities that grow on, or in, the soil. The

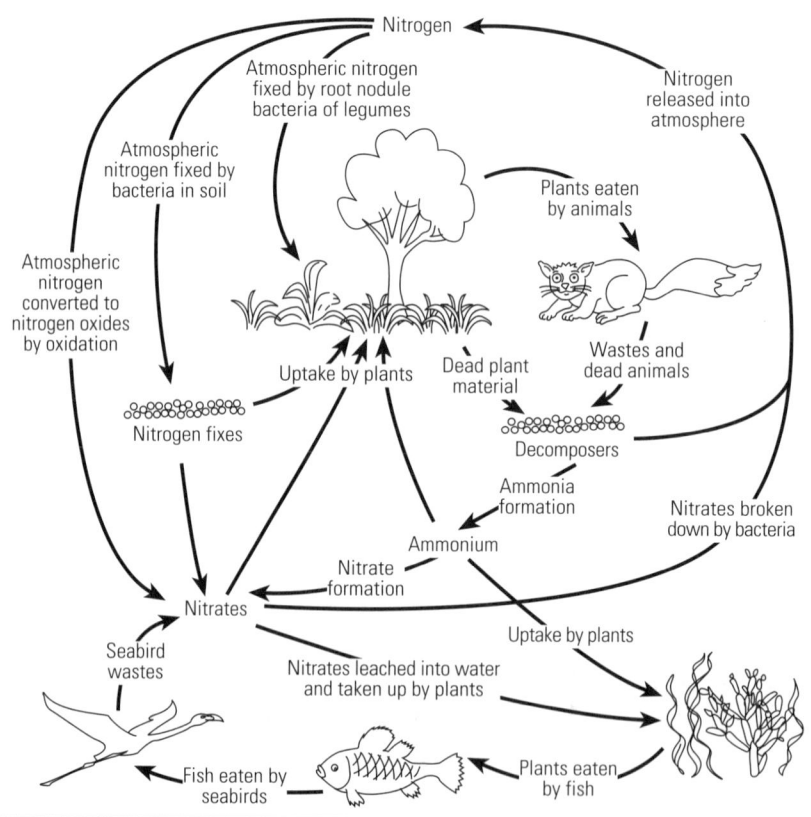

Figure 3.3.10 The nitrogen cycle.

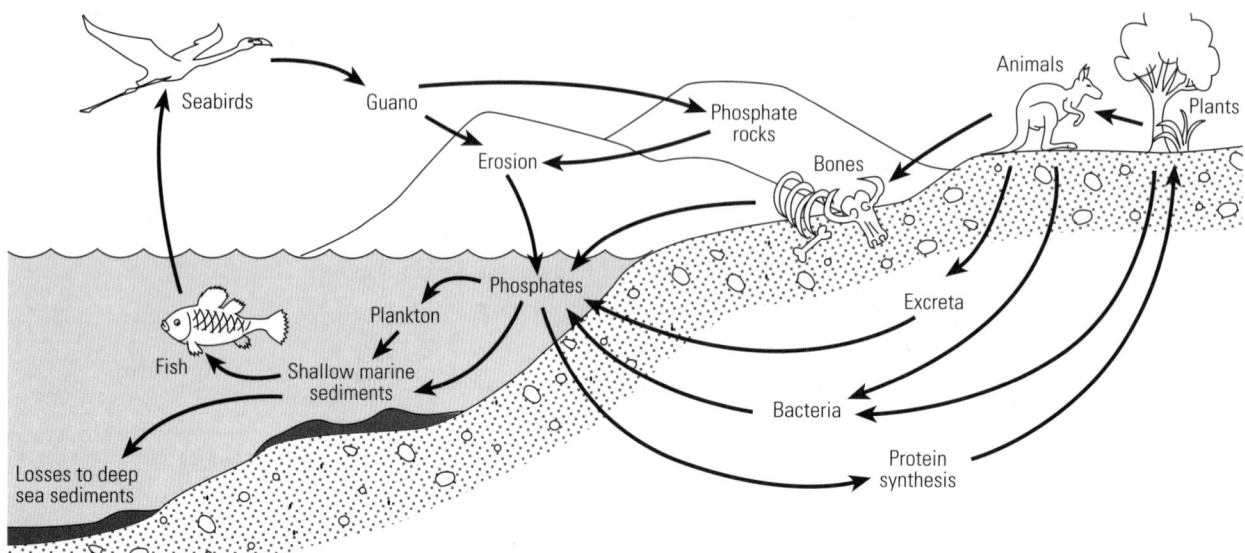

Figure 3.3.11 The phosphorus cycle.

nitrogen and phosphorus cycles are shown in Figures 3.3.10 and 3.3.11, respectively. Calcium is usually found as calcium carbonate. The interaction of calcium and carbon cycles is shown earlier in Figure 3.3.6 (page 152). Calcium carbonate is easily tested for using a few drops of acid. The production of rapid bubbling indicates the presence of calcium carbonate.

Nitrogen and phosphorus can be assessed using special tests. If your school has the material to do these tests try to assess the topsoil and subsoil of a soil. Nitrogen in soil is found in the form of nitrates. High levels of nitrates in soils are unusual and may indicate the addition of fertilisers. In many Australian soils, phosphorus is quite rare. Some plants, such as certain banksias, have special adaptations to obtain phosphorus.

Water in soil contains dissolved salts. Perhaps the most common salt is sodium chloride. Salinity refers to salts dissolved in water in the soil. High levels of salinity affect plant growth because it can damage plant cells and makes it hard for roots to absorb water. Later in this text we will examine soil salinity in more detail. (See page 241.) Another aspect of sodium in soils is what is called sodicity. Sodicity refers to the sodium held on the surface of clay particles. In sodic soils, clays swell and undergo dispersion. This leads to the formation of surface crusts and scalds. Scalds are areas of smooth, bare ground. Seedlings cannot force their way through the clay crust and water cannot infiltrate the ground. Sodic scalds are reasonably common in western New South Wales along the floodplains of the Macquarie, Murrumbidgee and Darling Rivers.

Sodium can be tested in a soil using silver nitrate solution. Caution is needed with this solution because it stains. Distilled water percolated through a soil will turn milky white when a couple of drops of silver nitrate are added if it dissolves chlorine. In the field a conductivity meter can be used to measure the presence of salts in soils or waters derived from them. High salt levels increase the conductivity of the soil.

REVIEW ACTIVITIES

1 Define the term 'texture'.

2 Discuss the relationship between a soil's texture and its physical properties.

3 Evaluate the importance of soil structure to plant growth.

4 Summarise the relationship between porosity, water and air in a soil.

5 Outline the chemical properties of a soil that are traditionally studied.

EXTENSION ACTIVITIES

6 Explain how soil biota (living things) increase the porosity of a soil.

7 Draw a diagram to show the nitrogen and phosphorus cycles as they exist in your local environment. Can you quantify (give a number to) the amounts of materials in your diagram?

8 Two students noticed that grass near a tree was not growing as well as grass further from the tree. One student said that the grass near the tree grew poorly because the soil near the tree was more acid than further away. The other student disagreed, saying that the grass near the tree grew poorly because students always sat under the tree and they had compacted the ground.
  a Identify the two hypotheses proposed by the students.
  b Design an experiment to test one of the hypotheses.
  c Identify three other possible explanations for the variation in grass growth.

SOIL PROFILES

A useful way of analysing and describing a soil is by studying its **soil profile**. In a cutting, or trench, one or more layers can be located above the unaltered rock on which a soil has developed. It is this succession of layers, or horizons, that makes up the soil profile. A diagram of a soil profile is shown in Figure 3.3.12.

Soils can show a great range of variation in their structures and it should be remembered that soil profiles like that in Figure 3.3.12 are models. Also, the boundaries between horizons are not always easy to observe. G. Northcotte proposed a useful classification of soils in 1960 based on the textures of the profiles. He recognised four ways in which a profile may occur. (See Table 3.3.3.) Bearing this in mind, let us look at the properties of horizons that may occur in soils.

Table 3.3.3 Four divisions of soils (after Northcotte, 1960)

Division	Description
Organic	Soils dominated by the presence of organic matter
Uniform	Soils with a uniform texture through the whole profile
Gradational	Soils that show increasing clay with depth but no sharp change in the profile
Duplex	Soils with a marked difference between the A and B horizons

O horizon

The O horizon is a layer dominated by organic material. This layer does not always occur. When an O horizon is present it may occur at the surface or at depth if buried. In soils developed in natural bushland the O horizon may contain recognisable layers, such as fresh leaf litter (sometimes labelled L), partly decomposed leaf litter (labelled F) and well-decomposed material in which original plant structures are not recognisable (labelled H). In waterlogged soils the O horizon may be very thick and made up of peat. This horizon may be referred to in some schemes as the A0 horizon. Review Figure 3.3.12 to see the ways in which these arrangements may occur.

A horizon

The A horizon contains organic material in the form of humus mixed with mineral grains. It is often the darkest horizon in the upper part of the soil. There is no evidence of the original material from which the soil formed. In this horizon you will find the greatest amount of biological activity and, possibly, evidence of human disturbance. In soils developed in very dry conditions the surface layer may be lighter than deeper horizons due to the absence of organic material.

E horizon

This layer is sometimes described as the lower part of the A horizon but when it occurs it is easily recognised. Like the A horizon, there is no evidence of the structure of the original parent material. It is a horizon in which clay, iron or aluminium, or some combination of these materials, has been removed. This leaves a lighter colour than the A horizon and it usually has a coarser texture than the B horizon below it. This horizon may be referred to in some schemes as the A2 horizon.

B horizon

The B horizon differs from the layers above and below it in structure and composition. It may contain concentrations of quartz, clay, humus, iron, aluminium, silicates, carbonates or gypsum either alone or, more commonly, in combination. Carbonate minerals may also be leached from the horizon and residual clays may give the horizon a granular, blocky or prismatic structure. This layer is a zone of accumulation and is termed an **alluvial horizon** because water has moved the oxides and clays to it from adjacent horizons. This process may result in the layer having a darker or redder appearance than the horizons above and below it.

soil profile
a vertical section through a soil showing differentiation and structures

alluvial horizon
a zone of mineral accumulation in a soil

definitions

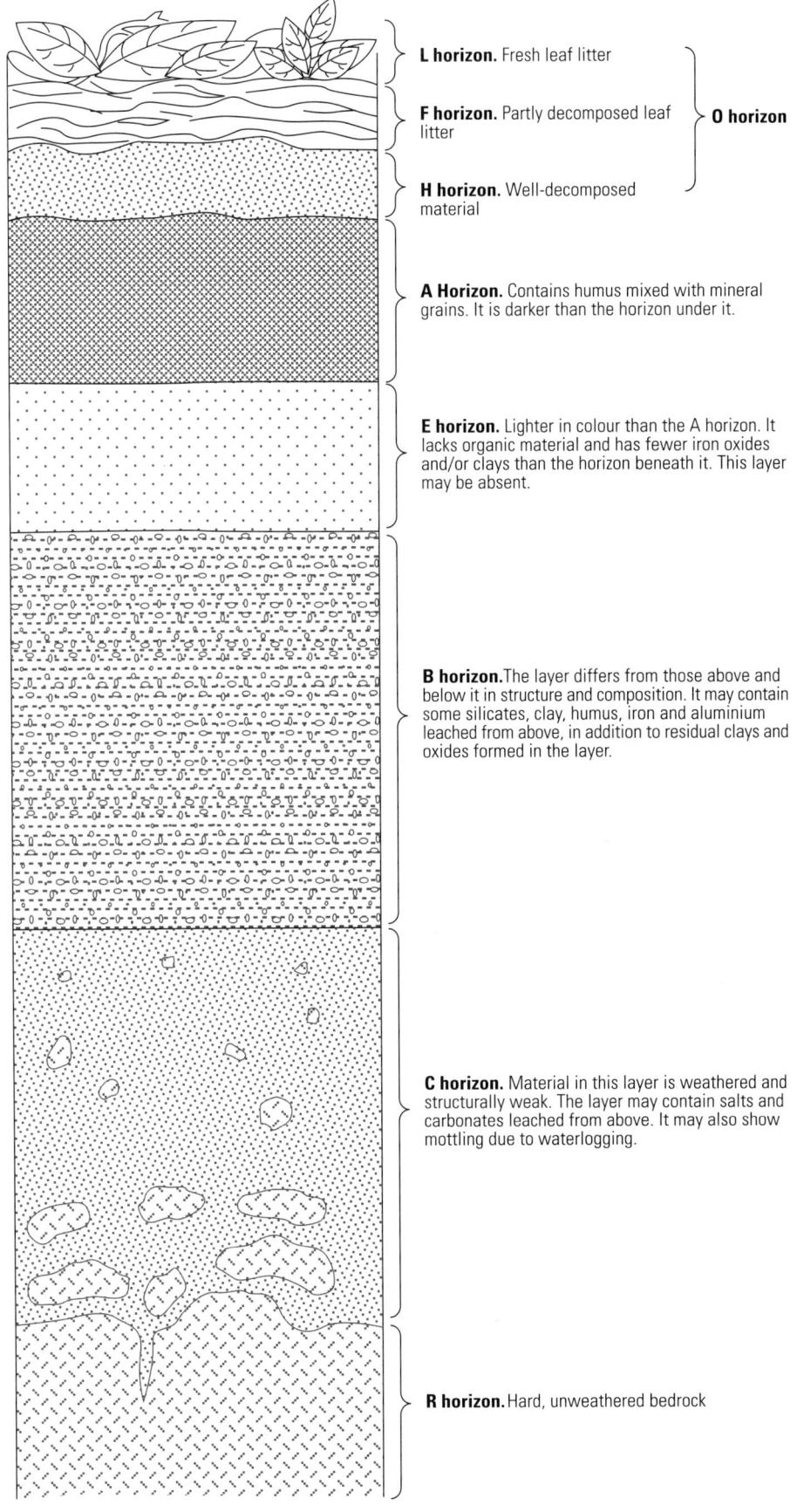

Figure 3.3.12 A soil profile. Some of the horizons shown here may not be present. In particular, L, F and H may be absent if there is little vegetation or the rate of decay is rapid. Horizon E may also be absent. The nomenclature is based on the FAO-UNESCO scheme.

C horizon

The C horizon contains evidence of its original parent material structure, although the material in it is weathered and structurally weak. This horizon contains weathered material but does not contain evidence of soil-making activities shown in the horizons above it. The layer may contain layers of salts and carbonates. It may also show mottling due to waterlogging.

R horizon

The R horizon is composed of the bedrock or alluvium on which the soil has formed. It is unweathered and hard.

REVIEW ACTIVITIES

1 What would the presence of an E horizon suggest about a soil?

2 Compare a B horizon with a C horizon. Under what circumstances would you expect them to be very similar?

3 In which parts of a soil do you expect to find organic matter and evidence of organic activity?

EXTENSION ACTIVITIES

4 Use information from your fieldwork to draw up a soil profile.

5 Compare the features you would expect to find in a tropical soil and a desert soil.

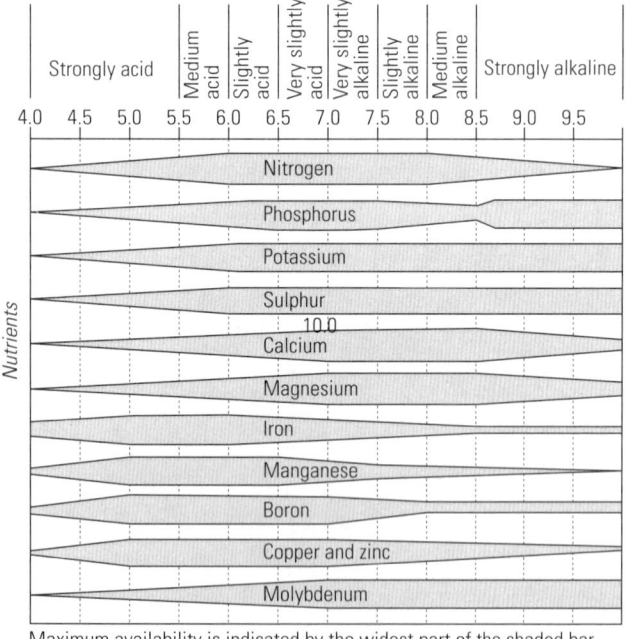

Figure 3.3.13 Nutrient availability and pH.

SOIL FERTILITY

The fertility of a soil is a reflection of its physical, chemical and biological properties. Plants and bacteria absorb nutrients from the soil. (The nutrients are in the form of ions, charged atoms and groups of atoms.) To do this, living things have to expend energy because the concentration in a cell may be less than the concentration of the ions in soil moisture. Once the ions are absorbed they must be replaced before the plants can absorb more.

Nutrients exist in four different forms, or reservoirs, in soil. They exist in the minerals making up the soil. Calcium, for example, is found in plagioclase minerals weathered from a granite. Nutrients also exist dissolved in the water in a soil. Nutrients such as sulfates and nitrates may exist in this form. Organic material contains nutrients and may bind some metal ions to them. Clays are also able to bind nutrients to them.

The presence or absence of these different nutrient reservoirs will affect the availability of nutrients for living things. Soils that have a long history of leaching may have the clay situated at a depth where the nutrients are unavailable to plants. The constant movement of water may also have removed important elements from the soil. A soil that lacks organic matter will not only lack nutrients but may not have a structure that allows plants to grow well.

The pH of a soil is important in terms of nutrient availability. The availability of important nutrients (such as nitrogen, phosphorus, calcium and magnesium) is highest in a pH range of 6 to 8. (See Figure 3.3.13.) Outside this range the pH causes nutrients to be unavailable and mobilises elements that are toxic. Zinc and aluminium are two potentially toxic elements that become soluble and available in acid conditions. The pH of a soil also affects the micro-organisms in the soil. The number of bacteria and fungi decrease rapidly as pH falls and, in soils with a low pH, fungi are more common than bacteria. As a consequence of reduced micro-organism numbers, decay rates are lowered and nutrient availability is further reduced.

In general, factors that affect a soil's water, air, nutrient or physical structure will influence its fertility. The addition of chemicals, such as salts in irrigation water or pesticides, can reduce soil fertility. Salts may limit plant growth by damaging cells or, as we have seen, affect the soil structure so that seeds cannot germinate. Pesticides kill organisms in the soil and reduce the decomposition of organic matter. This will result in fewer nutrients being available for plant growth.

REVIEW ACTIVITIES

1 Summarise the major factors affecting a soil's fertility.

2 Explain the importance of pH in nutrient availability to plants.

EXTENSION ACTIVITY

3 Some Australian plants are adapted to living on very infertile soils. Research some of the adaptations that such plants possess.

PROCESSES FORMING SOILS

If we are to conserve environments or seek to restore diversity to an area we cannot avoid paying attention to the soil. The nature of the soil determines what can grow in an area and, consequently, has a large affect on the animals found in the area. In terms of the biology of an area we should never forget the large variety and number of organisms that inhabit the soil. The structure and composition of the soil and the material from which it develops has a profound affect on the natural environment of an area. In the following section we will examine the factors that result in soil development and look in more detail at some soil types and the particular factors that have influenced their development.

The production of soils involves two sets of processes: the weathering of parent material and processes that modify the weathered material. Weathering involves the alteration of rocks by physical or chemical processes. It is dealt with in more detail on page 226. Weathering produces rock fragments, clays and salts. Soil-making processes redistribute and further modify these materials to form soils.

Weathering produces a layer of altered rock debris known as the regolith. The term 'regolith' literally means 'a rock blanket' and, apart from rock outcrops, the regolith covers most of the land surface. In a soil profile the regolith includes the A, B and C horizons. It ranges in thickness from a few centimetres to tens of metres. The regolith may consist of material that has weathered where it is or it may consist of materials transported by water, wind or ice. The nature of weathered material is most clearly shown by the C horizon.

The salts that are formed by weathering occur as ions in the soil water. Clays and organic compounds may bind these salts, or the salts may be moved as the water they are dissolved in moves. Leaching is the name given to the process of washing out a soluble material. Water may carry salts, together with clays and humus, down into the soil where they may be deposited in the B horizon. In dry conditions, water evaporation from the surface of a soil may draw water upward, carrying dissolved salts with it. (See Figure 3.3.14, page 160.)

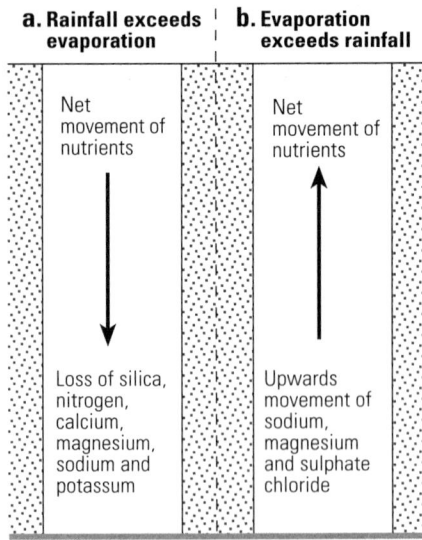

Figure 3.3.14 Water movement through soil.

Some rocks are more soluble than others. Limestone, while a hard rock, is dissolved by rainwater because rainwater is slightly acidic. Silicate minerals are much more resistant than the carbonate minerals that make up limestone, but different silicate minerals weather at different rates and provide soluble salts.

Factors important in the formation of soils

Soil formation is influenced by a number of factors. The most important of these factors are:
- climate
- topographic relief
- parent material
- living organisms
- time.

CLIMATE

Climate is the most important factor affecting soil formation. In particular, temperature and water play an enormous role in making and modifying soils. Water is necessary for the chemical reactions of weathering and it also moves materials away from the site of weathering. This allows new material to be attacked. Temperature is important because chemical reaction rates are affected by temperature. Higher temperatures mean faster reaction rates. Temperature variation is also important, particularly for physical weathering processes.

Figure 3.3.15 shows how latitude, climate and soil formation are related. Note that the depth of soils is related both to the amount of vegetation and the rainfall. The diagram also shows that the nature of the clays, reflecting different weathering reactions, is also related to rainfall. It is important to note that this diagram ignores topographic effects such as slopes and the altitude at which weathering occurs.

TOPOGRAPHIC RELIEF

The shape of the Earth's surface, or **topographic relief**, influences the formation of soils in a number of ways. High-altitude areas, such as the Snowy Mountains, are cooler than adjacent low-altitude areas. The mountains may also receive more rain than nearby areas. Areas near the Snowy Mountains and Mount Canobolas near Orange receive less rain because it has fallen on the high-altitude areas. Such areas are said to be in a rain shadow.

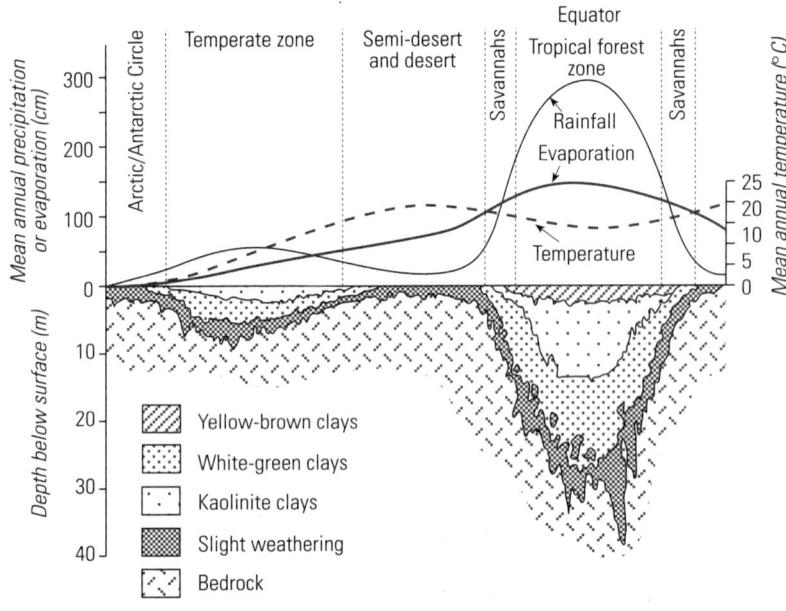

Figure 3.3.15 Climate and soils.

The slope of an area affects soil formation because it influences water movement and soil movement. (See Figure 3.3.16.) When rain falls, the water will either run along the surface or infiltrate the soil and move down the slope as ground water. The rate of movement depends on the steepness of the slope. On steep slopes the water moves away reasonably quickly, leaving a relatively small amount for soil formation. In flat areas, however, the water may remain in the soil for a relatively long time and be available for weathering and soil formation. Surface run-off on steep slopes causes more erosion than water running down gentle slopes. This, together with the motion of loose materials due to gravity on steep slopes, leads to soils on crests and steep slopes being shallow compared with soils on gentler slopes and in creek lines. (See Figure 3.3.17.)

topographic relief
variation in the shape of the Earth's surface; the set of landforms found in an area

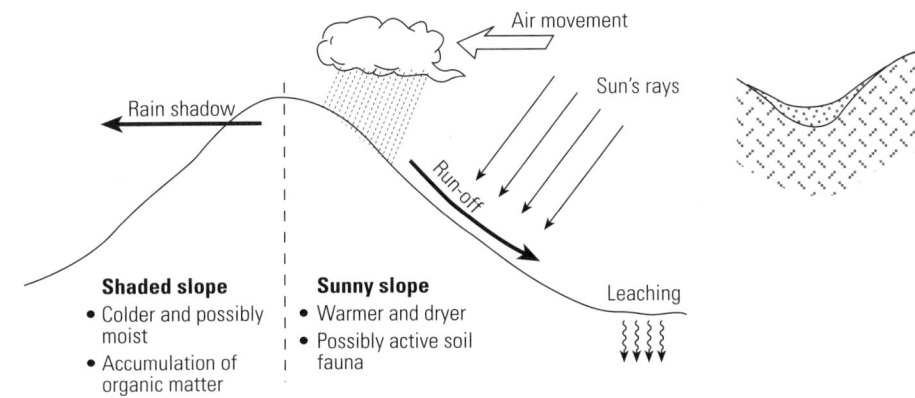

Figure 3.3.16 Topography and soil-forming processes.

Figure 3.3.17 Soil depth and topography.

The aspect of a slope—the direction a slope faces—may also affect the types of soils formed. The aspect determines the amount of heating the slope receives from the Sun and the amount of rainfall it receives. North-facing slopes receive more direct sunlight than south-facing slopes. This means that, other factors being the same, north-facing slopes are warmer and dry more rapidly than south-facing slopes. This has both costs and benefits for soil formation. Warmth encourages weathering and organic activity, but increased evaporation may lead to the soils being drier. The issue of rainfall then becomes important in determining how rapidly the soil will form. Because the south-facing slope has a different set of conditions from the north-facing slope, the nature of the soils formed may be different.

Remember that aspect also affects the vegetation growing in an area. This will affect the nature of the organic processes involved in the soil-making process.

Relief affects soil erosion. The two types of soil erosion you should know are called sheet erosion and gully erosion. (See Figure 3.3.18.) Sheet erosion is the removal of surface soil by water or wind. This leads to the removal of the fertile A horizon of a soil. Planting vegetation and controlling run-off can prevent this. Gully erosion begins when water running over the surface forms small channels called rills. Rills can enlarge and form gullies. These are hard to deal with because the sides of the gully may collapse and enlarge the area affected by the problem. Planting and fencing are required to prevent the formation or enlargement of gullies.

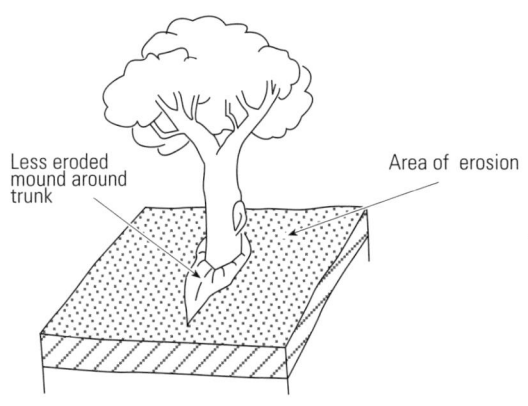

PARENT MATERIAL

Rocks weather at different rates depending on their composition and structure. Minerals such as olivine weather very rapidly compared to minerals such as quartz. Due to these differences a coarse, basic rock (such as gabbro) will weather faster than a granite under the same conditions. Grain size affects weathering rates and the rate of soil formation too. Large crystals have relatively large surfaces on which weathering can occur and the rock may break down faster than rocks of a similar composition but smaller grain size. Structures such as joints and faults will also affect soil formation by allowing water and air access to areas deeper in a rock body. This will increase the depth of the regolith.

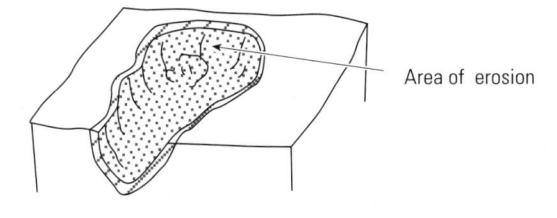

Figure 3.3.18 Sheet and gully erosion.

LIVING ORGANISMS

The importance of living things to soils has already been discussed. To summarise their influence, living things:

- allow water and air to enter the soil and cause soil formation
- provide organic material that affects soil structure and holds water and nutrients in the soil
- affect the pH and availability of nutrients
- prevent erosion by covering the soil and holding it together through root growth
- increase soil development by mixing soil materials and providing deep root penetration.

TIME

The longer a soil has to develop without disturbance the deeper it becomes. The effects of weathering and organic activity increase over time and the characteristics of parent material are removed from a soil. Certain minerals take time to develop or build up. Kaolin is a resistant white clay that builds up in old soils when other minerals are removed or altered. The redistribution of materials by leaching or upward water movement also takes time, particularly in areas of low rainfall.

Some soils in New South Wales are extremely old and, as a result, they lack certain plant nutrients due to prolonged leaching. Their red, yellow and orange colours may indicate prolonged periods of contact with air. The depth of the regolith also indicates their age in many areas. Along some rivers it is possible to find soils developing on other soils due to flood deposition. Such processes occur over relatively long periods of time.

REVIEW ACTIVITIES

1 Define the term 'leaching'. Explain its importance in affecting soil fertility.

2 Summarise the five factors that affect soil formation.

3 Describe how water movement through a soil transports materials.

4 Explain why many old Australian soils are relatively infertile.

5 Describe how the parent material on which a soil forms determines the type of soil that results.

6 Discuss why time is important in the formation of a soil.

EXTENSION ACTIVITIES

7 Suggest how a soil profile in your area would change if the rainfall in the area was to increase.

8 A palaeosol is a fossil soil.
  a Which parts of a soil might be fossilised?
  b How might a soil be fossilised?
  c Discuss the information a palaeosol may provide about conditions in the past.

BIOLOGICAL COMMUNITIES, CLIMATES AND SOILS

Native biological communities are adapted to the soils on which they grow. They are similarly adapted to the climate, or microclimate, within which they live.

Sample communities

Community adaptation can be illustrated by reference to some of the communities found in the Blue Mountains, west of Sydney. The three communities described below are relatively close to each other geographically but the materials on which the soils have developed, and the prevailing climates, have led to the evolution of three very different plant communities. The first is a community growing on a soil developed on basalt: tall open forest. It is reasonably protected from wind but receives less rain than the second community: a heath. The third community is an open woodland. It receives slightly less rain than the tall open forest but receives much less wind than the heath.

TALL OPEN FOREST

Rich, fertile soils have developed on basalt caps in the Blue Mountains outside Sydney. On these soils are found tall open forests. Within the open forest are patches of closed forests. These are dark, moist forests where the canopy is closed. The closed forest is found on sheltered slopes where moisture is more abundant.

The open forest communities contain trees over 30 m tall and the canopy covers 0–70% of the ground. Common trees include the ribbon gum (*Eucalyptus viminalis*) and Blaxland's stringybark (*Eucalyptus blaxlandii*). Magnificent brown barrels (*Eucalyptus fastigata*) and Blue Mountains ash (*Eucalyptus oreades*) occur together with a variety of understorey species. These include a variety of ferns and plants, such as blackthorn (*Bursaria spinosa*), native indigo (*Indigofra australis*) and groundsel (*Senecio species*).

The soils have a dark reddish clay loam A horizon up to 50 cm thick. These contain granular peds, have a slightly acidic pH (6.5) and evidence of a great deal of organic material. Roots and evidence of invertebrates are common in this horizon. The B horizon is also reddish brown but has a light clay texture and a neutral pH (7.0). The horizon shows a well-developed blocky structure and is often more than 1 m thick. Under this horizon is weathered basalt. The total thickness of the soils ranges from 50 cm to 3 m.

HEATH COMMUNITY

Two flat plateau ridges extending towards the south are Kings Tableland, near Wentworth Falls and Narrow Neck Plateau, near Katoomba. Both contain open heath communities developed on very thin soils. The communities are diverse. (See Plate 16.) Members of the Proteacea family are common and include the heath banksia (*Banksia ericafolia*), hakeas such as *Hakea teritifolia* and *Hakea dactyloides*, together with members of the genera *Petrophile*, *Isopogon*, *Conospermum* and *Grevillea*. Stunted she-oaks (*Allocasuarina nana*) and tea-trees (*Leptospermum* species), together with heaths (*Epacris* species) and boronias, are common. Eucalypts are not as diverse as in other communities. The most common is the Blue Mountains mallee (*Eucalyptus stricta*).

The soils are thin—rarely more than 20 cm—and rock outcrops are common. A thin layer of leaf litter is often present when the soil surface is protected from the wind. The soils are developed on quartz sandstones of the Narrabeen Group. Fertility is very low. The A horizon is often a white to light grey loose, coarse sand. It frequently contains quartz pebbles. The sand is moderately acidic (6.5–5.0) and very permeable. The B horizon may not be present under the sand. When it is it may be a brownish black loam sand. It is also very permeable and moderately acidic.

The thin soils do not provide good anchorage for plants and the area receives frequent, strong winds. The free-draining soils and low fertility mean that the plant communities have to be highly adapted to survive and thrive as they do.

OPEN WOODLAND

Near Blackheath, soils are again developed on Narrabeen Group sandstones. Low hills with slopes of 3–18° occur and some thin claystone layers are found among the quartz sandstones that outcrop. Dark, organic-rich sands occur and these are sometimes covered by layers of leaf litter up to 10 cm thick. The sand is quite acidic (pH 4.5) and contains abundant rock fragments and pebbles. Under the sands is a B horizon, which is quite different. It is a red-brown to yellow-brown clay sand or sandy loam. The pH is less acid than the A horizon (pH 5.0–6.0) and is less permeable than the sands above it. It is weakly pedal with angular, blocky peds. The depth of this horizon varies but may be 60–80 cm thick in places.

The vegetation in an open woodland is dominated by a variety of eucalypts that have approximately 20% canopy cover. (See Plate 17.) These include black ash (*Eucalyptus sieberi*), Blue Mountains ash (*Eucalyptus Eucalyptus*), scribbly gums (*Eucalyptus rossii* and *Eucalyptus sclerophylla*) and peppermints (*Eucalyptus radiata* and *Eucalyptus oblonga*). An understorey of shrubs occur (including members of the genera *Acacia, Persoonia, Leptospermum, Hakea, Banksia, Lomatia, Isopogon, Petrophile, Boronia, Daviesia, Monotoca* and *Telopea*), together with a number of grasses and mat-rushes (*Lomandra* species).

Interrelationships between soils and life

Changes to any of the soils described above will affect the communities living on them. Similarly, changes to the communities will affect the soils. The fertility of the basalt soils makes them very attractive for agriculture. Large areas have been cleared for orchards and farms, and cultivation changes the soil. In the case of the heaths, disturbance of the vegetation leads to severe erosion. Water and wind easily erode the sands. Once soil is removed, the weathering of the bedrock is extremely slow and weathered material is prone to erosion. The third community is relatively stable because it is not disturbed to any great degree. In some areas the building of houses disturbs parts of the area and erosion occurs readily once vegetation is removed.

The variations in plant communities are paralleled by variations in the animals that inhabit each community. Both invertebrates and vertebrates are present in varying numbers throughout the year. Their numbers change according to the temperature and other climatic conditions. The presence of different bird species varies due to migration through the area and due to changing food supplies as different plants flower or produce fruit. The tall open forest provides nesting areas for birds and mammals that are not found in the heath community. Reptiles are more abundant in the heath than in the tall open forest, and the open woodland appears to contain more carnivorous marsupials than the heath. In the more closed areas of the tall open forest, rotting logs contain the intriguing Peripatus, or velvet worm. They have similarities with worms and insects and have a fossil record dating back to the Cambrian period.

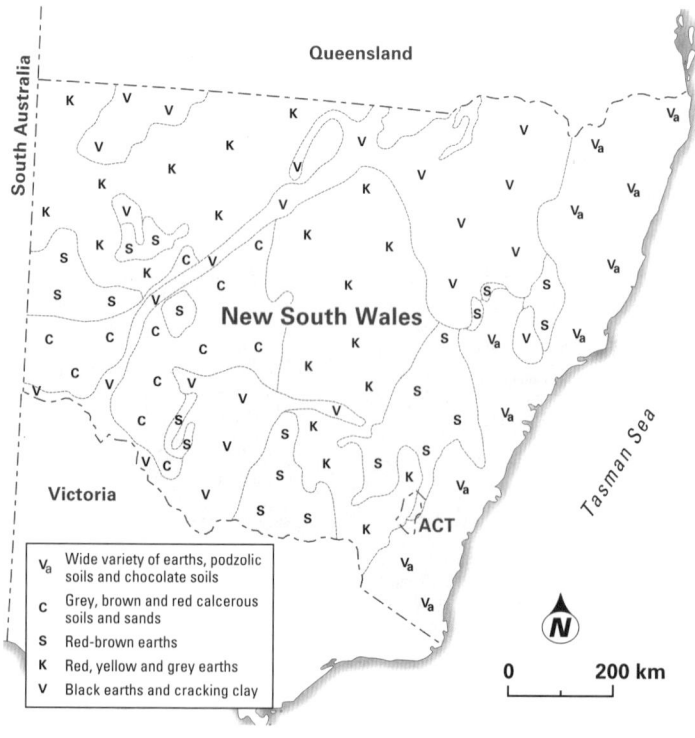

Figure 3.3.19 Soils of New South Wales.

Each animal within a community requires particular things from its environment. Food, water and shelter are obvious examples. Less obvious, perhaps, are appropriate climatic conditions such as a tolerable temperature range and humidity, which is important for some invertebrates. Other animals also affect the presence of a particular animal in an environment. Competition and predation play a big role in shaping the numbers of an organism in an area. More will be explained about such relationships on page 172.

It is important to recognise the close relationship between soils, life and climate. If humans alter any one of these three things the other two will also change. Examine Figure 3.3.19, which shows the soil types and their distribution in New South Wales. What are the types of soils found in your local environment?

REVIEW ACTIVITIES

1 Use the information in the section above to compare the three environments.

2 Assess the idea that the rock type on which a soil forms play a major role in the types of communities that live on the soil.

3 Summarise how human behaviour can affect soil formation.

EXTENSION ACTIVITY

4 Use secondary resources to identify a mammal, an invertebrate and a bird native to the area where you live. For each, identify the physical factors and biological factors that affect how the animal lives.

SUMMARY

- Soils are complex systems formed by weathering and containing mineral, organic, gaseous and water parts.

- The components of a soil determine the soil's properties.

- The physical properties of a soil include texture, structure, porosity and aeration.

- The chemical properties of a soil include organic content, pH and the abundance of elements, such as calcium, nitrogen, phosphorus and sodium.

- Soils can be analysed and described using soil profiles.

- A soil profile gives information about the conditions under which a soil forms.

- The fertility of a soil depends on the soil's physical, chemical and biological properties.

- There is a close relationship in an area between the soil and the biological community that exists on and in it.

- Soils are formed by weathering, biological activity and the redistribution of weathered material by water movement.

- Factors that affect soil formation include climate, topographic relief, parent material, living organisms and time.

PRACTICAL EXERCISE
The biota in your area

The Syllabus for the Earth and Environmental Science Course asks you to identify, gather and process first-hand or secondary data to identify the dominant plants and animals in your area. In this exercise we will examine a hypothetical area so that you can consider some of the factors to include in your fieldwork.

The collection of first-hand and secondary data

In studying the vegetation in a park near their school, students carried out the following procedures.

They ran a tape measure down a slope in the park and collected data about the plants along the transect. They did not know the names of the plants so they recorded as much information as possible, including photographs, about the major types of plants they encountered. Some of the students measured the shape of the slope using an inclinometer. Other students collected information about the pH, moisture content and leaf litter of the soil.

Back in the classroom, books from the library were used to help identify the plant species. Information about the local plants was also researched in the local council's State of the Environment Report and from a soil landscape map and report for the area. Some of the students also asked questions of members of the local Landcare group.

The fauna of the park presented some problems in terms of gathering data. Birds in the area were described and their numbers recorded. Insects were described when they were encountered but no-one systematically gathered any insects. Some of the leaf litter was placed onto a white sheet and sorted using gloves. The animals found were recorded. Back in the classroom, reference books were used to classify the insects into groups and attempts were made to classify other animals observed.

The results

The results of the fieldwork were recorded in a transect diagram (see Figure 3.3.20) and tabulated (see Tables 3.3.6 to 3.3.8). In Table 3.3.8:

- moisture is on a scale of 1 (dry) to 3 (saturated)
- leaf litter is measured in cm depth
- animals are by type, and the number observed is given in brackets.

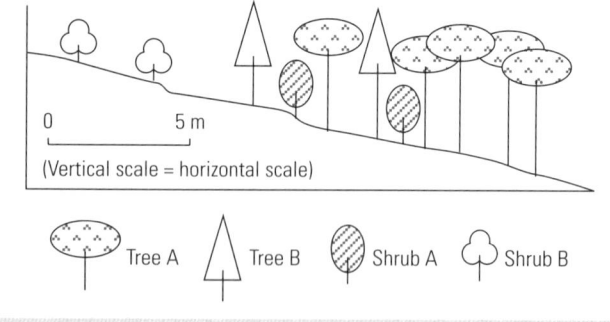

Figure 3.3.20 Transect diagram based on fieldwork.

Table 3.3.6 Plant species identified

Plant label	Scientific name
Tree A	*Eucalyptus crebra*
Tree B	*Eucalyptus punctata*
Shrub A	*Acacia buxifolia*
Shrub B	*Persoonia linearis*

Table 3.3.7 Animals identified and their location

- See Table 3.3.8 for insects in the leaf litter
- Young leaves of the Eucalypts had been eaten (possums?)
- Small insect-eating bird observed but we could not identify it
- Two red wattle birds observed in the Acacias

Table 3.3.8 Soil properties and invertebrates observed

Location	pH	Moisture*	Leaf litter	Animals
1	5.5	1	3 cm	Spiders (5) Beetles (2) Sucking insects (0) Millipedes (0) Centipedes (1) Moths and butterflies (3)
2	5.5	1	5 cm	Spiders (3) Beetles (6) Sucking insects (6) Millipedes (0) Centipedes (1) Moths and butterflies (5)
3	6.0	2	23 cm	Spiders (3) Beetles (5) Sucking insects (8) Millipedes (3) Centipedes (1) Moths and butterflies (0)
4	6.0	2	30 cm	Spiders (4) Beetles (7) Sucking insects (18) Millipedes (4) Centipedes (2) Moths and butterflies (0)
5	6.5	3	30 cm	Spiders (2) Beetles (4) Sucking insects (11) Millipedes (5) Centipedes (0) Moths and butterflies (0)

*Scale 1–3 (dry–wet)

ACTIVITIES

Analysis

Carry out the analysis of the data by completing the following activities:

1
Are the plants distributed randomly down the slope?

2
Is the type of plant in an area related to its position on the slope?

3
Graph pH against leaf litter depth. Describe the relationship you observe.

4
Is the distribution of millipedes related to the distribution of something else measured by the students?

5
Do any other relationships occur between invertebrate types and plant type or soil characteristics? If so, show how the relationships can be presented.

6
Can you suggest how the assessment of the plants could be improved?

7
How may the assessment of animals in the area be improved?

Conclusion

8
Assess whether the data in this exercise adequately describe the relationship between abiotic and biotic parts of the environment.

9
How may the data gathering be improved?

Extension

10
If you have the resources available, find out where the plants listed occur. (You could start with the Environment Australia website listed at the end of the section, page 203.) Can you identify the area where the students did their fieldwork?

PRACTICAL EXERCISE
Areas of environmental importance in your area

The Register of the National Estate is a database compiled by the Australian Heritage Commission. The database has been developed since 1976 and now contains information about more than 12 000 places of natural, historic and indigenous significance in Australia. In addition it contains information on all places that have been assessed by the commission, as well as those in the process of assessment.

In this exercise you will use the Internet to search the database. At the end of the exercise you will have identified places in your local environment that are regarded as being of environmental importance.

You may wish to follow up this exercise by using some of the search tools discussed on page 249 to locate sites of environmental importance in your area or to identify significant areas of biodiversity near where you live.

ACTIVITIES

1
Use a web browser, such as Microsoft Internet Explorer or Netscape, to visit the Register of the National Estate. You can practise your search skills or use the following address: www.environment.gov.au/heritage/register/

2
Click on the link to the Easy Search.

3
Enter your location into the location box and press the button marked 'Find' or click on the map.

4
A list of sites close to you should appear. Read about each one and try to identify the features that have led to it being placed on the register. Record the names of the sites and their significant features.

5
How many of the sites near you are natural sites and how many are part of our human heritage? Why are buildings or streets regarded as significant parts of the National Estate?

PRACTICAL EXERCISE
Analysing soil data

In this exercise you will use information recorded on a field trip to develop a summary and interpretation of a soil.

Background
The soil is formed on quartzites, impure limestones and hornfels. The soil is on a relatively steep slope and there were examples of sheet erosion near where the data (see Table 3.3.4) were collected.

The plant community where the soil was studied is an open woodland dominated by gum trees with an understorey of shrubs and grasses. There was 6 cm of leaf litter where the soil section was studied.

The farmer who owned the land provided chemical test results for the soil in the area sampled. The results are shown in Table 3.3.5.

Table 3.3.4 Data summary

Property	A horizon	B horizon
Colour	Brown to dark reddish brown	Yellowish brown
Texture	Fine, sandy loam	Silt loam
Structure	Massive	Weakly pedal
pH	6.0–6.5	6.5–7.0
Permeability	High	Moderate
Thickness	47 cm	87 cm
Organic matter	High	Low

Table 3.3.5 Chemical data from local farmer

Property	A horizon	B horizon
Depth of sample (cm)	30	80
pH	4.6	4.2
Exchangeable calcium (%)	37	18
Exchangeable magnesium (%)	11	24
Exchangeable potassium (%)	6	10
Exchangeable sodium (%)	2	3
Exchangeable aluminium (%)	3	17
Available phosphorus (ppm)	8	1
Organic carbon (%)	3.49	0.48

ACTIVITIES

1 Use the information provided to draw a diagram of the soil profile.

2 Compare the texture and structure of the horizons. Can you suggest a reason for the differences?

3 The exchangeable amount of the elements reflects the amount available to plants growing in the soil. Graph the information in Table 3.3.5 and summarise the elements that seem to be in short supply when the acidity is high (pH is low).

4 Farmers use the exchangeable aluminium percentage as an indication of potential aluminium toxicity. High levels of aluminium can lead to stunted roots and reduced growth in sensitive plants.
  a Predict whether you think the native plants in the area are likely to be sensitive to aluminium in the soil. Explain the reasons for your prediction.
  b Do the limited data presented here support the idea that low pH and available aluminium levels are related? If so, how?

5 Why would a soil like that described here be prone to erosion? Use the resources available to you to suggest some of the ways by which erosion could be avoided on this land.

6 What sorts of rocks has the soil developed on? Would these rock types weather slowly or rapidly?

3.4 Adaptations and survival in disturbed habitats

OUTCOMES

At the end of this chapter you should be able to:

- relate the presence of particular animals in the local environment to their requirements within the local environment

- explain how habitat disturbance from soil degradation can advantage introduced species of plants and lead to the reduction or elimination of native flora and fauna species in affected areas.

All living things possess adaptations. An adaptation is a feature of a living thing that improves its chances of survival and successful reproduction. Adaptations may be structural, physiological and behavioural and they give organisms advantages over similar living things that lack adaptations to a particular environment.

ADAPTATION AND NATURAL SELECTION

The word 'adaptation' may also be used to describe a process. The process of adaptation may occur in an individual during a short period of time or in a population over a series of generations. When a human spends time at a high altitude, changes occur in their body so that they can better cope with the amount of oxygen in the air. An example of these changes is an increase in the amount of red blood cells. In a closed forest, plants may respond to changes in light levels by producing larger leaves, more leaves or leaves with greater amounts of chlorophyll. An individual's ability to adapt to changes in the environment is fairly limited. Our adaptations are determined by the genetic information contained in the nuclei of our cells. Populations of individuals can respond to more drastic changes in their environments.

In a population, adaptation to an aspect of the environment may occur over a long time. Familiar plants such as the Eucalypts, members of the family Myrtaceae, and members of the family Proteaceae are plants that have relatively hard, leathery leaves. The leaves may also be spiny or reduced. This characteristic is called **scleromorphy** and plants with this adaptation are known as sclerophyllous. Such leaves are adaptations to relatively dry conditions. The adaptations are thought to have developed on the edges of rainforests when rainforest plants had to cope with a changing climate, fire and poor soils. The conditions inside the rainforest are quite stable, with nutrients and water being efficiently recycled. On the outer edge of the rainforest, however, conditions would have been quite different.

The process by which the frequency of an adaptation changes is called **natural selection**. Plants in a population on the edge of the rainforest that had leaves that allowed less water to escape and resisted wilting would have coped with poor soils and dry conditions better than plants that lacked these features. The better adapted plants would have reproduced and passed on the characteristics to their offspring so that over time the population would have come to contain a majority of plants showing

scleromorphy
having the anatomy and characteristics of scerophyllous plants, that is plants with tough leaves that help to reduce water loss

natural selection
a process that changes the frequency of genes and traits in a population

scleromorphy. These populations would have adapted further and given rise to the groups of plants that today dominate the dry conditions and poor soils of much of New South Wales and Australia.

ADAPTATIONS THAT ALLOW SURVIVAL IN CHANGING CONDITIONS

An important point to bear in mind is that some of the population already had the features that gave an advantage in the new environment. As the climate became drier during the middle Tertiary period, plants with the adaptations survived and reproduced more effectively and the make up of their populations changed. As a result, the whole population became adapted to the new conditions.

Adaptations within populations in different environments led to changes that meant the different populations could not interbreed. In this way, new species developed.

Plants and animals are adapted to their environments in a variety of ways. Plants are adapted to many factors, including soils, water availability, reproduction, predation and competition with other living things. Animals, too, are adapted to particular aspects of their environments. The factors include the way they find, eat and digest food as well as factors related to the need to compete with other animals and reproduce successfully.

The particular adaptations needed by an organism depend on the nature of habitat change. Figure 3.4.1 shows the environmental factors that control the nature and distribution of plant communities. What the diagram does not show is the affect on the composition of the community of the removal of particular species from it. Plant and animal species that depend entirely for their survival on other species will not survive their removal. In general, non-specialised species—those that are not highly adapted to a specific environment—stand a greater chance of surviving habitat change. Plants that grow on a variety of soils or animals that eat a wide variety of foods will stand a higher chance of survival than those organisms that are highly specialised.

Removing one or more species from an environment, even for a short time, may lead to an adjustment or redevelopment of the environment in which there is no place for the species removed. Relatively unspecialised species that tolerate a range of conditions may be the first to recolonise an area. The paradox of the unspecialised species surviving is that a healthy, stable environment is characterised by a high degree of biodiversity. In this situation many organisms are highly specialised to specific niches within the environment. They are not adapted to compete well in conditions caused by destruction of their habitat.

SUCCESSIONS

In the case of plants, habitat destruction may be followed by recolonisation of the environment so long as reservoirs of the plant exist in other areas and so long as seed dispersal and germination can proceed. Slow-growing plants may not compete effectively in disturbed areas as introduced weeds and other species may grow faster and exclude the original species.

The study of ecological successions gives us some indication as to how living things cope with habitat change or destruction. An ecological succession refers to the changes in the biological composition of a community over time. In general, plant species invade an area when conditions are favourable and disappear when the conditions change and become unfavourable. Each stage in a succession is

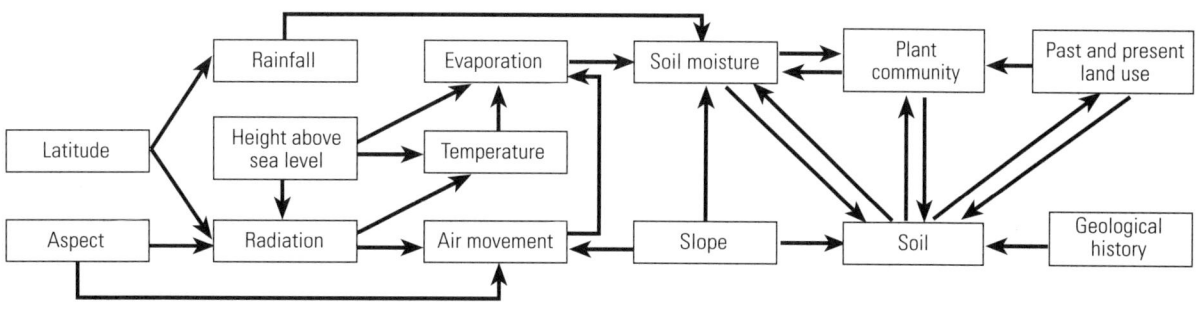

Figure 3.4.1 The environmental factors that control the nature and distribution of plant communities.

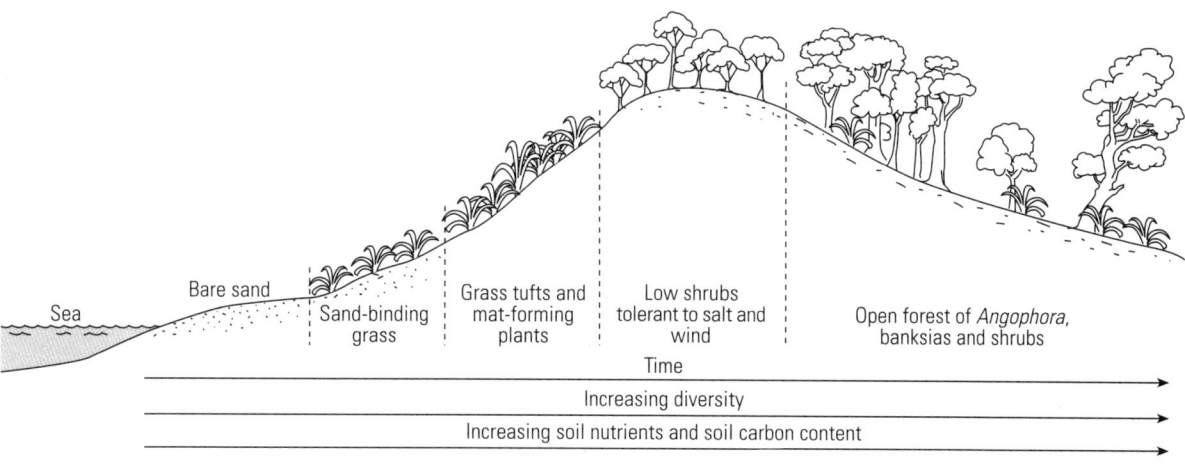

Figure 3.4.2 Succession on a sand dune.

characterised by the dominance of particular plant species and, often, by particular soils and fauna.

Successions may be classified as either primary or secondary. Primary successions are those that commence on bare surfaces. These may be new volcanic rocks or coastal sands. In areas disturbed by humans, the surfaces may be bare rock exposed by erosion or landslides on steep slopes. Secondary successions are those that develop on cleared areas. An area destroyed by fire or part of a forest that has been cleared by logging are places where such successions occur. Farming land that has been reclaimed by nearby bushland is another example.

Successions can be found on sand dunes (see Figure 3.4.2), on silted up waterways and lakes and in logging areas.

REVIEW ACTIVITIES

1 Discuss the meaning of the term 'adaptation'.

2 Explain how natural selection occurs.

3 Outline the characteristics of a succession.

EXTENSION ACTIVITY

4 Identify the types of ecosystems that exist in your local environment. For one of them, identify the physical factors that characterise the ecosystem.

DESCRIBING PLANT COMMUNITIES

Before we can describe and discuss the way humans have changed the make-up of vegetation in New South Wales we need to understand the way scientists describe natural systems. Scientists develop a range of technical terms to describe things precisely. Understanding these scientific terms is important. If you are to follow the arguments put forward in any scientific debate you must understand the concepts being used. If you do not, you risk not understanding the issues and being excluded from discussions about them.

All living things are part of ecosystems. An ecosystem is the system comprising the organisms found in a place and their environment. The environment is the sum of the living and non-living things that surround an organism. The living things interact both with other organisms and with their environment. Ecosystems can be of varying sizes and are named according to their environment or habitat types. Using the environment in which they occur, we may refer to ecosystems as terrestrial, freshwater or marine. These types can be further subdivided into smaller ecosystems. (See Table 3.4.1.)

To survive in a particular habitat an organism must obtain certain things from its environment. The physical factors of an environment include:

- light
- temperature
- soil type
- water quality and availability
- nutrient availability
- shelter.

Table 3.4.1 Types of ecosystems

Ecosystem category	Examples of subdivisions of ecosystems
Terrestrial	Alpine, temperate, arid, semi-arid and tropical
Freshwater	Creek, river, pond, swamp, lake and wetland
Marine	Reef, rock platform, littoral, shelf, pelagic and deep sea

A habitat is a place within an ecosystem in which an organism lives. An ecosystem contains many habitats and some organisms can inhabit a number of habitats while others are restricted to only one. Some waterbirds, for example, exist not only on beaches and in coastal wetlands but may also be found in the wetlands formed by flooding in the far west of New South Wales, or in places such as Lake Eyre in South Australia. Some organisms, on the other hand, are found in a very specific habitat. Sometimes this is due to the changes that have restricted the environment to very small areas. Examples of these plants and animals are found in the alpine areas of the Snowy Mountains.

The living things in the environment may play different roles. The roles may include being:

- competitors from the same species who compete for nutrients, places in a habitat and reproductive opportunities
- competitors from different species who compete for nutrients and particular places in the habitat
- predators, which use other organisms for food
- part of symbiotic relationships in which
  - both organisms benefit (mutualism)
  - one organism benefits while the other is unaffected (commensalism)
  - one organism benefits while the other is harmed (parasitism).

Living things make up communities. A community is the living part of the ecosystem and we may describe various parts of the community as animal or plant communities. A community may also be described in terms of the part of the ecosystem it occupies. In an open woodland ecosystem, for example, a soil community will contain different organisms from those found in the understorey of the woodland.

Plant communities are very important in describing different environments. The types of plants, and the ways in which they are arranged, are due to the physical characteristics of the environment in which they occur. Plants are able to modify the physical environment and thus influence the particular animal communities that inhabit them.

Plant ecologists (scientists who study the ecology of plants) describe plant communities according to their structure. An important part of the structure are the types of plants making up the tallest part of the community. These plants are described as being 'dominant'. This is not because they are necessarily the most common plants, but because their relative height makes them easily recognised. The dominant plants may be trees, shrubs, herbs, ferns, grasses and, in rare environments, mosses and liverworts. A second important characteristic of the structure is the foliage cover provided by the highest layer of the community. The uppermost foliage of a tree is called the crown and the crowns form a layer referred to as the canopy. The amount of sky that the canopy blocks out is called the cover or canopy density. A widely used classification scheme using these two structural characteristics is shown in Table 3.4.2.

Table 3.4.2 Classification of Australian vegetation by structure (adapted from Specht's 1970 classification)

Life form of tallest layer (stratum)	Symbol for tallest layer type	Foliage cover density of the tallest layer (stratum)				
		Dense 70–100% Crowns interlocking (4)	Mid-dense 50–70% Some crowns interlocking (3+)	Mid-dense 30–50% Crowns touching (3–)	Sparse 10–30% Up to one crown separation (2)	Very sparse Less than 10% More than one crown separation (1)
Trees[a]: greater than 30 m	T	Tall closed forest	Tall forest	Tall open forest	–	–
Trees: 10–30 m	M	Closed forest	Forest	Open forest	Woodland	Open woodland
Trees: less than 10 m	L	Low closed forest	Low forest	Low open forest	Low woodland	Low open woodland
Shrubs[b]: greater than 2 m	S	Closed scrub	Scrub	Open scrub	Tall shrubland	Tall open shrubland
Shrubs: 0.25–2 m						
Sclerophyllous	Z	Closed heathland	Heathland	Open heathland	Shrubland	Open shrubland
Non-sclerophyllous	C	–	–	Low shrubland	Low shrubland	Low open shrubland
Shrubs: less than 0.25 m						
Sclerophyllous	D	–	–	–	Dwarf open heathland	Dwarf open heathland
Non-sclerophyllous	W	–	–	–	Dwarf shrubland	Dwarf open shrubland
Hummock grasses[c]	H	–	–	–	Hummock grassland	Open hummock grassland
Herbaceous layer[d]						
Grasses	G	Closed (tussock) grassland	(tussock) grassland	(tussock) grassland	Open (tussock) grassland	Very open (tussock) grassland
Sedges	Y	Closed sedgeland	Sedgeland	Sedgeland	Open sedgeland	Very open sedgeland
Herbs	X	Closed herbland	Herbland	Herbland	Open herbland	Very open herbland
Ferns	F	Closed fernland	Fernland	Fernland	–	–

[a] Trees are woody plants, usually with a single stem
[b] Shrubs are woody plants, usually with many stems arising near base
[c] Hummock grasses are grasses that form mounds standing above the surface of the ground
[d] The herbaceous layer consists of ferns, grasses, herbs, sedges and rushes

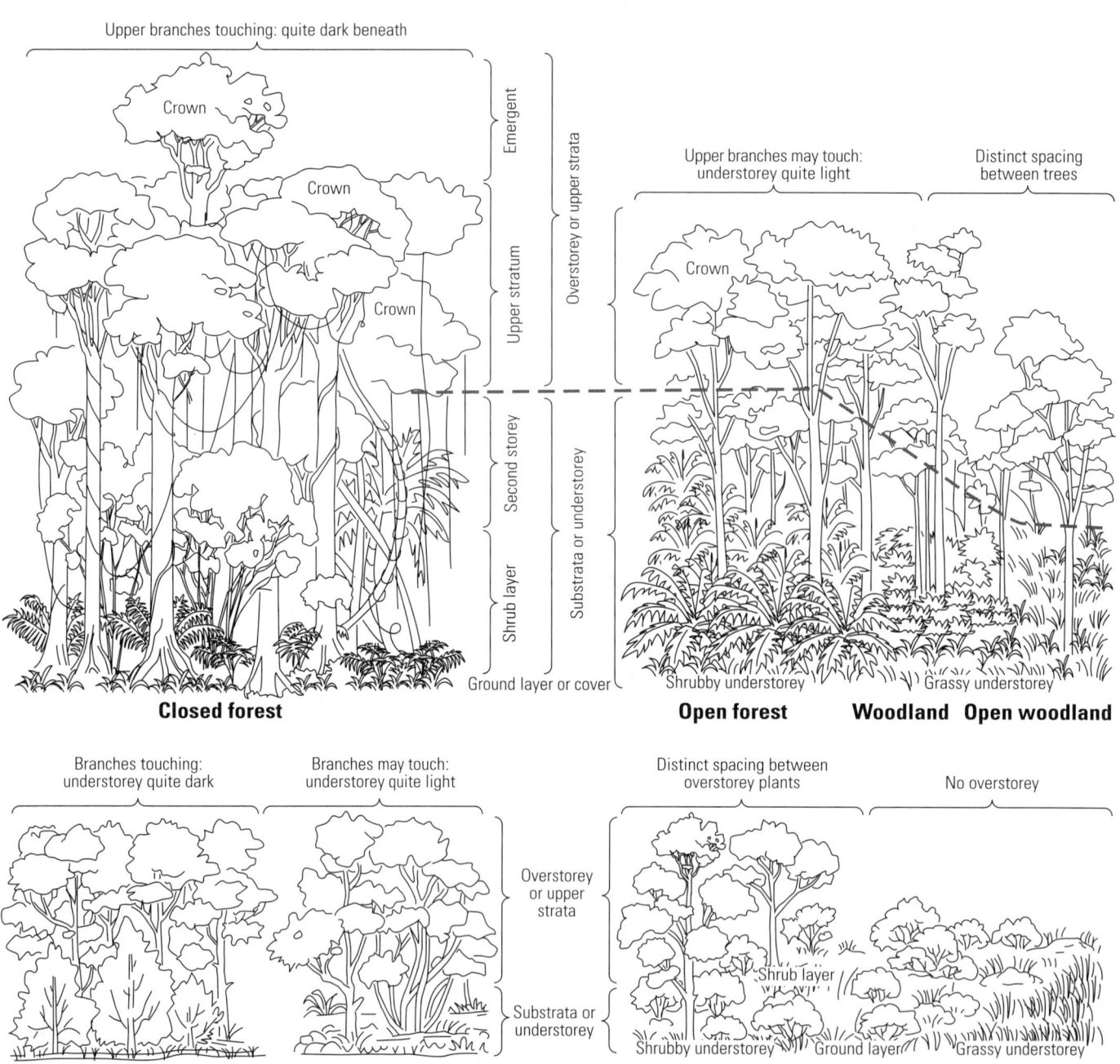

Figure 3.4.3 Different community types.

> When scientists repeat the name of a plant frequently they may abbreviate it so that the genus name is a letter. For example *Eucalyptus fibrosa* becomes *E. fibrosa*.

Plant communities (see Figure 3.4.3) can also be described in other ways beside structure. A richer definition of a plant community may involve using the name of the dominant plant. In the Sydney Basin, open forests occur in different areas. They occur on different soils and the dominant trees in the various open forests reflect the different conditions in which they occur. An example from two areas west of Sydney can be used to illustrate the naming of communities based on the dominant plant.

On the Cumberland Plain west of Parramatta the open forest areas are dominated by the ironbark (*Eucalyptus fibrosa*). The broad-leaved ironbark, as it is commonly called, is found from near Bodalla on the south coast to Rockhampton in Queensland. It is a tall tree that does not branch much below half its height. It also has an easily recognised grey bark that appears to be composed of flat, slightly flaky strips due to furrows running down the trunk. The forests occur on alluvial soils and they may be termed *Eucalyptus fibrosa* open forests.

Further to the west, in the Upper Blue Mountains near Wentworth Falls, the forests grow on soils derived from quartz sandstones. The dominant species here are also Eucalypts but the species are different. The Sydney peppermint (*Eucalyptus piperita*) and the black ash (*Eucalyptus sieberi*) are not restricted to the area. The Sydney peppermint is one of the most common Eucalypts in the Sydney Basin and is found from Bulahdelah in the north to the Illawarra, south of Sydney. The black ash occurs from near Melbourne up to Wyong on the central New South Wales coast. Near Wentworth Falls the two species, together with a rich understorey, form a readily identifiable community. It may be described as an *E. piperita*/*E. sieberi* open forest or, if the black ash is more common, as an *E. sieberi*/*E. piperita* open forest.

The species richness of an area is an excellent way of describing a community in detail. If you attempt to describe a natural community as part of a local environment report be sure to describe the composition of different layers of the community, such as the understorey and canopy. Plant classification is not easy and you may need to seek some assistance in order to classify plants to the species level. The changing nature of communities can be also described in a quantitative (using numbers) way through the use of quadrats and transects.

The distribution of communities can also be described in terms of vegetation maps. Such maps are particularly useful for analysing the effects of soils, geology, topography or water availability. Distribution maps, like vegetation maps, can also be used to describe the occurrence of animals within areas you may study.

REVIEW ACTIVITIES

1 What are the characteristics used to define a plant community?

2 Recall the roles played by an organism in an environment.

EXTENSION ACTIVITIES

3 Apply the classification system in Table 3.4.2 (page 173) to an ecosystem in your local environment. Assess how the system works.

4 Consider an area that has been partly cleared to build a house. Describe the characteristics an animal must have in order to survive the disturbance and exist near human activity.

SUMMARY

- All living things possess adaptations.
- Adaptations arise as a result of natural selection.
- Some adaptations allow survival in changed environments.
- Communities exhibit successions over time.
- Living communities can be described in a variety of ways.

PRACTICAL EXERCISE
Examining adaptations

The purpose of this exercise is to familiarise you with the significance of adaptations and the role they play in a species becoming endangered.

ACTIVITIES

1 Identify two organisms that occur in your local environment. If possible, identify two that are endangered in your area—a State of the Environment Report for your area would be a good place to start. Insects are good specimens for this exercise and so are plants.

2 Describe, using words and drawings, the characteristics of the organisms you have identified. A labelled drawing may be appropriate, and a map may show its distribution.

3 Summarise three characteristics of each organism's structure, functioning or behaviour and try to identify how the features act as adaptations to the environment.

4 If your organisms are endangered, try to determine why they have become endangered and what adaptations no longer promote the organisms' survival.

PRACTICAL EXERCISE
Describing a local environment

The purpose of this exercise is to give you practise in gathering data to describe a local environment. The things you learn in this exercise will help you in your field study for this chapter.

ACTIVITIES

1 Identify an area that has different plants distributed within it. The area may be your backyard of part of your school. Be sure to identify any possible safety risks in the area before you begin and ensure you have permission to be in the area.

2 Use a tape to record the vegetation along two or three transects across the area. Record the occurrence of plants along the lines and the habits and heights of the plants.

3 On a piece of paper, draw in the transects on an outline representing the area. Use the information on the transects to make a map of types of vegetation in your area.

4 Identify the dominant (most characteristic) species in your area. Try to identify the five most common types of plants present.

5 Examine the area and try to relate the plant distribution to the physical factors (such as light, water and soils) that are present.

6 Write a short summary describing the characteristics of your area.

176 EARTH AND ENVIRONMENTAL SCIENCE: THE PRELIMINARY COURSE

3.5 Human impacts on the environment

OUTCOMES

At the end of this chapter you should be able to:

- summarise and assess the changes in the local environment in the last fifty years in terms of:
  - vegetation cover and diversity
  - animal diversity and abundance
  - water flow and quality

- explain why different groups in the local society have different views of the impact of human activity on the local environment

- gather and process information from secondary sources to describe changing vegetation cover, plant and animal diversity and abundance, and water flow and quality in the local environment over the last fifty years

- use available evidence to assess current human impact on the local biotic and abiotic environment

- recall strategies used to balance human activities and needs in ecosystems with conserving, protecting and maintaining the quality of the environment

- assess the impact of human alterations to the environment, including land clearing, in terms of some specific consequences, such as increased run-off, increased soil erosion, changes in river flows, in-stream sedimentation

- describe, using examples from the local environment if possible, ways in which artificial structures can disrupt natural surface processes.

THREE FACTORS AFFECTING HUMAN IMPACT ON THE ENVIRONMENT

Three factors that are known to influence the degree of human impact on the environment are population, **affluence** and **technology**. Population refers to the number of individuals living in an area. Affluence reflects the demand on resources produced by a single person. Technology is a measure of our ability to cause environmental change through the use of tools and knowledge. These factors can produce both positive and negative effects.

Large populations may draw on more resources than smaller populations. However, if they contain a diversity of skills and the wealth to care for the environment, the environmental impact caused may be less than that produced by a smaller population.

In Australia most of us are quite affluent compared to people in other parts of the world. We use large amounts of energy, minerals and food. Often, we consume more than we need and we produce a great deal of waste. Being poor, however, can also produce environmental problems. If a community does not have the resources to

definitions

affluence
having material wealth

technology
materials, techniques and knowledge arising from the application of science and art

prevent or repair environmental damage its impact on the environment will be large and increase over time.

The use of technology can also produce positive or negative effects. Land clearing was once done using ringbarking and partial clearing. While effective, this method did not occur at the rates produced by the chainsaw and bulldozer. Yet, the use of satellites and other technologies to monitor the environment and research into preventing pollution are helping to reduce the impact of humans on the environment.

HOW WE LIVE AND ITS EFFECT ON THE ENVIRONMENT

Australia is a highly urbanised country. Over time, settlement patterns in New South Wales have shown three strong trends. One has been to increase the population along the coast. Another is to concentrate the population into large towns and cities. A third trend has been the decline in the population of rural centres. This is not only true of New South Wales but of Australia as a whole.

Concentration of population

By 1990 some 85% of Australians lived in towns and cities with populations of 10 000 or more. Between 1970 and 1990 Sydney's population increased by 31%. At the same time the city has been growing, with the creation of new suburbs and the urbanisation of areas in the city that were once used for industry or agriculture.

A town's or city's influence on the environment extends beyond its geographical limits. To function and grow, a population centre must draw resources to it, such as minerals, fuels and food. By using these resources, it produces wastes that must be removed. As a town or city grows, we see both an increase in consumption and an increase in waste. These have a large impact on the environment.

Coastal development

Coastal development has produced particular problems for the local environment. Like cities, coastal towns require resources to grow. Building on slopes and adjacent to areas sensitive to change produces problems. In some areas adverse effects have been caused by the rates of population growth outstripping the available infrastructure and services. High levels of unemployment may result and this influences the ability of communities to fully fund environmental programs. This is due to the diversion of funds to deal with unemployment and the resultant lowering of the community's overall wealth.

Many coastal centres discharge their **effluent** and its pollutants into oceans and tidal estuaries. The success in treating sewage is a function of population size. Small volumes of sewage can be treated more effectively than large volumes. However, the 1996 Australian State of the Environment Report reported that very small communities may still discharge untreated sewage into the oceans.

effluent
liquid discharged as waste

Decline in rural population

Inland towns may also struggle with environmental problems. Lower average incomes and declining population numbers affect the ability of communities to fund environmental rehabilitation.

Rural communities must also work with soils that have been damaged by unknowing mismanagement in the past and other consequences of inappropriate land clearance, such as salinity and dieback. In central New South Wales, land clearing has reduced the amount of native vegetation to less than one-fifth of its original coverage. Land clearing has led to problems associated with soil management and water management.

REVIEW ACTIVITIES

1 Describe the three factors that influence the degree of human impact on the environment.

2 Explain why a town's influence on the environment extends beyond the town boundaries.

3 Outline some of the environmental problems generated by coastal development and inland towns.

EXTENSION ACTIVITY

4 Identify the principle environmental problems in your area. If a local State of the Environment Report is available, compare your list of local environmental problems with those contained in the State of the Environment Report.

HABITAT DISTURBANCE SINCE 1788

During the last 220 years European settlement has substantially modified Australia's landscapes and ecosystems. For example since European settlement in 1788 some 61% of native vegetation in New South Wales and the Australian Capital Territory has been cleared, thinned or significantly disturbed. We had, and still have, a wonderful heritage in our plants and ecosystems. There are some 53 000 species of plants recorded from New South Wales, not including mosses, liverworts and algae. The plants, and the animals that live with them, occur in a wide range of communities. They range from the alpine herb fields of the Snowy Mountains to the semi-arid shrublands and treeless grasslands of the north-west of the state.

The greatest changes to Australia's plant communities and ecosystems have been due to human settlement, particularly urban development and agricultural practices. The main causes of change in the environments west of the Great Dividing Range escarpment since European settlement have been:

- land clearing
- grazing by stock
- the introduction of weeds and feral animals
- the changes in such things as fire patterns and the number of macropods.

In other parts of Australia we can add to this list logging, soil degradation, pollution and natural climate variations. The alterations we have caused to river systems in Australia have also produced changes. These changes have not only affected the water-based environments but also the terrestrial environments adapted to the behaviour of the river systems. Let us look at the causes of environmental change in a little more detail.

definition

bioregion
an area defined on the basis of the characteristic plant communities found there

Land clearing

Between 1788 and 1921 some 44% of New South Wales was wholly or partially cleared. This was done using ringbarking and burning. The most intense period of land clearing and modification occurred between 1893 and 1921. This mainly occurred on the Western Slopes, Central Plains and in the Riverina. Clearing of the northern floodplains has occurred mainly during the last two decades to make land available for cropping. During the 1980s, satellite imaging technology was used to estimate that 50 000 ha of vegetation, with more than 20% canopy cover, was being cleared annually. At the same time, sparse woodland and grasslands were being cleared at the rate of approximately 15 000 ha per year.

Maps of cleared or thinned vegetation by **bioregion** (see Plate 19) and modification of native vegetation since 1788 shows that the most modified areas of

New South Wales occur in the central area of the Western Slopes and the eastern parts of the Western Plains. There is a strong relationship between the quality of soils, topography and habitat disturbance. Land suitable for cropping and grazing, particularly if it is fertile and flat, is the most likely to be cleared. Rocky, infertile areas, such as the areas covered by National Parks along the Great Dividing Range, have been less subject to clearing. It is in these areas that a lot of the state's **remnant communities** are to be found. While the popular perception among some people in the state is that it is the rainforests and tall forests that are most at risk, it is the woodlands and very sparse vegetation areas that have been modified most.

Besides the central part of New South Wales, the other areas of the state that have been highly modified are the coastal alluvial valleys. The Bega, Clarence, Hunter, lower Shoalhaven, Richmond and Tweed Valleys together with the Cumberland Plain have all been extensively cleared. In many of these areas, unique communities occurred on the rich alluvial soils and land clearance has either removed them or left remnants at risk. The Big Scrub lowland rainforests of the Tweed and Richmond Valleys were reduced from 75 000 ha to some 300 ha (0.4% of the original area) between 1788 and 1900. Plant communities around Sydney have been similarly reduced. The Eastern Suburbs Banksia Scrub, Cumberland Plains Woodlands and Castlereagh Woodlands are of very limited extent compared to what they once occupied.

Some areas are more prone to disturbance than others. Coastal areas, particularly on the north coast, are under pressure due to population growth. The provision of housing, roads, pipelines and transmission lines all contribute to the demand for land clearing. Areas in which valuable building timbers occur have also been significantly cleared.

The clearing of land for crops and grazing has removed certain habitats, altered some and fragmented others. Fragmentation of habitats produces a number of problems. A small area of habitat has a large boundary relative to the area it occupies. The large boundary means that the area is more likely to be affected by events such as grazing by animals from outside the area, the invasion of weeds and fire. Small, isolated areas may not be able to support certain animals, particularly large carnivores. Territorial species, in particular, may be affected. Individuals control large areas, leading to low densities of animals. Small areas may not support populations large enough to cope with change.

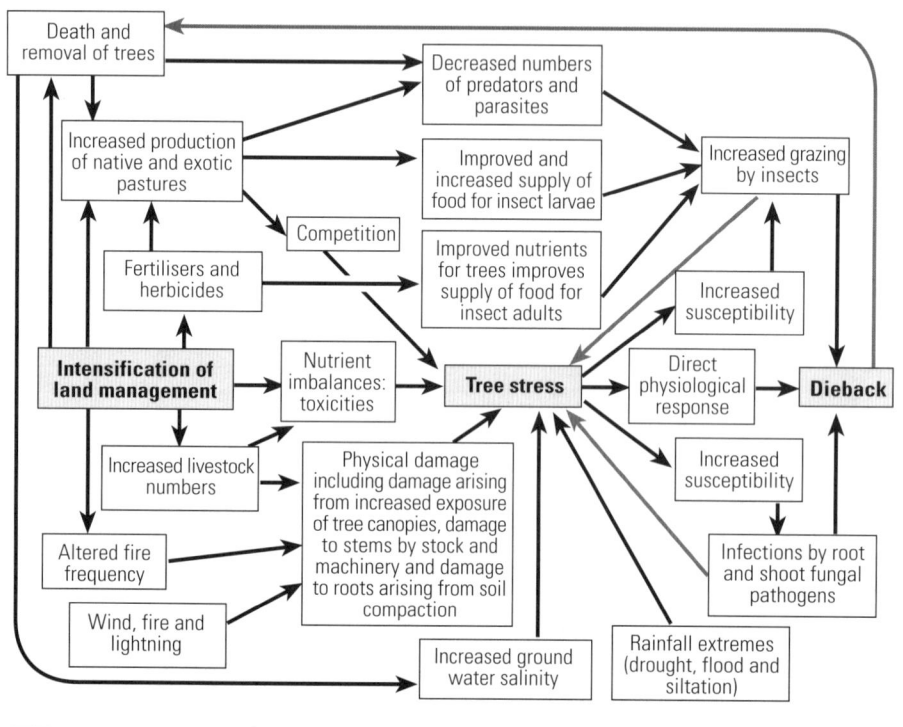

Figure 3.5.2 Factors involved in dieback.

Dieback

Other stresses that small, isolated areas of native vegetation are prone to include **dieback** and low rates of plant regeneration. Dieback is the term used to describe the long-term decline in the health of rural gum trees. It is widespread in the New England area of New South Wales and is due to a number of factors. Figure 3.5.2 shows the relationship between land management and tree stress. Overclearing, together with fertiliser poisoning, fungal pathogens and insect attack all act to stress the trees and cause increased mortality in seedlings. Changes to the structure of the plant community also contribute to the ill health of the plants. Clearing of acacias in the understorey removes both animal habitat and a source of nitrogen for the plants in the community.

The relationship between insects, land management and dieback shows the way in which a change to one part of an environment can affect other parts of it. Scarab beetle larvae have benefited from the development of pastures on cleared land. They live in the upper part of the soil and feed on the roots of the grasses. When they become adults, however, they feed on the leaves of gum trees. While the pastures sustain a large number of larvae, the number of trees is such that the beetles defoliate many of the trees, killing those that are already stressed. Often the vegetation that remains undergoes structural changes, removing habitats for insect-eating birds. As a result, the vegetation lacks the protection that such birds provide. At the same time, a diversity of other insect types are feeding on the leaves, sucking sap and contributing to the stress on the trees.

Table 3.5.1 Droughts in Australia since 1864

Drought period	Extent	Effect
1864–66	Severe through all of Eastern Australia	–
1888	Affected all of Australia except the Northern Territory	–
1895–1903	Regarded as probably Australia's most severe drought since written records kept	Large dust storms
1911–16	Whole country affected for some of this period	–
1918–20	Whole country affected	–
1939–45	Whole country affected for some of this period	Severe wind erosion
1958–68	Second in severity to the 1895–1903 drought	–
1972–73	Affected mainly Eastern Australia	–
1982–83	Severe in South-Eastern Australia	Dust storms decrease in severity, possibly due to an increase in woody weed cover
1997–98	Widespread throughout Australia	–

Natural change

Climate change has also affected our relationship with the land and understanding of vegetation changes. Table 3.5.1 shows the occurrence of droughts in Australia since 1864. Caused by the **ENSO** climatic pattern, these droughts are not only correlated with periods of erosion but also with intense bushfires. In native communities such times are stressful but can be overcome. The presence, however, of introduced animals and plants may cause changes to an ecosystem. Weeds may germinate and grow faster than native plants, altering the ecosystem diversity. Grazing by rabbits after fires may also affect plant regeneration. Similarly, fragmentation of a plant community may leave the remnants less able to survive periods of climatic stress.

Grazing by stock and introduced animals

The nature of the plants and the animals eating them affects the plants' survival and abundance. While understorey vegetation is affected both by burning and grazing, the frequency of grazing can have a greater affect than burning on biological communities in many areas. Grazing is carried out by a variety of animals, some being natives and others introduced species. Grazing reduces plant abundance and health because plants are eaten and also because of the trampling that occurs as hard-hoofed animals move around an area. The trampling not only damages plants but compacts the soil, affecting germination and growth.

The behaviour of the animals living in a plant community has an affect on plant communities and the other animals that rely on them. Not all animals prefer to eat the same plants. This means each animal species has a different affect on the plants in a community. Introduced animals, such as sheep, bite off plants very close to the ground. In areas where their grazing is sustained for a prolonged time, their feeding behaviour and hard hooves may remove all the groundcover in an area. Persistent

definitions

remnant communities small areas containing what remains of once widespread communities

dieback the long-term decline in the health of rural gum trees

ENSO El Niño-Southern Oscillation

grazing by feral rabbits has a similar affect on some, but not all, plants. If a previously affected area is fenced, preventing rabbits from entering the area, the growth of acacias and other natives naturally occurs. In non-fenced areas these same native plants do not survive long after germination.

The nature of a plant determines the degree to which grazing affects its survival. Woody plants, such as herbs and shrubs, grow from buds on the tips of stems. Once eaten, the buds have to be replaced by dormant buds further down the stem. While this can benefit some plants by increasing the amount of foliage, continuous grazing can kill the plant as growth points are removed. Other plants, such as some grasses, grow mainly from the base of the leaf and are less affected by some grazing animals. It is the palatable, more edible plant species that are affected most. Their loss relative to other, less palatable species leads to changes in community composition. A well-known example is the disappearance of *Thelmeda* grass species and the increased occurrence of the low-nutrition, unpalatable *Hetropogon* grass species.

The effects of grazing on animals in an area can take a number of forms. Competition of introduced animals with native herbivores for the available food supply can lead to smaller populations of the native animals or their disappearance. The removal of habitat and food sources can also affect native animals. Some species of seed-eating birds, for example, have been displaced from grasslands due to overgrazing and land clearing.

Groundcover removal and disturbance also leads to increased sheet and gully erosion. The loss and degradation of soils by erosion reduces the areas in which native organisms can live and the repair of such areas can take a great deal of time and effort. Remember, however, that erosion occurs in natural environments from time to time and habitats can be re-established. It is sustained and extensive erosion that presents long-term impediments to bush regeneration.

PREVENTING DAMAGE BY GRAZING

A number of steps can be taken to ensure grazing does not lead to adverse affects on native ecosystems. Some of these measures are discussed below.

Areas can be fenced to exclude grazing animals, such as relatively undisturbed sites, revegetation areas, steep slopes and the edges of waterways. Fencing of these habitats protects the vegetation and also maintains better water quality.

It is important to manage grazing in a responsible way. Poorly managed grazing prevents vegetation recovery and leads to some species losses. Crash grazing (the short-term, intense grazing of an area) can allow an area to recover better than those that are grazed continuously over a long period of time.

Careful property planning in both rural and urban areas can protect native vegetation and environments of value. The importance of planning and laws to conserve and protect the environment are discussed in chapter 3.6.

INTRODUCTION OF WEEDS AND FERAL ANIMALS

A weed is a plant that interferes with our use of land for a particular purpose. In a vegetable garden or a garden bed, a weed is a plant that we do not want there. In another part of the garden, however, it may not be considered a weed because it is not 'out of place'. In Australia, it is estimated that 10–15% of plant species are introduced from other countries. In Sydney, as many as 500 of the 2000 known species are introduced.

In areas of native vegetation, introduced plants are weeds because they upset the interrelationships that exist in the environment. Some introduced plants can dominate quickly in areas that have been disturbed. They have rapid germination rates and few predators so they compete very well with native plants. As adults they may exclude other plants and remove food sources for animals.

Feral animals are, in a sense, animal weeds. An introduced animal, such as a cat or goat, has a recognised and valued place in a particular environment but in natural ecosystems these animals upset the interrelationships and abundance of native species. Carnivores (such as cats, dogs and foxes) prey on native animals. Small marsupials can be eliminated from areas by feral and domestic cats. Herbivores (such as goats, horses and rabbits) compete with native animals for food, and hard-hoofed animals produce soil erosion.

Altered fire regimes

Mention has already been made of fire and its affect on the Australian environment. Frequent fires will not affect all species in the same way. Some plant species may be adapted to fires of a particular frequency and their abundance will be affected when the fire frequency changes. Animal species are also affected by the frequency of fires. For example fires can affect the population numbers of some birds by removing nesting sites.

It is important to remember that with regard to fire there are both adapted and unadapted species of plants in Australia. Unadapted plants are found in rainforests, alpine areas and saltbush communities. These communities are often unique and may be remnants of plant communities more abundant in former times. Their protection is particularly important because they are refugia and not replaceable.

The difficulty of measuring change

While the details related earlier paint a picture of environmental change, caution should be applied to the numbers quoted in many publications. The data collection and recording methods used over 200 years ago are very different from those used today and the changes caused by humans and those caused by natural events are sometimes hard to distinguish.

Our knowledge of the distribution of plants in 1788 has largely been derived from the written records of early settlers and explorers. This information has allowed scientists to prepare maps that are accurate down to areas of approximately 25 km^2. Modern technology, using satellite images, allows us to generate maps of much higher accuracy. Satellites give detail on areas measured in hundreds of square kilometres. When we compare the results, changes that are noted may be a consequence of the differences in accuracy being used.

REVIEW ACTIVITIES

1 Outline the changes that human activity has caused to the Australian environment.

2 Summarise the effects of land clearing, grazing and fire on native habitats.

3 Evaluate the criteria a plant should have in order for it to be a good weed.

4 Account for the changes in the natural environment caused by changes in fire regimes since 1788.

5 Explain the problems involved in measuring environmental change.

EXTENSION ACTIVITIES

6 Refer to Table 3.5.1 (page 181). Match up the periods of drought with events in Australian history since 1864. Consider Federation and the First and Second World Wars. Assess whether there is a relationship between large droughts and these historical events.

7 Consider the statement 'A feral animal is a weed' as a metaphor. Metaphors are important in science because they allow scientists to compare phenomena they know with new ideas. In what ways is a feral animal like a weed? How is it different? Is there another metaphor that could be made using a feral animal?

HUMAN IMPACTS ON ECOSYSTEMS

Logging and mining

Intensive and frequent logging has significantly changed the environment in some parts of Australia. Between 1788 and 1995, 43% of forests in New South Wales were cleared. Communities containing valuable timbers are particularly vulnerable to logging and older methods of forestry would not only remove most of a particular tree species but cause enormous disturbance to other plant and animal species. Logging and mining can both lead to a change in the age structure and composition of a community. While natural succession, if undisturbed, will lead to a complex community being formed it is unlikely to replace exactly the community that existed prior to disturbance.

Mining directly affects very small areas of New South Wales. The regulations that relate to mining and the environment, together with the fact that most miners live close to the areas mined, mean that the industry today takes seriously its responsibilities to protect and conserve the environment. In the past, however, some mines failed to ensure that toxic materials did not enter ecosystems or that areas were properly revegetated after mining ceased. Even today, the disturbance caused by mining may not lead to the re-establishment of exactly the same communities in an area. Accidents do happen and very careful consideration of possible environmental impacts has to be given to planning of mining and processing operations.

> **definition**
>
> **urbanisation**
> the building of towns and cities, replacing agricultural land and native bushland

Urbanisation and recreation

Urbanisation has an enormous effect on an environment. It replaces a particular environment with another and makes a significant impact on environments adjacent to the area of urbanisation. Some of the impacts may be destruction of vegetation, weed invasion, siltation of creeks, nutrient pollution of waterways, introduction of animals (such as cats), industrial pollution and problems associated with water management, such as dams and sewage treatment.

Recreation as an environmental problem

Associated with the problems of urbanisation are those caused by recreation. When we use parks, drive through forests, waterski, visit the snowfields or go to the beach we risk altering the environment. Ground compaction by walkers or vehicles may stop regeneration and produce erosion. Pollution by exhausts, litter and sewage are problems that need to be managed. In the case of water sports, overfishing and the erosion due to boats are issues needing consideration.

More about water quality, its measurement and ways of conserving water are explained in section 4.

Figure 3.5.3 Industry can affect both local and regional environments.

REVIEW ACTIVITIES

1 Explain how logging and mining have altered parts of Australia.

2 List some of the problems arising from urbanisation.

3 Why is recreation a potential environmental problem?

EXTENSION ACTIVITIES

4 Research the possible effects of mining on the environment. What precautions do mine operators use to minimise environmental impacts?

5 Identify the impacts of urbanisation given in the text. Describe the nature of each impact and suggest a way in which each may be minimised.

6 Compare the environmental impact of tourism and mining. Which resource use has the greatest long-term impact on the natural and human environments?

IMPACTS OF HUMANS ON WATER SYSTEMS

Water is one of our most valuable resources. Because Australia is the driest continent in the world (other than Antarctica), and our rainfall is unpredictable, Australia has the highest water storage capacity per person in the world. In New South Wales, water resources are concentrated; ten of the reservoirs contain 90% of the storage capacity. Seventy per cent of the water resources are used for irrigation, with the next biggest use being for domestic purposes. In New South Wales 1580 million L of water is stored for every square kilometre under irrigation. Sydney stores over 930 000 L of drinking water per person, which is more than three times that of New York and more than five times that of London.

To store water we use a combination of dams and ground water. Ground water is stored in aquifers—porous rocks in the ground. Sixty per cent of Australians depend on water in the ground for everything except drinking water, which is sometimes called potable water. Drinking water is of importance to everyone, but in urban areas we put it to a wide range of other uses. In 1991 each person in Sydney, on average, used 200 L of drinking water per day for washing, irrigation and other purposes.

To say we 'use' water is misleading. What we do in most cases is dirty it and then return it to the environment. The way we use water is of interest because by understanding its use we can decide how we may conserve it better. Data, again for Sydney, suggests that during the early 1990s our use of water in the garden, toilet and bathroom accounted for a little more than 70% of water use. The kitchen accounted for 10% and the laundry 15%. Swimming pools accounted for 2% of water use.

The importance of catchments

Both dams and ground water rely on recharge from catchments and water movement through the ground. Catchments are areas where rainfall and ground water movements are converted to stream flow. Catchments can be affected by land uses that change water flow and cause pollution. In some areas changes in farm practices have led to an increase in small dams, which alter the amount of water reaching streams and creeks. While this presents problems for managing water resources, it has provided a large number of small, semipermanent water habitats for animals and plants. Pollution can arise from sediment run-off due to clearing, nutrients (such as nitrogen and phosphorus) and pesticide residues.

Our demands for water can produce problems for ground water reservoirs when we extract water faster than it can be recharged or cause changes that increase the recharge rate. In the Namoi Valley in northern New South Wales, during the 1980s water was extracted from the ground faster than the natural recharge rate. Water is also extracted from the Great Artesian Basin faster than it can be replaced by the natural recharge rate.

When land clearing or the establishment of irrigation alters the surface, water may build up too quickly. Tree removal changes the rate at which water is removed from the soil. Irrigation artificially adds water to the soil. Together these processes can lead to a possible increase in waterlogging and salinity and an alteration in soil life. Alternatively, the draining of areas may lower the watertable and change the structure of plant communities as plants that rely on shallow water disappear from the area.

Changing water flow

Urbanisation and land clearing both alter water flow in the environment. The changes may be to the flow behaviour of the water or to the quality of the water. Factors such as storing water in dams, draining natural water stores and creating hard surfaces, all of which change infiltration and run-off, affect the flow behaviour of water systems. Two factors that affect the quality of water are the pollution caused by nutrient run-off and the nutrients in sewage added to water systems.

The effects of drainage systems and dams

Here we will look at the effects on the natural environment of two artificial structures: drainage systems and dams. Both alter the movement of water and both, as we shall see, alter the environments in which they occur as well as those outside their local areas.

DRAINAGE SYSTEMS

The amount of water we use in urban areas outside the house can have a significant effect on the environment. Water used for gardens and parks, together with water used for cleaning cars and yards, adds significant amounts of water to the environment. In some places this may represent as much as 30% extra water over that provided by rainfall. Some of the extra water is stored in the soil and some runs off into drainage systems, both natural and artificial.

By altering the surface of the ground for urban or agricultural purposes we also change the way rainwater is carried within a catchment. Hard surfaces (such as roads, roofs and concrete surfaces) do not allow absorption by the ground and the water runs off into drainage systems. When water runs over artificial surfaces it often flows more rapidly than it would over a natural surface. As a result, the erosion caused by the water cleans the surfaces, carrying chemicals and litter into the drainage systems. Because less water is absorbed, the volume of water running off an area is also increased. The water produced by rainfall rapidly accumulates in drains and watercourses and is referred to as stormwater.

Land clearing and the construction of channels can result in significant changes to water movement. Artificial channels with hard surfaces and steep sides behave in different ways from natural channels. (See Figure 3.5.4.) They increase the speed of the stormwater they carry and the rate at which the stormwater volume increases. Natural areas, swamps in particular, absorb water and release it slowly. Clearing of such areas removes habitats for animals and plants and causes the water to flow across the area and build up more rapidly. If drains cannot cope with the build-up, the area may flood. It is in the upper parts of the catchment where minor flooding is worst.

Sediment produced by land clearing or agriculture can also alter the flow of water courses. Sediment can build up in creeks, changing their water-carrying capacity.

In the upper parts of creeks and rivers, deep-flowing areas can be replaced by shallow, stagnant pools. In estuaries the build up of sediment may lead to changes in the flow of the tide and affect the flushing of the river by tidal changes. Mining of sediment along rivers also produces environmental changes. Alterations to the flow patterns may result in erosion further upstream and where the channel has been changed.

Water carried into drains and waterways produces environmental effects. In a creek or river, deposition and erosion are processes that occur at the same time. With stormwater, the increased speed of the water produces greater erosion where it comes into contact with soil. The erosion produced is worse because the water does not carry a lot of sediment and so there is nothing to be deposited where material is eroded.

DAMS

Dams have a significant effect on the environments where they are built. They alter the characteristics of water entering and leaving them. Dams also affect soil, the landscape, climate and the living things in the environment.

Dams are sinks: they are places where things accumulate. Into a dam run sediments and nutrients, which can alter the chemistry of the water. Temperature plays a role too. Dams can contain a huge volume of water and much of it is cold. The temperature has important consequences for life in the dam and where the water flows out of it. Because organisms have particular ranges of temperature within which they function, cold water can change the distribution of living things downstream.

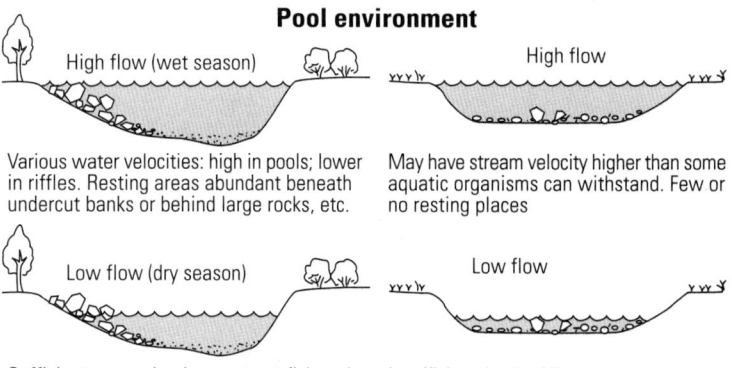

Figure 3.5.4 A comparison of natural and artificial channels.

Dams not only change the properties of water in the dam but also alter the flow of water in the catchment downstream. Dams have a large surface area and so they lose a great deal of water due to evaporation. This, together with water being held back, means that the water flowing downstream is much less than it may be. Seepage into the rocks of the floor and sides of the dam may also alter ground water levels near the dam.

Soil downstream of the dam may suffer because silt that once travelled to the area is now held behind the dam wall. Soil upstream of the dam may also be affected. Changes to watertables may lead to salinity changes. Flow changes may result in greater rates of deposition of sediment adjacent to the dam.

Changes to the landscape may be caused by sediment being withheld. As was explained earlier, water that does not contain sediment causes greater erosion than water from which the sediment can be deposited. On the coast, processes may increase erosion due to changes in the sediment being deposited at deltas and in estuaries. Changes in the gradient above the dam may also lead to landscape changes as the creeks and rivers alter their behaviour to suit the new conditions.

Around the dam the climate may be altered. Comment has already been made about increased rates of evaporation. This may lead to changes in the humidity and also to changes in the rate of precipitation, depending on the size of the dam. Water, being a great heat sink, may alter air temperature around the dam.

Living things are affected by flooding of habitats, blocking of fish migration, changes to water composition and temperature, and creation of new habitats for water-living organisms. In some areas a dam may provide a permanent source of water that allows a species to survive during periods of drought when it previously might have died out of the area. On the other hand, small floods downstream of a dam, which trigger breeding in some species of native fish, are reduced.

REVIEW ACTIVITIES

1 List the ways in which potable water is used in Australia.

2 Explain why the care of catchments is important for the quality of water we drink.

3 Describe how irrigation can alter the native vegetation of an area.

4 Contrast the characteristics of natural channels and artificial channels.

5 Describe how land clearing can alter water movement in the environment.

6 Summarise the effects of dams on river systems.

EXTENSION ACTIVITIES

7 Assess the affect that the roof of your house has on water movement in the area around your house.

8 As large cities continue to grow, local flooding can become an increasing problem in some areas. Analyse possible reasons for this and suggest ways of minimising such flooding.

SUMMARY

- Population, affluence and technology influence the degree of impact that people have on the environment.

- Concentration of population affects the natural environment.

- Since European settlement in 1788 the Australian landscape has been altered.

- Major causes of environmental change have been land clearing, grazing, introduction of weeds and feral animals and changes to fire frequencies.

- Logging and mining have changed parts of the Australian environment.

- Urbanisation and recreation have a very large affect on the natural environment.

- Other than Antarctica, Australia is the driest continent on Earth and Australians have the highest water storage capacity per person in the world.

- Water is stored in ground water aquifers and dams.

- Catchments are important for recharging water storage and can be affected by other land uses.

- Water use in urban areas can affect ground water and surface flow.

PRACTICAL EXERCISE
Dams and surface processes

In this exercise you will examine some effects of a dam on the natural environment. You will analyse data and assess the effects that the dam described has had on its environment.

Background
A dam was constructed on a river over a four-year period. Data about the sediment being carried in the river were collected before, during and after the construction of the dam at two sites on either side of the dam.

Table 3.5.2 Suspended load in the river upstream and downstream of the dam

Period	Load upstream (tonnes/year)	Load downstream (tonnes/year)
Pre-dam		
1974	1 710	1 790
During dam building		
1975	4 700	4 300
1976	7 300	8 770
1977	8 900	1 940
1978	5 100	1 600
Post-dam		
1979	13 200	480
1980	4 300	209
1981	11 600	2 110
1982	4 900	80

ACTIVITIES

Processing the data

1 Calculate the mean (average) for the sediment load during and after dam building for the locations upstream and downstream.

2 Calculate the percentage difference for each year between the sediment load upstream and downstream of the dam.

3 Graph the sediment load for each site against time.

Analysing the data

4 Describe what your graph shows for the sediment load downstream of the dam.

5 During the dam's construction the data for 1976 was unusual compared to other years during its construction. What possible reasons can you suggest for the high sediment load downstream of the dam?

6 After the dam's construction the sediment load measured upstream of the dam varies from year to year. Does the sediment load downstream show the same variation? Suggest a reason for any differences you observe.

7 Calculate, using the means, the average amount of sediment being trapped in the dam from 1979 to 1982. What effect will the sediment accumulation have on the volume of water in the dam?

8 Suggest why the sediment load upstream of the dam increased after the dam was built. If the nutrients entering the dam showed a similar change to the sediment load, what could you predict about the water plant growth around the edges of the dam?

9 Suggest how the dam would affect:
a fish populations downstream of the dam
b water evaporation along the whole length of the river
c salinity of river waters downstream of the dam.

Summary and conclusions

10 Summarise the effects of the dam on the river upstream and downstream of the dam wall. What do the data show and what can you infer from the data?

11 If dams affect the natural environment in adverse ways, why do we build them? What possible alternatives are there to dams as a source of water for human use?

PRACTICAL EXERCISE
Current human impacts in your area

One way to assess human impacts on your local environment during the last fifty years is to compare visual records of the local area dating from that time with the features of the environment that exist now. Photographs, paintings and maps are all useful for measuring the changes that have occurred.

ACTIVITIES

1 List the sources of information about your area fifty years ago. Consider local libraries (school and council), the local historical society and people who have lived in the area for a long time.

2 Consider what aspect of the local environment you will assess. Vegetation cover may be a good place to start. Another aspect you could assess is whether any major changes have been made to the water services in your local environment.

3 Collect information from the sources you have identified. For each piece of information record the following: its origin; the time it represents (fifty years ago or today); the information contained in the source; and the reliability of the source. Try to collect at least four pieces of information from the past and four from the present.

4 Compare each piece of information you have from fifty years ago. Does each tell the same story? What variations are there between the data you have collected? Repeat the comparison using the information you have from the present.

5 Summarise the differences you have identified between now and fifty years ago. Are there things that you did not find out about conditions in the past? Does this mean the information does not exist or was it unavailable to you?

6 Write a short summary of what you learnt about gathering and analysing data by doing this exercise.

3.6 Balancing development and conservation

OUTCOMES

At the end of this chapter you should be able to:

- identify one environmental issue that requires some government regulation or management, such as:
  - sustainable development
  - exploration
  - mining
  - environmental planning
  - air and water quality management
  - land use and rehabilitation

- identify an appropriate local environmental document that aims to address one of these issues (for example environmental impact study, catchment management plan)

- gather, process and analyse information from secondary sources to identify and discuss the scientific basis of the issues in the chosen local environmental document.

WHEN IS AN ENVIRONMENTAL CHANGE A PROBLEM?

The preceding chapter should have indicated to you that human changes to the environment produce changes to other parts of the environment. Some of those changes are unexpected and some of them not only affect other living things but harm the quality of people's lives. In this chapter we will examine why the environment needs to be managed and some of the ways by which governments at different levels try to regulate the way we affect the environment.

Change, as we have seen, is the driving force that shapes life. Long before humans began to change their environment living things appeared, changed and became extinct as they adapted, or failed to adapt, to environmental changes. Changes, both natural and those caused by people, occur today. So why should we be concerned when we see the effects of change in the natural environment?

An environmental situation is only a problem when someone sees it as a problem. In the 1930s few people regarded land clearing as a problem. While some thoughtful people saw what it did to the soils on their farms, other farmers continued to clear land to improve their productivity. Indeed, they were encouraged to do so by government subsidies—money was given to them to help them clear their land. To understand that something is a problem requires two prerequisites. First, we need to understand the environment to the extent that we know what is healthy and what is not. Secondly, we need to be able to monitor changes and understand the meaning of the changes we see.

ENVIRONMENTAL AND RESOURCE MANAGERS: DIFFERENT PRIORITIES?

The problem facing us is that we do not know enough about the natural environment: either the things that make it up or the processes that shape it. We really do not know how many living things are to be found in Australia. The smaller they are, the less we know about them. For the living things that have been named, we know a relatively small amount about their biology. We have less than 200 years of written information about the processes that shape Australian environments.

As a result of our lack of knowledge, those who are charged with managing the environment must be cautious. They have to understand and maintain a widespread and complex system. They need to acquire information on which to base their management and ensure that they conserve what is already here. Once extinct, species cannot be resurrected or exactly replaced. Such an approach must eventually bring the environmental manager into conflict with others who use the environment because other people have different priorities with regard to the environment.

Competing environmental uses

Environmental managers will not always have the same priorities as people who manage resources. A resource is something that we use. Resources include minerals, food, water, energy sources, timber and soil. Resource managers (such as farmers, foresters and miners) are responsible for providing our society with the materials we need. They must also try to provide the resources in a profitable way. A farmer or miner who uses water to produce the resource they need may be most concerned about having enough water to achieve their goal. The effect of water removal from natural areas, or its effect on another community's drinking water supply, will not necessarily be of immediate concern to the resource manager.

It is important, however, that resource managers and those who must share the environment's resources understand that changes they cause may have long-term effects on their wellbeing. An example of the importance of this is to be seen in the salinity problems faced by those who live along the Murray. Land clearing and water use practices have led to a situation where the jobs and wellbeing of people who live along the Murray have been put at risk.

Part of the problem with sharing and managing our environment is our affluence. We live in an affluent society because we have relatively easy access to resources. In one sense we find resources to be cheap. Two problems with easy access to resources are that we often do not appreciate their real value and, because of this, historically we have not worked to conserve the resources. Overuse of a resource is not sustainable. Sooner or later the resource will become scarce or unattainable. A belief that something is cheap leads us to use it without concern for its replacement and may lead to it being unavailable to our descendants.

ENVIRONMENTAL REGULATION

Education and legislation are both needed to ensure that resources are managed well. Education helps people to see the competing needs of resources and to understand the value of things they would otherwise take for granted. Legislation is needed to ensure that those who disregard the needs of the wider community and future generations do not abuse the available resources.

The laws or legislation that regulate our use of the environment aim to promote the following:
- *Sustainable use.* This means that resources are not used to the extent that they are exhausted.
- *Efficient resource allocation.* This means that resources are not wasted and that the best use is made of the limited resources we have.
- *Equity.* This idea aims to ensure that minority groups and those with few resources have the ability to make their views heard even if other interests are more powerful.

- *Environmental quality.* This principle implies that nature has a right to exist in a state that does not depend on our sense of its usefulness. As we understand that the quality of the environment affects the quality of our life, the need to preserve and maintain the environment for itself is becoming more important. It is interesting to note how recent laws have dealt with this aspect of the environment.
- *Public interest.* This principle recognises that society as a whole has special needs with regard to the environment and that special treatment should be afforded the environment in all our interests.

These five aims are known as the principles of ecologically sustainable development. (See Figure 3.6.1.)

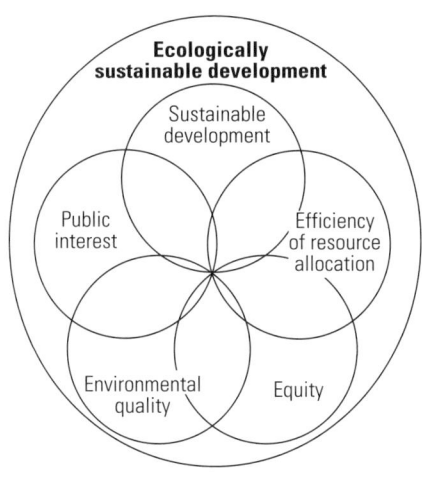

Figure 3.6.1 The principles of ecologically sustainable development.

Direct and indirect regulation

The law can act to protect the environment either directly or indirectly.

Examples of direct intervention by government include laws that prohibit certain actions, such as polluting. The need to obtain permits, such as those for mining, are associated with particular requirements that protect the environment. Mining, for example, requires agreement about conservation and rehabilitation before mining proceeds. At the local government and state government level, planning controls (such as building permits and developers being required to produce environmental impact statements) regulate people's interaction with the environment. With direct regulation there are usually sanctions, such as fines or imprisonment, with which the regulations can be enforced.

Indirect regulation involves the use of regulations that indirectly affect the economics of a development. The Federal Government can regulate the mining of minerals, such as uranium, with its international trade powers. The laws do not necessarily prevent a company mining the uranium but they stop it being sold overseas. Without a market in Australia, such legislation effectively makes the mining unprofitable. Taxation can be used in a similar way.

The imposition of particular taxes can make the difference between a resource being harvested at a profit or a loss. Subsidies can work in the opposite way. Subsidies for recycling or revegetation make it worthwhile, in an economic sense, for people to act in a way that benefits the environment.

International, federal, state and local regulation

Besides looking at legislation in terms of its direct or indirect effects we can also look at it in terms of its scale. There are four scales to be aware of: international, national, state and local. All levels of government enact regulations that affect how we treat the environment.

INTERNATIONAL REGULATION

The Federal Government, on behalf of us all, signs international conventions that Australia must honour. Over the last half century, people have become aware that environmental issues are global in their scope. Pollution of the oceans, for example, affects many countries besides the one that is the source of the pollution. Federal laws are generally indirect. When the Australian Constitution was written one hundred years ago, the environment was not an issue. As a result, the Constitution does not give the Federal Government powers to directly legislate on environmental issues in preference to the states.

FEDERAL REGULATION

Federal laws that have been used to regulate environmental issues include powers relating to trade and commerce, external affairs, corporations and finance and taxation.

On 16 July 2000 the Federal Government enacted new legislation called the *Environmental Protection and Biodiversity Conservation Act*. This act replaced a number of Commonwealth statues, namely:
- *Environmental Protection (Impact of Proposals) Act 1974*
- *Endangered Species Protection Act 1992*
- *National Parks and Wildlife Conservation Act 1975*
- *World Heritage Conservation Act 1983*
- *Whale Protection Act 1980*.

Some environmental issues are likely to produce a significant impact on matters of national significance. The new act aims to ensure that such issues are assessed in a rigorous way and that research and conservation of biodiversity occurs. The matters covered by the act that trigger the need for government assessment are related to:
- World Heritage properties
- Ramsar wetlands
- nationally threatened species and ecological communities
- migratory species
- Commonwealth marine areas
- nuclear actions, including uranium mining.

Assessment by the Environment Minister may occur in a number of ways. These include:
- assessment of preliminary documents
- through a public environment report (PER)
- through the consideration of an environmental impact statement (EIS)
- by public inquiry
- an accreditation process on a case-by-case basis that is delegated to a state agency.

In coming to a decision, the minister must consider a number of issues. These include not only environmental impacts but social and economic matters. The minister must also take into account the principles of ecologically sustainable development, environmental impact assessment reports and relevant comments from other ministers.

STATE REGULATION

The State Government plays an important role in regulating how we interact with the environment. In New South Wales the principle agency responsible for environmental regulation is the Environmental Protection Authority (EPA).

With the *Protection of the Environmental Operations Act 1997* (POEO Act) the State Government was able to set out explicit protection of the environment policies (PEPs). PEPs are instruments for setting environmental standards, goals and guidelines. The PEPs provide both a framework for government decisions affecting the environment and a means of adopting national environment protection measures set by the National Environment Protection Council. Before a PEP can be made, the POEO Act requires public participation and an analysis of the economic and social impact of the PEP.

In particular, the POEO Act allows the State Government to deal with pollution. The act replaced a number of prior pieces of legislation that dealt with water, air, noise and other forms of pollution.

In New South Wales the government also regulates how we deal with the environment through such measures as environmental impact assessments, the protection of natural and cultural heritage and the protection of specific places, including marine areas.

LOCAL GOVERNMENT REGULATION

Local government has responsibilities to the state Environmental Protection Authority (EPA) and to its residents. Local governments must consider any PEPs and enforce them through the use of council regulations, licences and development applications (DAs). Councils are also required to prepare local environment plans (LEPs).

A DA is required in order to obtain a development consent. Such consents are approvals for development and they apply to subdivisions, building applications and developments. A development may include short-term uses of land, such as concerts or parties. A development consent is needed whenever it affects an environmental planning instrument (EPI). An EPI may be a local environment plan (LEP), a regional environmental plan (REP) or a state environmental planning policy (SEPP). Relevant EPIs affecting your area are available for inspection at your local council chambers.

In order to carry out development work a person must lodge a DA with their local council. Depending on how their activities may affect the environment they may be required to obtain a licence from the EPA or the council. In considering DAs a council must take into account any threatened species or community and associated habitats. If such DAs are to be approved, the agreement of the Director-General of the NSW National Parks and Wildlife Service is also required.

In deciding significant development proposals, councils may require an environmental impact assessment to be undertaken. Such assessments aim to inform both the public and government decision makers about the development. The document that reports an environmental impact assessment is called an environmental impact statement (EIS). An EIS is required whenever a development is classified as a designated development (a 'high-impact' development) or the development is thought likely to significantly affect the environment.

An EIS must include:
- an analysis of possible alternatives to the development plan
- a detailed analysis of the likely environmental impact
- a justification of the development, with regard to the principles of ecologically sustainable development.

Councils are also required to publish management plans and an annual State of the Environment Report for their area. In the report the council must assess the environmental impact caused by council activities and detail the steps being taken to minimise the impact of the activities.

Councils control building and land use. They must consider when deciding to grant an approval such matters as:
- the principles of ecologically sustainable development
- protection of the environment
- protection of public health, convenience and safety
- items of cultural and heritage significance that may be affected.

Councils also play a role in pollution control. They do so through development decisions, such as zoning.

If it needs to, a council may approve a development but impose what are called consent agreements, which outline the conditions under which the development can proceed. By contributing to regional waste plans, issuing notices ordering pollution control measures and being involved in the removal of noxious weeds, councils play a significant role in environmental protection.

Figure 3.6.2 Who is responsible for developing large projects, such as dams?

REVIEW ACTIVITIES

1 Discuss the reasons for different groups of people seeing environmental issues in different ways.

2 Explain why legislation is needed to protect the environment.

3 List the principles of ecologically sustainable development.

4 Contrast direct and indirect methods of environmental regulation.

5 How does the Federal Government regulate environmental issues?

6 List the ways in which councils regulate or control environmental impact.

EXTENSION ACTIVITIES

7 What does Ramsar refer to? What are Australia's international obligations with regard to wetlands?

8 Discuss the importance of enforcing environmental legislation. What are the consequences of a regulatory authority lacking the resources to monitor environmental impacts?

SUMMARY

- Environmental problems are sometimes hard to recognise and often have complex causes.

- Competing demands by resource users and environmental managers can lead to conflict.

- Environmental regulation is needed to ensure that some people do not disregard the needs of others in the community and to preserve the environment for future generations.

- Ecologically sustainable development involves sustainable use, efficient resource allocation, equity, environmental quality and public interest.

- The law can protect the environment using both direct and indirect regulation.

- Different levels of government are responsible for different aspects of environmental regulation.

PRACTICAL EXERCISE
Regulation of environmental impact

This exercise will involve you in the use of secondary sources to determine how human impact on the environment is being regulated in your area. At the end of this exercise you should have identified and examined a local environmental document that is being used to regulate or manage an environmental issue in your area.

A good place to start in identifying issues in your area that are being regulated is by examining the State of the Environment Report produced by your local council. Councils are required to produce such reports regularly and identify changes that are causing environmental impact. Does such a report exist in your school library or local library? If you have access to the Internet, can you locate a copy of the report on the council's website? Use the report to identify the council's response to issues highlighted in the report.

Councils use environmental planning instruments (EPIs) to determine whether a planning permission is required. EPIs include local environment plans (LEPs), regional environmental plans (REPs) and state environmental planning policies (SEPPs). Relevant EPIs affecting your area are available for inspection at your local council chambers and are sometimes identified in State of the Environment Reports.

An environmental impact statement (EIS) is required whenever a development is considered likely to significantly affect the environment. Because such statements are intended to inform the public as well as managers they are usually on display at the council.

ACTIVITIES

1
Use the resources available to you to locate and examine a local government document in one of the categories outlined here.

2
Write a summary of the document. Include its title, its subject, what type of document it is and the measures described in the document that seek to regulate environmental impact.

3
Review the ecologically sustainable principles in the text. At the end of your summary, discuss the ways in which the document you have examined supports these principles.

3.7 Biodiversity

OUTCOMES

At the end of this chapter you should be able to:

- use examples to describe and explain what is meant by biodiversity
- outline the potential effects of a loss of biodiversity in destabilised ecosystems
- discuss the importance of refugia in conserving biodiversity
- gather information from secondary sources, including the Register of the National Estate (ERIN) or other databases to identify significant places of environmental importance in the local area.

definition

biodiversity
diversity of life forms, genes and biological systems.

Biodiversity refers to the diversity of life forms—not only plants and animals but micro-organisms too. The term also applies to the variety of genes living things contain and to the ecosystems in which life exists. The term 'biodiversity' is a contraction of the term 'biological diversity'. It is a relatively new term that you are unlikely to find in dictionaries published more than ten years ago. As our understanding of the importance of other living things has increased, so has the use of this term.

TYPES OF BIODIVERSITY

It is common to find biodiversity being described at one of three different but interrelated levels, species diversity, genetic diversity and ecosystem diversity.

Species diversity

Species diversity refers to the variety of living things in an area. It is perhaps easier to measure than the other types of diversity and it can be described in a number of ways. We can describe species biodiversity in terms of the number of species present. This is a measure of species richness. We can also describe species diversity in terms of species abundance, which is the number of a particular species present. This can provide important indicators as to the health of a species. Lastly, we can describe species diversity by using classification and looking at the occurrence of closely related species.

A problem in assessing species diversity is that we do not know enough about the species that exist. We have named relatively few and know, in detail, the biology of even fewer. It tells us something about our priorities that the organisms we know most about are often the pests that affect our ability to feed ourselves. Many of the organisms that benefit us are much less understood.

The number of species that have been named around the world is a little more than 1.5 million. The number of species that actually exist has been estimated at around 5 million, although some researchers have estimated the number as high as 100 million. It is thought that in Australia there are probably about 1 million species of living things but only about 15% of them have been named. Organisms such as invertebrates, micro-organisms and lower plants (mosses, liverworts and algae, for example) are the least understood, but it is from these organisms that we are beginning to obtain chemicals for use as drugs. Such organisms also play important roles in maintaining soil fertility and the health of other ecosystems.

Genetic diversity

Genetic diversity refers to the variety of genetic information contained within and between populations. Genes are the information that produces the characteristics of living things. They are part of the material called DNA (deoxyribonucleic acid), which makes up the chromosomes found in cells. In multicelled animals, protozoans and plants the chromosomes are found in the nucleus of the cell. In eukaryotic cells, those with sub-parts built of membranes, genetic information is also found in the mitochondria and chloroplasts of the cells. This occurrence has been used as evidence to support the idea that eukaryotic cells arose from symbiosis between primitive cells.

Two processes—mutation and genetic mixing—cause the diversity of genes in a population of living things. Mutation produces new types of a gene. Some mutations are beneficial and some are harmful. Others are either beneficial or harmful but may confer a benefit under changed environmental conditions. Genes vital to survival, such as those concerned with respiration or photosynthesis, show little variation between different organisms but other, specialised genes show a higher degree of variation. Mixing, or recombination, occurs when the genes of two organisms are combined during sexual reproduction.

Sexual reproduction provides an advantage to those species of organism that use it. Some organisms do not reproduce using sexual reproduction but many invest large amounts of energy in the process because it provides a survival advantage. Earlier we learnt about the process of natural selection. It is the variation of genes in a population on which natural selection works. If a population of living things contains variation, some of the variations may allow individuals to reproduce and survive in changed environmental conditions. Populations in which little variation occurs, such as some plant crops, are at risk because a change in the environment that affects one individual will affect all the other members of the population in a similar way.

When a population becomes small the amount of variation in the population decreases. This means that the population may not be able to adjust to change and is at risk of extinction. It sometimes means that the frequency of bad genes is relatively high. They make up a greater proportion of the total gene number and so breeding produces more offspring with the characteristics that are disadvantageous.

A guide to the diversity in the genetics of a species is the number of subspecies present. If a biologist can distinguish a number of distinct types within a species it indicates the presence of a number of different genes, coding for a particular characteristic.

Ecosystem diversity

Ecosystem diversity describes the broad differences between different ecosystem types. An ecosystem is defined not only by the living things it contains but also by the physical conditions and the processes that occur within it. In naming ecosystems it is usual to use a combination of the habitat type and the climatic conditions in which the habitat occurs. A temperate dry forest and temperate grassland, for example, are characterised by different habitats in similar climatic conditions.

The boundaries of ecosystems are sometimes hard to define. This is because ecosystems are dynamic. As the conditions around the ecosystem change, so do the boundaries. Organisms within the ecosystem may also alter the environment and produce successions that gradually alter the habitat. In addition to natural environmental change, ecosystems are experiencing changes brought about by human impacts.

The principal impacts on ecosystems by people include:
- population growth and changes in patterns of settlement
- economic factors due to decisions by those who derive income from natural resources
- lack of awareness by those who live in or adjacent to an ecosystem.

These impacts produce a range of pressures that can be identified in different environments. They include:
- habitat loss
- habitat fragmentation
- competition from introduced species
- predation by introduced species
- introduction of diseases
- lack of sustainable practices
- illegal collection of organisms
- pollution
- climate change of varying scales
- lack of education.

The importance of ecosystem diversity is that ecosystems interact with one another and provide a buffer to change. Mangrove ecosystems illustrate the benefits of such buffers. Natural systems are generally quite stable when subject to minor changes in the environment. Major changes, however, can have an adverse, possibly detrimental effect.

REVIEW ACTIVITIES

1 Describe what is meant by the term 'biodiversity'.

2 Identify the types of biodiversity that exist. Name an example of each type.

3 Discuss which type of biodiversity is hardest to measure.

EXTENSION ACTIVITIES

4 Identify an endangered species in your local area and the pressures that have led to it being at risk.

5 Describe, using examples, how organisms alter their environment.

THE IMPORTANCE OF BIODIVERSITY

The importance and value of biodiversity to humans is due to a number factors. We depend on biodiversity for important resources. Humans derive their food, many medicines and industrial products from domesticated and undomesticated biodiversity. Wheat, for example, is bred to be resistant to fungal disease using genetics from wild grasses.

There are also aesthetic and ethical reasons for protecting biodiversity. Rich, complex environments are great places to experience. We are also coming to understand that life does not exist only for our use. Other living things have a right to exist and we are obliged to ensure that they do not disappear through our actions.

Biodiversity is essential for the maintenance of human life on earth. Forests provide oxygen and slope stability, control water flow and modify climate. Biodiversity also provides a model for an ecologically sustainable society. If we can learn to use the techniques of waste minimisation that have evolved in the natural world over billions of years we stand a better chance of living sustainably on Earth.

Biological diversity is the result of billions of years of evolution. The diversity of organisms is the result of populations, communities and ecosystems adapting to changes in the environment. We know that over time new types, or species, of organisms appear and that they also disappear. The disappearance of a species is known as extinction. Once extinct, populations cannot be resurrected.

Humans are part of the biological world. Like other organisms, the environment we live in has shaped our characteristics and our wellbeing depends on the living things around us. If we fail to maintain the natural environment we risk our health.

THE IMPORTANCE OF REFUGIA

A refuge is a region where certain types or groups of organisms are able to survive during periods when their original distribution is reduced due to environmental change. The term 'refugia' refers to a number of refuges. Refugia can be produced in a number of ways.

The northward movement of Australia and the Pleistocene ice age resulted in the increasing aridity of the environment. As organisms adapted to the new conditions, rainforests and temperate vegetation communities became restricted to areas that suited them. In arid areas it is possible to find plants that were once widespread living in sheltered areas near permanent water. On the western edge of the Great Dividing Range some rainforest remnants exist in dry forests because they are situated in sheltered gullies or valleys.

In the Snowy Mountains animals and plants exist that are adapted to the very cold conditions that occur there. During the last ice age such organisms evolved and had a larger distribution than they do now. As conditions have changed they have become restricted to increasingly smaller areas.

Refugia can also be places where organisms retreat to during short-term crises, such as droughts and floods. Once the crisis is over the organisms can move back to the areas they previously inhabited.

It can be seen, then, that refugia are important because some contain unique communities and species that no longer exist in the wider area, or anywhere else. The preservation of such unique places protects our biodiversity and allows us to see what parts of Australia were like in the past.

Some refugia allow native organisms to cope with the pressures of human activity. Clearing is an example of an activity that has produced refugia. In remnant vegetation areas, animals survive that can provide a source of organisms while adjacent areas are being revegetated or rehabilitated.

REVIEW ACTIVITIES

1 Outline the possible effects of allowing biodiversity to decline in an area.

2 Explain what a refuge is.

3 Describe two reasons why it is important to maintain refugia.

EXTENSION ACTIVITIES

4 Use the secondary resources available to you to identify and describe a biosphere reserve. What is the aim of the reserve?

5 Should humans be allowed access to areas that are refugia? Discuss this question using evidence from secondary sources. In your answer try to identify at least two areas of New South Wales where human activity occurs close to a refuge.

SUMMARY

- Biodiversity refers to diversity in life forms.
- Biodiversity can be described in terms of species diversity, genetic diversity and ecosystem diversity.
- Major impacts by people on the environment are due to population growth and changing patterns of settlement, economic decisions and lack of awareness.
- Refugia play an important role in protecting organisms threatened by changing environments.

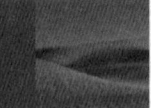

PRACTICAL EXERCISE
A field study report

During your study of the local environment you will undertake at least one field study. The purpose of this exercise is to help you organise the information you gather into a report of your study. While the outline below will help you organise the information you have gathered, it will be to your advantage to consider the issues discussed below before you conduct your fieldwork.

It is suggested here that your report should include seven important parts. Your teacher may ask you to present the information in a different way from that outlined here, but the information and evidence of analysis discussed here will be the same for a wide variety of areas that may be visited.

Title and introduction
At the beginning of your report you should outline the purpose of the field study. Was it intended for you to study the relationship between geology and landscape? Was the purpose to study the relationship between geology, landscape and vegetation or was it to investigate the impact of humans on the local environment? In addition to stating the purpose of your study you may record the date on which the study took place and the weather conditions you worked under. Such conditions affect what you see and are able to study. Don't forget to acknowledge the people you work with. Good field studies are usually the result of a team effort.

The study area
A detailed description of the area in which the study occurs is an essential part of the report. Ideally your description should include a map. The map may be a simple sketch map but it should record the locations you visited. Don't cram too much information into your map. Other maps or diagrams may be used later to present data or your interpretations.

Background information
Ideally you will do some preliminary research before you visit the area. Resources you could study include topographic maps, geology maps, mine reports, soil landscape maps, information about flora and fauna, historical information and computer databases dealing with biodiversity. Such research is not a substitute for what you record in the field. Rather, it prepares you for what you may see and helps you identify issues that may require your attention.

A summary of the data
During your fieldwork you may study a variety of things: soils, rocks, landscape, plants (including weeds) and the effects of human impacts on the environment. How you summarise this information will make all the difference to the quality of your report. Consider the use of tables, diagrams, maps, photographs and drawings. Be careful to distinguish between what you are recording and your analysis of the data.

Analysis of data
In working out what your data mean, remember the purpose of your study. Focus on the information that is relevant to the purpose of the fieldwork. Your analysis may take the form of written text or it may make use of maps and diagrams. Be careful to consider cause and effect relationships in the data you recorded.

Human impact in the area
The effect of human actions on the natural environment is an important part of this course. Whether you make a discussion of this issue a separate section or part of the analysis it should be in your report. Humans affect so many things in the environment, including soils, vegetation, landscape and water. This is a part of your report that should not be lacking in detail.

A bibliography
In preparing your report, if you use information from texts, maps or other reports you need to acknowledge them in this section.

Use these headings to organise your report. When you have finished drafting it, go back and check it. Make sure it fulfils the purpose of the study and that your writing makes sense.

RESOURCES

Adapting to human impacts

Environment Australia: Environmental Education Network
www.environment.gov.au/education/aeen/

Environment Australia: Plants and Animals
www.environment.gov.au/bg/plants/plants.htm

Oceans Alive *www.abc.net.au/oceans/alive.htm*

Australian environments

Australian Physical Environments: Evolution, Status and Management
www.arts.monash.edu.au/subjects/ges/ges1020/LECT20A.HTM

Erosion *www.netc.net.au/enviro/fguide/soil1.html*

National Landcare Program
www.landcare.gov.au/landcare/pub/guide/keyindex.html

The Pedosphere and its Dynamics *www.pedosphere.com/*

Balancing human needs and conservation

Australian Rangelands
www.ea.gov.au/land/bushcare/publications/rangelands/

Best Practice Environmental Management
www.environment.gov.au/ssd/

Environment Australia Online *www.ea.gov.au*

North Coast Region Total Catchment Management Home Page
www.scu.edu.au/sponsored/tcm/index.html

NSW Environment Protection Authority
www.epa.nsw.gov.au/index.asp

Ramsar Convention on Wetlands *www.ramsar.org/*

Remnants: Birds as Indicator Species
www.birdsaustralia.com.au/remnants/

UN Framework Convention on Climate Change
www.unfccc.de/resource/beginner.html

Biodiversity

Australian Biodiversity *www.ea.gov.au/biodiversity/*

Biodiversity and its Value
www.ea.gov.au/biodiversity/publications/series/paper1/index.html

Community Biodiversity Network
www.cbn.org.au/links/NBC_Statement.html

Conference on Biological Diversity
www.biodiversity.environment.gov.au/biocon/natrep/context.htm

Conservation in the Australian Alps
www.environment.gov.au/bg/protecte/alps/nature.htm

CSIRO: Biodiversity *www.csiro.au/page.asp?type=faq&id=BiodiversityAndEndangeredSpecies*

Refugia for Biological Diversity in Arid and Semi-arid Australia
www.ea.gov.au/biodiversity/publications/series/paper4/bio15.html

The Centre for Biodiversity and Bioresources
www.bio.mq.edu.au/kcbb/

Threatened Species and Ecological Communities
www.biodiversity.environment.gov.au/plants/threaten/

Built structures and natural processes

Background to Damming the Fitzroy River
www.sinclair.org.au/fitzroy/background.html

Ecological Sustainable Development: A Concept Map for Sustainable Development *www.origen.com.au/esd.htm*

Freshwater Ecology *enterprise.canberra.edu.au/WWW/www-crcfe.nsf*

International Rivers Network: Rivers and Dams
www.irn.org/basics/basic.shtml

Natural Resources and Climate *www.agric.nsw.gov.au/reader/16*

Stormwater Pollution from Building Sites (EPA)
www.epa.nsw.gov.au/envirom/stormwater.htm

Urban Stormwater Program (EPA)
www.epa.nsw.gov.au/stormwater/

Water Facts (Publications)
www.wrc.wa.gov.au/public/waterfacts/index.html

Effects of human impacts

Caring for our Natural Resources
www.dlwc.nsw.gov.au/care/index.html

Extinction of Australia's Fauna
online.anu.edu.au/Forestry/silvinative/kk/1intro.html

Landcare Australia *www.landcareaustralia.com.au/*

National Conservation Council of NSW *www.nccnsw.org.au/*

Threatened Species and Ecological Communities
www.biodiversity.environment.gov.au/plants/threaten/information/overview/overview.htm

Environmental law

The *Environment Protection and Biodiversity Conservation Act 1999* (Cwlth): Overview *www.environment.gov.au/epbc*

Environmental Defenders Office NSW
www.edo.org.au/edonsw/edonsw.htm

NSW Environmental Protection Authority *www.epa.nsw.gov.au/*

Farrier, D. *The Environmental Law Handbook: Planning and land use in NSW*. 2nd edition, Redfern Legal Centre Publishing, Sydney 1998

Geology and landscape

Geomorphology from Space
daac.gsfc.nasa.gov/DAAC_DOCS/geomorphology/GEO_HOME_PAGE.html

Humans and the environment

Caring for our Natural Resources
www.dlwc.nsw.gov.au/care/index.html

Catchment Communities in Victoria
www.nre.vic.gov.au/catchmnt/conditn/soils/index.htm

Environment Australia: Plants and Animals
www.environment.gov.au/bg/plants/plants.htm

NSW Environment Protection Authority *www.epa.nsw.gov.au*

State Forests of New South Wales *www.forest.nsw.gov.au/*

Other influences on ecosystems

Australian State of the Environment Report 1996: Human Settlements *www.environment.gov.au/soe/soe96/key-findings/key-human.html*

CSIRO's Urban Water Program *www.dbce.csiro.au/urbanwater/*

Threatened Species and Ecological Communities
www.biodiversity.environment.gov.au/plants/threaten/information/overview/overview.htm

Rock types

Geoscience Australia *www.agso.gov.au/*

Igneous Rocks Homepage
seis.natsci.csulb.edu/basicgeo/IGNEOUS_TOUR.html

Igneous, Sedimentary and Metamorphic Rock Basics from University of British Columbia
www.science.ubc.ca/~geol202/petrology/rock.html

Lynn Fichter's Metamorphic Rocks
csmres.jmu.edu/geollab/Fichter/MetaRx/index.html

Lynn Fichter's Sedimentary Rocks
csmres.jmu.edu/geollab/Fichter/SedRx/index.html

Petrography Homepage
sorrel.humbolt.edu/~jdl1/petrography.page.htm

Scheibner, E. *The Geological Evolution of Australia: A Brief Review*. Geological Survey of New South Wales, Sydney, 1999

Volcano World *volcano.und.nodak.edu/vw.html/*

Soils

Australian Soil Classification *www.cbr.clw.csiro.au/aclep/asc.htm*

The Australian Soil Environment
cathar.tesag.jcu.edu.au/~JLULY/Soils.html

NSW Department of Land and Water Conservation
www.dlwc.nsw.gov.au/

Soil Erosion *www.epa.nsw.gov.au/soe/97/ch2/7.htm*

Soil Erosion in Agricultural Systems
www.msu.edu/user/dunnjef1/rd491/soile.htm

Soil Sodicity *www.science.org.au/nova/035/035key.htm*

Systems

Basic Concepts of the Systems Approach
pespmc1.vub.ac.be/SYSAPPR.html

Other useful resources

Australian State of the Environment Committee. Australian State of the Environment Report 2001, CSIRO Publishing, Melbourne, 2001

Bradstock, R. A., et al (eds). *Conserving Biodiversity: Threats and Solutions*. NSW National Parks & Wildlife Service, Sydney, 1995

Branagan, D.F. & Packham, G.H. *Field Geology of New South Wales*. 3rd edn. NSW Department of Mineral Resources, Sydney, 2000

Buchanan, R. Bush Regeneration: *Recovering Australian Landscapes*. TAFE Student Learning Publications, Sydney, 1989

Clark, I.F. & Cook, B.F. *Perspectives of the Earth*. Australian Academy of Science, Canberra, 1983

Department of the Environment, Sport & Territories (DEST), Biodiversity Unit, 1994–96 Biodiversity Series, Papers 1–8. DEST, Canberra

Goudie, A. *The Human Impact on the Natural Environment*. 5th edn. Blackwell Publishers, Oxford, 2000

Recher, H.F., Lunney, D. & Dunn, I (eds). *A Natural Legacy: Ecology in Australia*. 2nd edn, Pergamon Press, Sydney, 1986

White, M. E. *Listen—Our Land is Crying: Australia's Environment; Problems and Solutions*. Reed Books, Sydney, 1999

Young, A. *Environmental Change in Australia Since 1788*. 2nd edn. Oxford University Press, Melbourne, 2000

Water issues

4

The abundance of water on Earth and the way water has shaped the Earth's surface makes our planet stand out from other parts of the solar system. This section examines four important aspects of water's relationship to the environment. It begins by examining where water is found and how it moves between different subsystems of the Earth. The importance of water to the Australian environment is studied and the role of dissolved gases, salts and pollutants described. The third aspect covered in this section is the role water plays in the breakdown of rocks. This leads to the last aspect, which is the water resources of Australia, past and present, and some of the problems facing people in their attempts to conserve our water resources.

CONTENTS

Chapter 4.1	Water movement on the Earth	206
Chapter 4.2	Water and Australian environments	218
Chapter 4.3	Water and weathering	226
Chapter 4.4	Water resources: Past and present	232
Resources		245

4.1 Water movement on the Earth

OUTCOMES

At the end of this chapter you should be able to:

- recall some impacts of natural events (including cyclones, volcanic eruptions and earthquakes) on the atmosphere, hydrosphere, lithosphere and/or biosphere

- recall the distribution of the atmosphere, hydrosphere, lithosphere and biosphere on planet Earth

- outline an estimate of Earth's total water budget and the percentage available for terrestrial organisms

- identify factors (including geographic position, climate and topography) that determine the present distribution of water on the planet

- identify the interrelationship between sea levels, ocean currents, global temperatures and ice deposits.

EARTH SYSTEMS AND THEIR DISTRIBUTION

Relative to the size of our planet, humans and other life inhabit a relatively thin layer of the Earth. From space, the atmosphere appears as a thin blue layer. The densest part of the atmosphere is its lowest 12 km. The highest mountains are 8.5 km above sea level. The average depth of the oceans is approximately 4 km below sea level and the average height of the land less than 1 km above sea level.

Life inhabits four major subsystems: the atmosphere, lithosphere, hydrosphere and biosphere. Each subsystem affects, and is affected by, the other subsystems as it receives and transfers matter and energy.

The atmosphere

The **atmosphere** is the outermost layer of the four subsystems. The extreme outer limit of the atmosphere is 30 000–40 000 km above the surface. Most of the gas exists in a layer approximately 120 km thick and contains four recognisable layers. (See Figure 4.1.1.) We inhabit the **troposphere**, a layer approximately 12 km thick. The motion and behaviour of this layer directly affects life through such climatic events as wind and storms. Above the troposphere is the **stratosphere**. This layer extends up to 50 km above sea level. It contains the ozone layer at approximately 25 km, which restricts the amount of ultraviolet light reaching the Earth's surface. Within the stratosphere, the temperature increases from about –58°C to –3°C.

Above the stratosphere is the **mesosphere**. The boundary between these two layers is marked by a sudden decrease in temperature. The temperature of the mesosphere falls to about –90°C at the boundary with the **thermosphere**, which is the outermost layer. The thermosphere extends to approximately 120 km above sea level. Within this layer cosmic rays and ultraviolet light interact with gas molecules to produce the polar lights, such as the Aurora Australis. Two other layers are recognised by Earth

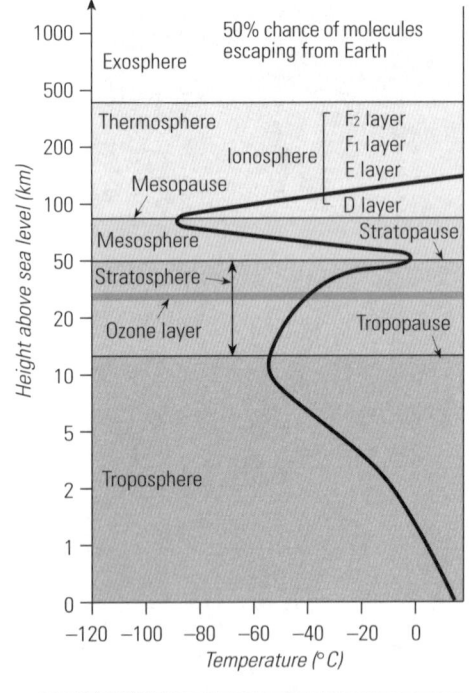

Figure 4.1.1 The layers of the atmosphere. Most of the gases making up the atmosphere are located in the lowest 120 km of the atmosphere.

scientists: the exosphere and the magnetosphere. It is worth noting here that although the magnetosphere is thousands of kilometres from the Earth's surface, changes within it affect weather at the surface.

The lithosphere

The **lithosphere** is the outer part of the solid Earth. It is composed of the crust, both continental and oceanic, together with the cold outermost mantle. Beneath the lithosphere is the **asthenosphere**, a **plastic** part of the mantle that flows when force is applied to it and allows the lithosphere to move.

The hydrosphere

The hydrosphere is the system of the Earth's water. An obvious part of the hydrosphere are the oceans, which cover 71% of the Earth's surface. The water in the atmosphere and in living things is also part of the hydrosphere, as is water in the ground and in the subducting crust.

The biosphere

Life is found where water exists. The **biosphere**—the zone of life—extends from the lower part of the troposphere down into the lithosphere. Bacteria known as entremophiles have been found in rocks kilometres below the surface of the Earth and we know of communities around submarine vents in the ocean that exist kilometres below the surface.

NATURAL PHENOMENA AND THE EARTH'S SYSTEMS

Within the **ecosphere**, natural events occur that affect a number of subsystems. The gas and ash emitted from an erupting volcano not only affect the lithosphere and biosphere, but the atmosphere too. Gases, such as sulfur dioxide, and fine ash from the eruption act as barriers to light radiation and can reduce the temperature at the surface of the Earth, while carbon dioxide and water gas can increase the atmospheric temperature. Such temperature changes can be global. When Mount Pinatubo erupted during the 1990s it caused a temperature fall of several degrees thousands of kilometres from the volcano.

Cyclones are atmospheric events but they affect the hydrosphere and biosphere. The wind and wave action they cause damages reefs and erodes coastlines. Winds also damage trees and other vegetation. The heavy rains can cause landslides on steep slopes and local flooding. The winds of the cyclone can also create waves that affect coastlines a long way from the location of the cyclone.

---

atmosphere
the layer of gases that surround the solid Earth

troposphere
the layer of the atmosphere closest to the surface of the Earth

stratosphere
the layer of the atmosphere above the troposphere

mesosphere
the part of the atmosphere where gas molecules have a 50% chance of escaping into space

thermosphere
the layer of the atmosphere above the mesophere

lithosphere
the outer rigid part of the Earth composed of the crust and cold outermost mantle

asthenosphere
the plastic layer beneath the lithosphere that allows the plates to move

plastic
changes shape when a force is applied to it

biosphere
the system consisting of the Earth's living things

ecosphere
the zone in which life exists

definitions

REVIEW ACTIVITIES

1
List the subsystems that make up the ecosphere.

2
Describe the features of the atmosphere.

3
Classify the layers of the Earth from the thermosphere to the asthenosphere according to the subsystems they contain.

4
List the effects of a volcano on the subsystems of the Earth.

EXTENSION ACTIVITIES

5
Discuss the effects on the subsystems of the Earth of a natural phenomenon not discussed above.

6
Analyse the effects of a recent natural phenomenon using information from media sources. Evaluate the science in the reports.

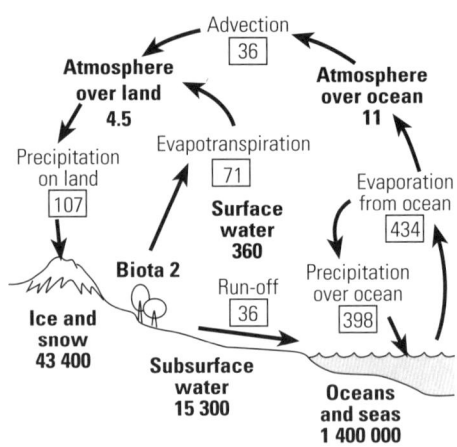

Figure 4.1.2 The global water cycle. The numbers set in bold are amounts in units of 10^{15} kg. The numbers in the boxes represent fluxes, or rates of exchange. These values are in units of 10^{15} kg per year.

THE EARTH'S WATER BUDGET

It is estimated that the atmosphere, hydrosphere and lithosphere contain 1.454 million km³ of water. Figure 4.1.2 shows the global water cycle and the water stored within parts of the hydrosphere. Worldwide, lakes and inland seas represent only 1% of the freshwater currently locked in glaciers and icecaps. Of the water in the ground (soil water), 6.7% is in areas where it can be returned to the surface. Water in soils constitutes only one six thousandth of the total water budget. The water in the atmosphere on which we depend constitutes one ten thousandth of a per cent of the total.

In all, there are nine storage areas for water. Water molecules will exist in the different areas for varying times. (See Table 4.1.1.) Estimates for **residence times** of water in different places continue to be refined by scientists. Good data in this area are important because residence times can be used to assess the effects of pollution and other damage caused to water storage areas. The information is also important in modelling climate change.

Terrestrial organisms depend on water from lakes, rivers, wetlands and the atmosphere. Plants may also gain access to some ground water. While some organisms have adaptations that allow them to drink water containing high levels of salt, most organisms require freshwater to live.

The largest storage of freshwater on the planet is ice, but living things have difficulty obtaining water from the cryosphere (the subsystem of the Earth that is frozen). Organisms use water as a liquid. They must expend energy to convert ice to water and, in environments where ice exists, energy is a very valuable commodity. To eat ice, or just thawed water, would drastically lower the body temperature of an animal. Frozen areas are deserts to most living things. Animals that live in the Antarctic, like some animals in the Simpson Desert, obtain water through the things they eat.

The residence time of water in parts of the hydrosphere (the subsystem of water on the Earth) depends on processes that move water. The rate at which water is moved from one part of a system to another is called a **flux**. Important fluxes in the water cycle include evaporation, advection, precipitation, convection, run-off, infiltration and percolation. Their causes are discussed below.

Table 4.1.1 Storage and average residence times of water

Storage	Average residence time of water molecules
Oceans and seas	4000 years or longer
Ice and snow	Tens to thousands of years
Ground water and soil (subsurface water)	Days to thousands of years
Biota	One week to years
Rivers, lakes and reservoirs (surface water)	Up to two weeks
Atmosphere	Eight to ten days

Evaporation is the conversion of liquid to vapour. Water molecules with a lot of kinetic energy escape from the surface of liquid water into the atmosphere. In doing so, they remove their energy from the water and the liquid remaining cools.

Advection is the process of moving matter or energy from one place to another in the horizontal stream of a gas.

Precipitation is the process of depositing water or ice from the atmosphere to the Earth's surface. Note that the same term is used to describe how dissolved substances are deposited from a fluid.

Convection is a process of heat transfer. It is not only important in the atmosphere but in the oceans too. Hot fluids are less dense than the fluids around them and the density difference causes them to rise. In doing so, the molecules in the air or water carry heat upwards.

Run-off describes the movement of water along the surface of the ground. It may simply run across surfaces or along channels.

Infiltration describes how water enters soil or rock. The term 'percolation' describes how water moves through the spaces within a soil or rock.

Condensation and crystallisation are not fluxes but they are processes that affect the way water behaves. Evaporation and convection move water vapour to cooler areas. **Condensation** occurs when the water vapour forms droplets of liquid water. The droplets are often small and light enough that the air around them stops them falling to Earth. If the droplets become large and heavy enough they may undergo precipitation. It the atmosphere is cold enough the water vapour may condense and then crystallise (form a crystal). Ice crystals may form as frost on surfaces or in clouds. The ice crystals in clouds may precipitate as rain, hail or snow depending on atmospheric conditions.

residence time
the average time a molecule exists in a certain place

flux
the rate at which material or energy flows

condensation
the process of forming a liquid from a vapour

definitions

REVIEW ACTIVITIES

1 Use the information in this section to calculate the percentage of the Earth's water budget that is available to terrestrial organisms.

2 Describe how an increase in advection from the marine to terrestrial atmosphere may affect the water available to terrestrial organisms.

3 Heat and gravity are both involved in the global water cycle. Identify the processes in the water cycle that involve heat and those that involve gravity.

4 Construct a diagram of the water cycle. Use information in this section so that your diagram contains more detail than the water cycle diagram in the text. (See Figure 4.1.2.)

EXTENSION ACTIVITIES

5 Analyse the processes involving water that would directly affect the water availability to a plant in your local environment.

6 Using the information contained in Figure 4.1.2, write a description of the processes that move and change water.

FACTORS AFFECTING THE MOVEMENT OF WATER ON THE EARTH

In the last section a number of processes were described that affect the movement of water around the Earth. In this section, we will look at other factors that determine the distribution of water on Earth. How these factors specifically affect the Australian continent will also be considered.

Geographical position

SOLAR HEATING

The location of Australia affects the water availability to organisms living on the continent. To understand why this is so we need to examine how solar heating drives the motion of the atmosphere and affects the average surface temperatures over the Earth.

Most of the energy available at the surface of the Earth comes from the Sun. Solar energy (or radiation) accounts for 99.98% of the energy that drives Earth processes, with the remainder being heat from within the Earth (0.018%) and gravitational tidal energy (0.002%).

The Earth's atmosphere controls the amount of solar energy absorbed by the surface. (See Figure 4.1.3, page 210.) Clouds, aerosols and gases reflect, or absorb,

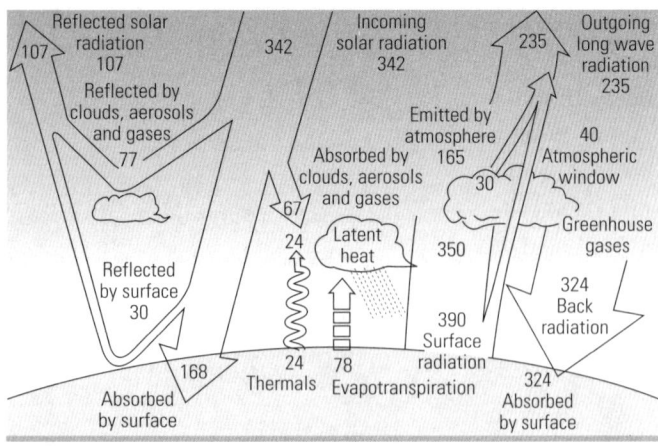

Figure 4.1.3 Energy paths into and out of the atmosphere. Note that radiation can be reflected, absorbed or re-radiated. All values given are in webers per square metre (W^{-2}).

energy before it reaches the surface. Of the light energy that reaches the surface some is directly reflected back into the atmosphere and the rest is absorbed. Once absorbed, the energy may return to the atmosphere as infrared radiation. Radiation from the atmosphere can also be absorbed by the surface. This is called back radiation.

The location and nature of a surface on the Earth will determine how much heating occurs. The reflectance, or **albedo**, of a surface describes how much light is reflected. It is expressed as a percentage, with an albedo of 100% meaning perfect reflectance and an albedo of 0% meaning total absorption. Snow and ice have an albedo of 40–95% and bare soils have an albedo of 5–10%. Plant communities have a higher albedo than soil, with different communities having an albedo between 5% and 30%. As a general rule, the lower the albedo, the greater is the surface heating.

Latitude plays an important role in heating. At high latitudes, near the poles, the Sun is at a low elevation. (See Figure 4.1.4.) Light travels through more atmosphere to reach the surface and so more light is absorbed or reflected. The light that reaches the surface is spread over a relatively large area so the amount of energy per square metre is low. This means that a relatively small amount of heating occurs. At the equator the path through the atmosphere is short and the high angle of incidence generates a relatively high level of heating.

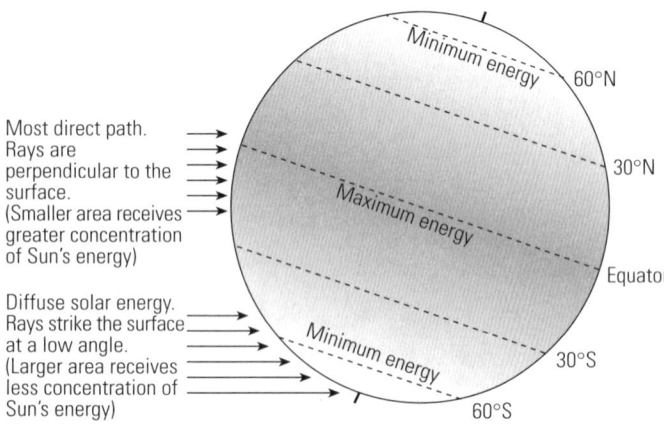

Figure 4.1.4 The Sun's energy is not evenly distributed on the Earth's surface. The amount of solar energy per square metre increases as the latitude decreases. In high latitudes, heating is less and more energy heats the atmosphere than at lower latitudes.

The net effect of variations in the Sun's heating is that surface temperatures fall as one moves from the equator to the poles. Another effect of the Sun's heating is that there is more energy to drive evaporation at the equator and more rain falls over the oceans there than at higher latitudes.

AIR CIRCULATION

Differences in heating cause the circulation of air. As warm air rises and cool air sinks, air streams move air along to fill the low-pressure areas that result. (See Figure 4.1.5.) The air streams are the high-level and surface winds that move through the atmosphere. These winds not only help circulate air in the atmosphere but they play an important role in the circulation of the oceans too.

The winds are deflected from travelling between the poles and equator by the spinning Earth. The **Coriolis effect** causes the winds to form spiralling convection cells. (See Figure 4.1.6.) A consequence of the Coriolis effect is that air circulates differently in the Northern and Southern Hemispheres, with the two hemispheres being separated by a zone of low pressure. This zone is called the intertropical convergence zone.

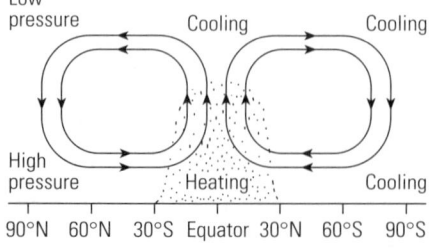

Figure 4.1.5 A simple model of atmospheric convection and circulation showing the effect of latitude on heating and cooling. The air of the atmosphere moves from place to place and so does the heat it contains. Heating occurs mainly in the lower latitudes and cooling occurs mainly in the higher latitudes. Surface evaporation (indicated with dots) is greatest near the equator.

The pattern of airflow in the atmosphere and of currents in the oceans are affected by the distribution of land. (See Figure 4.1.6.) During winter, landmasses cool and areas of high pressure form. During the northern winter, high-pressure areas within Asia force the intertropical convergence zone south of Papua New Guinea. The resulting winds bring moisture-laden air to Northern Australia and produce the wet season. Differences in the pattern of atmospheric circulation are shown in Figure 4.1.7. Note how the subtropical ridge is pushed southward during the summer. The change in air flow patterns that results causes changes to winds along the east coast and also affects rainfall patterns.

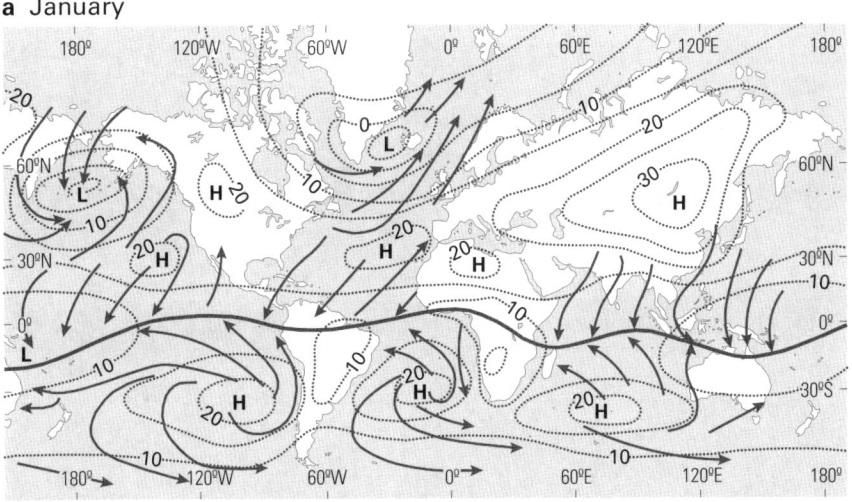

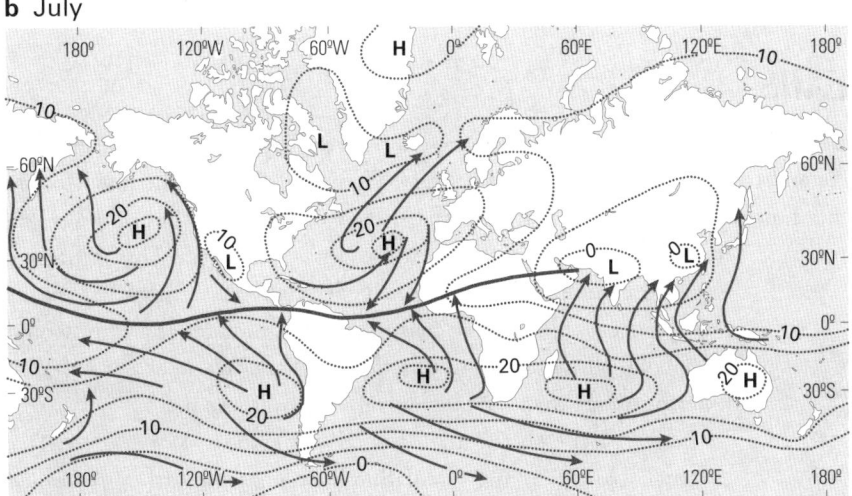

Figure 4.1.6 The distribution of land and sea affects the air circulation patterns and pressure of the Earth. Heating and cooling in the hemispheres causes the intertropical convergence zone to move during the year, changing the airflow around some continents. The intertropical convergence zone is indicated with a thick line.

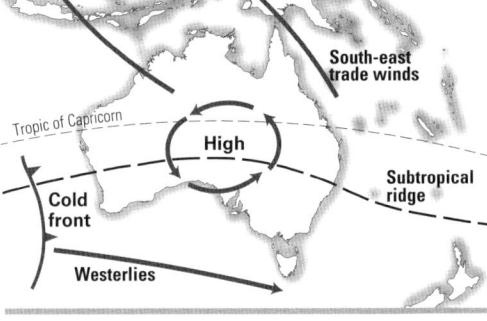

Figure 4.1.7 The atmospheric circulation over Australia during summer and winter. The high-pressure system is deflected south during summer as monsoons reach the north of Australia.

A feature shown in Figure 4.1.7 is that a cell of high pressure dominates the centre of Australia. This is a result of Australia's latitude and causes the centre of Australia to receive very little rain. The low rainfall, together with high evaporation rates, makes Australia the driest continent on Earth, after Antarctica. The dryness of Australia is not helped by it being so flat. The lack of mountains means that there are no barriers to cause moist air to produce rain.

Australia's size makes the centre an area of extreme climatic variability and also results in a wide range of climate zones. Plate 18 shows Australia's climate zones. They range from alpine and moist temperate zones in the south-east to moist tropical zones in the north. It is not a surprise to find most of Australia's population located along the east coast when one considers the relatively low variability of climate within the zones there. Compare the movement of winds around the winter high-pressure cell (see Figure 4.1.7) with the alignment of sand dunes in the arid zone. The dunes formed parallel to the wind directions when the last ice age was at its coldest.

Some of the dunes are still active. At Lake Mungo National Park, sheep grazing has caused the western side of ancient dunes to be more rapidly eroded. The sand has been deposited on the eastern side where the dunes are slowly moving eastward. A consequence of this process has been the uncovering of evidence of early human habitation in Australia.

albedo
a measure of the reflecting ability of a surface

Coriolis effect
the tendency of fluids to be deflected from straight-line flow; caused by the Earth's rotation

definitions

REVIEW ACTIVITIES

1 Explain how the latitude of Australia influences our climate.

2 Describe why high-latitude areas warm more slowly than areas near the equator.

3 Clarify the relationship between climate and latitude.

4 Explain why the centre of Australia is arid.

5 Explain why the lack of high mountain ranges in Australia reduces rainfall and surface run-off.

EXTENSION ACTIVITIES

6 Contrast the summer and winter atmospheric circulation shown in Figure 4.1.7.

7 Design an experiment to measure the albedo of different surfaces.

OCEANS

There has been a growing understanding of the relationship between Australia's climate and the oceans during the last century. Figure 4.1.8 shows the major ocean currents flowing around Australia. Note that major currents flow south along the west and east coasts, carrying warm water with them. This is unusual when compared with other continents, and the warm water means that the oceans along these coasts are not as productive as they would be if cold nutrient-rich waters flowed there. Warm water contains fewer nutrients than cold waters moving from the poles.

Figure 4.1.9 shows the general circulation of ocean surface currents. Note that the circulating currents form structures called **gyres**. The circulations of gyres in the ocean and of pressure systems in the atmosphere have similar patterns, as shown by a comparison of Figures 4.1.6 (page 211) and 4.1.9. This is because the winds are the major factor controlling the formation of the surface currents.

It is variations in the pressure of the atmosphere due to changes in ocean temperature that cause variations in Australia's rainfall. In years of normal rainfall, air pressure over the eastern Pacific Ocean is high relative to that near Darwin. Moist air rises in the low-pressure areas over the east coast of Australia, where rainfall is normal or above normal.

In some years the waters in the eastern Pacific Ocean warm and, as a result, the atmospheric circulation of the whole Asia-Pacific region is changed. Eastern Australia is a zone of relatively high pressure and experiences below average rainfall. Such years are also times of increased bushfire activity.

The changing nature of the ocean and atmosphere, which changes Australia's rainfall, is called the El Nino-Southern Oscillation (**ENSO**). (See Figure 4.1.10.) ENSO can act within a year, over a period of years or across decades. Scientists are beginning to understand how similar events in the north-eastern Indian Ocean also affect Australia's rainfall and weather.

Like the atmosphere, the ocean has a structure. Three layers can be recognised based on temperature and salinity. (See Figure 4.1.11.)

Surface waters are relatively thin (up to several hundred metres) and warm. The average temperature of the oceans is 3.6°C but the average surface temperature is 19°C. The composition of the surface layer varies with latitude. In the tropics and

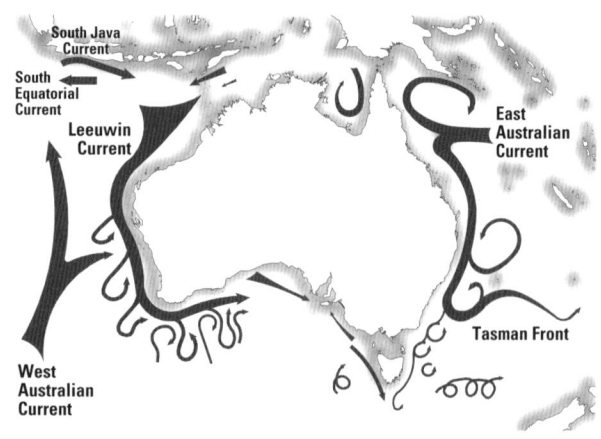

Figure 4.1.8 Ocean currents around the Australian coast. South-flowing currents along the east and west coasts affect the abundance within the marine ecosystems.

gyres
systems of wind-driven ocean surface currents

ENSO
the El Niño-Southern Oscillation

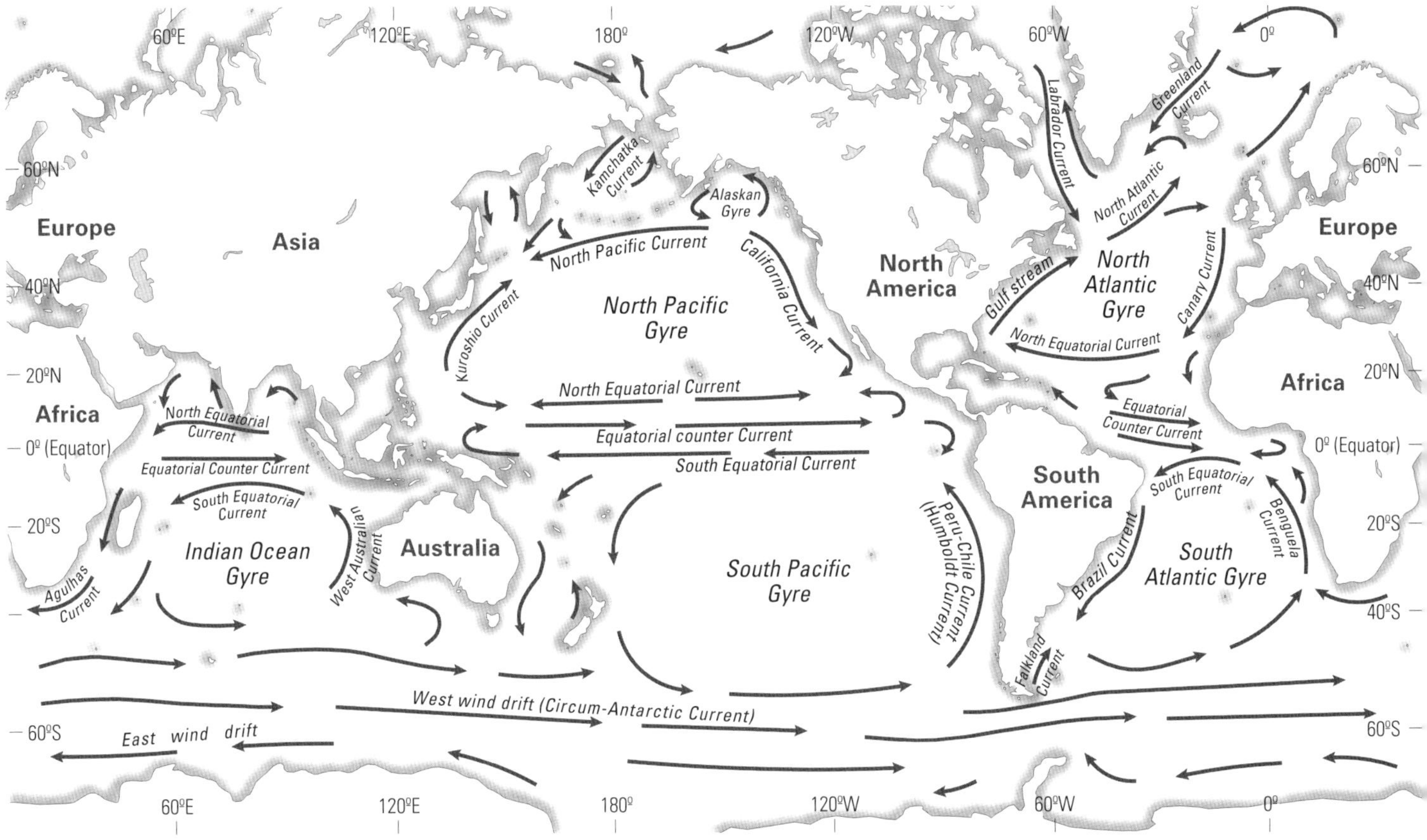

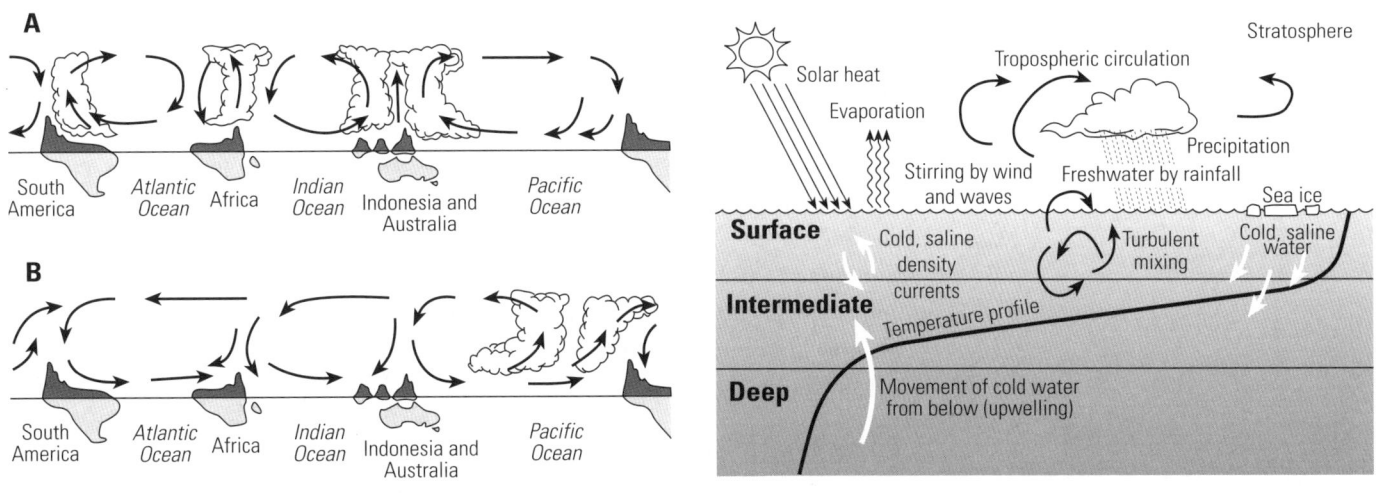

Figure 4.1.9 Major surface currents and gyres of the world's oceans. This pattern has changed during the Earth's history and the changes have affected climate.

Figure 4.1.10 A diagram showing how ENSO affects weather over Australia and Indonesia. (A) Rising air over Australia and Indonesia generates rain. (B) When the western Pacific Ocean is warmer than normal, descending air over Australia leads to less rain in Australia and Indonesia.

Figure 4.1.11 The ocean and atmosphere interact with each other. The deep ocean is isolated from the atmosphere by the surface layer, except at the poles.

polar regions, precipitation is greater than evaporation. This causes the surface water to be relatively stable and fresh. In mid latitudes evaporation dominates precipitation and the surface layer is more saline and unstable.

Deep waters are colder (–1°C to 5°C) and more saline than surface waters. There is relatively little variation in temperature. The water is cooled near the poles and rapidly sinks. It gradually moves towards lower latitudes. Near coasts it may move up to the surface. This phenomenon, called 'upwelling', is very important because the

WATER ISSUES 213

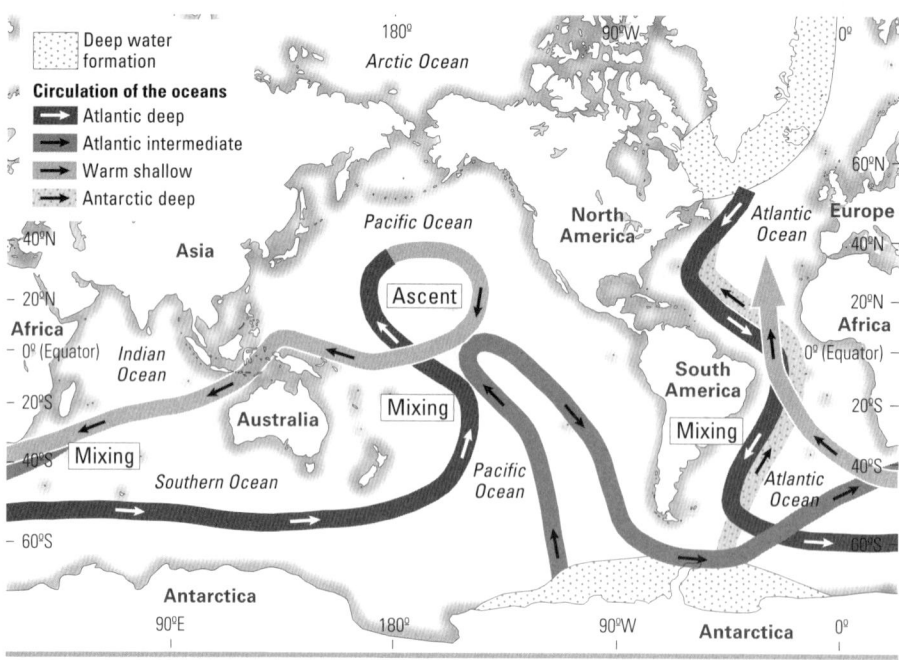

Figure 4.1.12 The deep circulation of the oceans. Cold water formed near the poles sinks and moves through the oceans. Over time it causes mixing of the ocean waters.

water contains nutrients. The nutrients are used by the plankton that support marine food chains. It is known that the deep currents and shallow currents are linked and that they slowly cycle water in the oceans. (See Figure 4.1.12.) It is thought that plate tectonics may alter the movement of such currents and affect climate, but this is not well understood.

An intermediate layer can be identified between the surface and deep layers of the ocean. This layer is absent at the poles where the deep layer forms. The intermediate layer is a transitional layer in terms of temperature, salinity and density. It extends from a depth of 200 m to about 900 m.

The importance of the ocean's structure is that the ocean acts to slow the rate of climate change. The surface layer acts as a source of heat and reduces seasonal temperature changes along coastlines. The deep layer stores a great deal of carbon dioxide, locking it out of the atmosphere. This tends to prevent warming in the atmosphere. It is also important to understand that the ocean and atmosphere interact with each other. Our knowledge of how they influence each other continues to grow but more needs to be understood before we really understand the dynamics of climate change.

Changes in ocean currents

The three diagrams in Figure 4.1.13 show how ocean circulation has changed over the last 60 million years. One change you will note is that waters at the equator could circulate around the equator. Note, too, that at that time no icecap existed on Antarctica.

By 30 million years ago the climate had begun to cool. Barriers to ocean currents stopped water circulating at the equator and currents had begun to circulate around Antarctica. Between 3 and 5 million years ago the Isthmus of Panama closed, changing circulation within the northern Atlantic Ocean. The current known as the Gulf Stream strengthened along the east coast of North America. It carried warm water north and precipitation began to increase in Europe and North America. By 3 million years ago icecaps had begun to develop in the Northern Hemisphere and the last ice age was under way.

What this history illustrates is the importance of ocean currents to climate change. The surface currents redistribute heat and influence precipitation. As the deep layer of the ocean developed it would have removed carbon dioxide from the atmosphere, causing temperatures to fall.

Falling temperatures meant precipitation was more often in the form of ice. As the ice deposits built up they reduced the amount of water in the oceans and sea levels fell. Increased land area would, in turn, have affected climate and the circulation of surface currents.

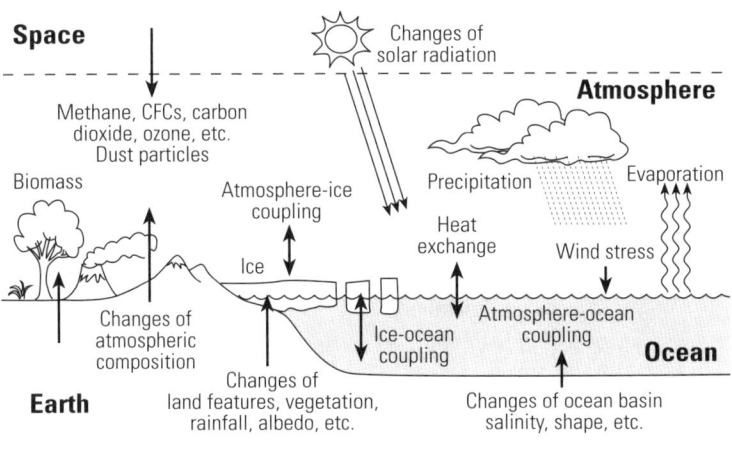

Figure 4.1.14 The physical processes that govern global climate.

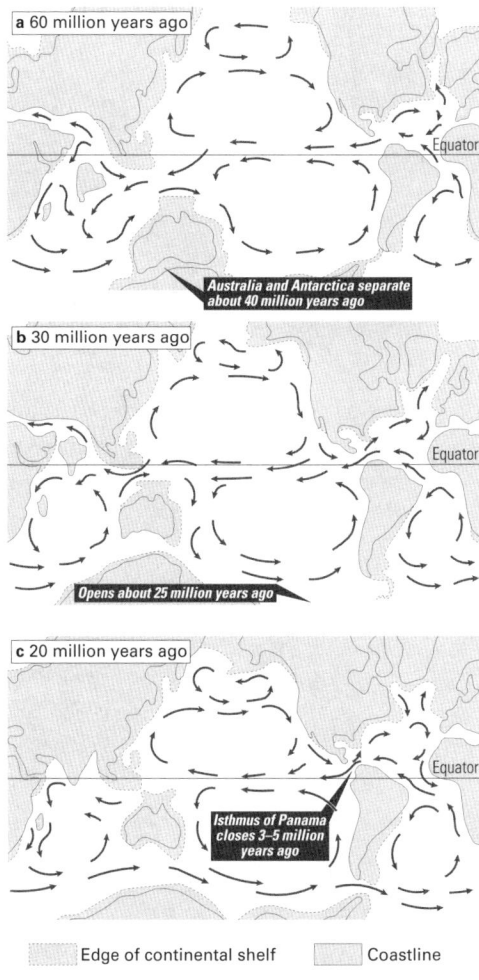

Figure 4.1.13 The evolution of the present ocean circulation. The closing of the Isthmus of Panama and the creation of the Circum-Antarctic Current have altered climate. The mechanism is explained in the text.

Currents, temperature, sea level and icecaps are all interrelated. (See Figure 4.1.14.) Changes in one area can cause changes in the way other areas operate. In each case the movement or distribution of water changes and those changes influence the living world and abiotic features of the environment.

REVIEW ACTIVITIES

1 Construct a diagram to show how ENSO changes rainfall in Eastern Australia.

2 Contrast the surface layer and deep layer of the ocean.

3 Use the information on pages 213–14 to construct a time line of changes leading up to the last ice age.

4 Describe how ocean currents can change temperature and ice deposits.

5 Make a diagram that identifies the interrelationships between sea levels, ocean currents, ice deposits and global temperatures.

EXTENSION ACTIVITIES

6 Evaluate the effect of sea level change on climate within the Australian continent. (Hint: Consider the effect of ocean currents on areas near the coast.)

7 Explain why farming in Europe is easier than farming in Nova Scotia, despite the fact that both areas have similar latitudes. (Hint: Consider the effect of ocean currents.)

8 Explain why periods of extensive glaciation are also periods of dryness.

9 Identify the proportion of time during the last 700 million years during which icecaps were either very small or non-existent.

SUMMARY

- The Earth can be studied as a set of four interacting subsystems: atmosphere, lithosphere, hydrosphere and biosphere.

- Each subsystem has a structure and processes that affect, and are affected by, other subsystems.

- Natural events, such as volcanoes and cyclones, produce short-term changes to some of the Earth's subsystems.

- The Earth's water budget is very large but the water on which we depend is a very small part of this.

- Water moves between parts of the hydrosphere and can exist in different areas for different lengths of time.

- A range of processes moves water within the hydrosphere. These include evaporation, advection, precipitation, convection and infiltration.

- The distribution of water on the Earth is affected by factors such as climate, topography and geographical position.

- Global climate is affected by a number of interrelated factors. Some of these factors include global temperatures, ice deposits, sea levels and ocean currents.

PRACTICAL EXERCISE
Where is the water?

In this exercise you will process information from the text to identify any relationships that exist between the locations of water within the hydrosphere and the residence time of water in those locations.

ACTIVITIES

1 Prepare a graph with mass on the vertical axis and time on the horizontal axis.

2 Mark the time axis in days, weeks, years, millennia (thousands of years) and tens of millennia. Your graph should look like the diagram below.

3 Plot the information in Figure 4.1.2 and Table 4.1.1 (page 208) onto your graph. Use lines to indicate ranges of time.

Mass of water (x 10^{15} kg)

Days   Weeks   Years   Millenia   Tens of millenia
Residence time

4 Name the largest reservoir, or storage area, of water on your graph. Calculate the percentage of the total water budget that this amount represents.

5 Name the smallest reservoir shown on your graph. Calculate the percentage it represents.

6 Does your graph show any relationship between residence time and mass of the reservoir? If so, describe it.

7 Discuss whether it is an advantage for living things to rely on water reservoirs in which water has a small residence time.

8 Summarise what you have learnt from this exercise.

4.2 Water and Australian environments

OUTCOMES

At the end of this chapter you should be able to:

- recall differences between biotic and abiotic features of the local environment

- explain the importance of water as a solvent in biological systems

- compare the relative solubility of oxygen and carbon dioxide in water and how the solubility of each changes with temperature

- choose equipment and perform first-hand investigations to gather first-hand information about the presence of dissolved oxygen in water at different temperatures using indicators or appropriate technology

- predict the potential impact of excessive water evaporation and subsequent increase in salinity on common terrestrial and inland aquatic organisms

- identify common water pollutants that can affect the growth of plankton

- plan and perform first-hand investigations to gather first-hand information that demonstrates the effect of varying salt concentrations on plant growth.

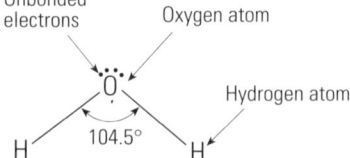

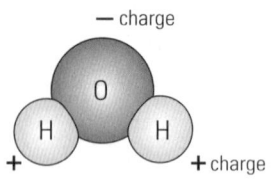

Figure 4.2.1 The structure of a water molecule. The two diagrams show different aspects of the same molecule.

Ecology is the study of organisms and where they live. We have already discussed the systems that interact at the Earth's surface: atmosphere, lithosphere, hydrosphere and biosphere. The site where these four systems interact is sometimes called the ecosphere, and it is a very complex system. Within it, water plays a vital role. In this chapter we will examine the properties of water and the behaviour of dissolved gases and salts. These subjects are important because water and the things it contains have a direct affect on our wellbeing and the health of the ecosphere as a whole.

FEATURES OF THE LOCAL ENVIRONMENT

In understanding the ecology of an area we can consider the environment as consisting of two parts. The non-living (**abiotic**) part consists of solar energy, liquid water, air, the mineral part of the soil, chemicals in the environment and aspects of the climate, such as wind, humidity and heat. The living (**biotic**) part involves living things, such as plants, animals, bacteria and fungi.

Properties of water

Water is something common to the biotic and abiotic parts of the environment. It is a remarkable substance, with properties that make it essential to living things.

Many of water's unique properties arise from the structure of the water molecule. A water molecule consists of two hydrogen atoms bonded to an oxygen atom. (See Figure 4.2.1.) Lines joining the centres of the atoms make an angle of 104.5°. In addition to its bent shape, the water molecule is charged. The oxygen atom attracts electrons more strongly than the hydrogen atoms and, as a result, the oxygen has a slight negative charge while the hydrogen atoms have a slight positive charge. Such molecules are called **polar**.

As a result of the electrical charges, water molecules form short-term bonds with each other. This causes that water to hold together; it shows **cohesion**. The molecules are also attracted to surfaces with electrical charges. We say the water is adhesive because it sticks to such surfaces.

Perhaps the most important property of water for living things is its role as a **solvent**. Many substances **dissolve** in water because the polar water molecules dissolve ions (charged atoms and groups of atoms) and other molecules that are polar. (See Figure 4.2.2.) The polar water molecules do not dissolve all substances. Non-polar substances, such as many carbon compounds, are not dissolved by water. Once dissolved, substances can easily move within, into and out of cells.

Cell membranes that prevent water movement through them can, as a result, control the dissolved substances moving across them. Such substances are the raw materials from which cells obtain substances for growth, repair and reproduction. Without water's ability to dissolve materials it is hard to see how life could exist.

Another important property of water is its heating behaviour. Water has a high **heat capacity**. This means that water absorbs a lot of heat for a relatively small rise in temperature. This is important to living things for two reasons. The high water content of cells means they gain and lose heat relatively slowly. Stable temperatures help processes in the cell to proceed at relatively constant rates. Water's high heat capacity is also important to life because cell processes create heat. Water absorbs the heat without a large change in temperature, allowing chemical processes to continue at relatively low temperatures. If the temperature in cells becomes too high, the enzymes that carry out cell processes are altered and the cells die.

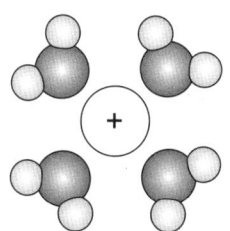

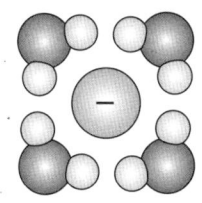

Figure 4.2.2 In a solution, water molecules surround and interact with charged atoms (ions). Note how the oxygen is attracted to the positive charges and the hydrogens to the negative charges.

Properties of oxygen and carbon dioxide

Solubility is a measure of how much material (**solute**) will dissolve in a certain amount of another material (solvent). The solubility of a gas depends on a number of factors. These include the temperature, the pressure and the type of gas involved.

Both oxygen and carbon dioxide dissolve in water but the amount that dissolves decreases as the temperature rises. This means that if water warms it loses oxygen and living things may die. If the water in the deep layer of the ocean was to warm by only a few degrees, the amount of carbon dioxide released would have a drastic effect on the Earth's climate.

The pressure of a gas above water affects how much gas dissolves. At a constant temperature the solubility of a gas is directly proportional to the pressure above it. This means that if the pressure of oxygen doubles, the amount that dissolves will double too.

Different gases have different solubilities. Carbon dioxide is more soluble than oxygen at a given temperature. Thirty millilitres of oxygen will dissolve in a litre of water at 30°C. This low solubility is enough, however, to support aquatic life and allow the breakdown of organic matter in water. A high temperature not only reduces the solubility of oxygen but also reduces oxygen in other ways. Some aquatic organisms are very sensitive to temperature changes. If the temperature rises and they die their bodies decay and this removes oxygen from the water. Increased temperatures also increase the metabolism (life processes) of surviving organisms. This will increase their use of oxygen and cause its concentration to decrease further.

ecology
study of organisms and where they live

abiotic
concerning non-living things

biotic
concerning living things

polar
describes a molecule that has an uneven distribution of electrical charges

cohesion
the property of something that makes its parts hold together

solvent
the substance into which something dissolves

dissolve
to become part of a liquid

heat capacity
the amount of heat needed to raise the temperature of a substance by 1°C

solubility
the amount of a material that will dissolve in another material

solute
the material that dissolves in something

Carbon dioxide behaves differently from oxygen when it dissolves. It is only slightly soluble in water but forty times more soluble than oxygen. About 1% of the carbon dioxide reacts with water to form carbonic acid or hydrogen carbonate ions. The rest stays in solution bound to water molecules. Carbon dioxide is also capable of forming what is called a supersaturated solution. At higher pressures, such as those in the deep ocean, a large amount of carbon dioxide would dissolve. When the cold water rises we say the water is supersaturated because it continues to hold the dissolved carbon dioxide in amounts it would not contain if the carbon dioxide dissolved into the water at the lower pressure. Carbon dioxide is added to soft drink to make a supersaturated solution. When the bottle or can is opened some, but not all, of the gas is released, forming bubbles. If the drink is stirred or shaken the carbon dioxide escapes rapidly until the amount dissolved is proportional to the air pressure.

Understanding the solubility of gases is not only important for understanding the survival of living things. Today scientists use such knowledge for engineering. Removing oxygen from water in contact with metal reduces rusting and saves money. Adding oxygen is important in speeding up processes in sewage treatment. In a similar way, understanding the behaviour of other dissolved substances has important economic implications and helps us understand the natural world.

REVIEW ACTIVITIES

1 Classify features of the environment as either biotic or abiotic factors. Use a table to arrange your classifications.

2 Explain why water's solvent properties are important to life.

3 Describe how the structure of water affects its solvent properties.

4 Compare the solubility of carbon dioxide and oxygen in water.

5 Identify the factors that affect the solubility of a gas.

EXTENSION ACTIVITIES

6 At 30°C 30 mL of oxygen will dissolve in 1 L of water. Calculate the solubility of oxygen in the water if the pressure is:
a doubled
b increased by 30%.

7 Contrast the importance of water's solvent properties and its heat capacity in living things.

Table 4.2.1 Dissolved substances in three naturally occurring waters measured in millimoles per litre (mmol/L)

Substance	Average seawater	Rainwater	Ground water
Sodium (Na^+)	456.50	0.280	0.40
Potassium (K^+)	9.70	0.010	0.05
Calcium (Ca^{2+})	10.00	0.025	0.70
Magnesium (Mg^{2+})	55.60	0.035	0.25
Ammonium (NH_4^+)	0.03	0.002	–
Nitrate (NO_3^-)	0.02	0.005	–
Chloride (Cl^-)	549.00	0.320	0.15
Sulphate (SO_4^{2-})	27.60	0.009	0.30
Hydrogen carbonate (HCO_3^-)	2.50	0.014	1.50

EVAPORATION AND SALINITY

Salts are present in all natural waters. As minerals weather they release dissolved substances that remain in the water until they are deposited. During the history of the Earth the concentration of salts in the oceans has risen as water derived from the land has carried dissolved substances to it. The concentration of dissolved substances in seawater, ground water and rainwater is shown in Table 4.2.1. Note that the rainwater is not pure. It contains salts that are held in small water droplets derived from the ocean. You should also note that the ground water and rainwater are both acidic while seawater is alkaline.

In many parts of New South Wales, large volumes of water are applied to fields during irrigation. Crops may require more water than the native plants to thrive and the amount of water added during irrigation may be as much as four times the natural rainfall for an area. Besides changing the height of the watertable, this large amount of water is exposed to evaporation, which changes the salinity of the water. The salinity of the water rises because evaporation occurs and dissolved substances are left behind. They crystallise as salts if all the water evaporates, but they may just become concentrated by the removal of some of the water.

Water is not only exposed to evaporation in the fields during irrigation. The construction of dams increases the amount of evaporation as the water moves sideways into the surrounding ground.

The salinity rise affects plants in a number of ways. Saline water around a plant cell will cause water in the cell to move outwards by a process called osmosis. This causes the cell to shrink inside the cell wall and the cell may die. In addition, the changing salinity in the soil may destroy the soil structure, affecting germination and growth.

REVIEW ACTIVITIES

1 Describe how and why saline water affects plant growth.

2 Evaluate the importance of irrigation as a cause of increasing salinity.

3 Different plants tolerate salt in different ways. Explain why research into salt tolerance in plants is important in dealing with salinity.

4 Increased water availability due to irrigation can lead to problems with weeds competing with native vegetation. Explain why this occurs.

EXTENSION ACTIVITY

5 Look at Table 4.2.1. In scientific work it is sometimes useful to recalculate data in order to find relationships more easily. Total the number of millimoles per litre for the nine substances in seawater. Next, calculate the percentage of the total for each substance. Repeat the process for the data about the rainwater. Compare the percentages and interpret any patterns you observe.

WATER POLLUTION

Human activity produces waste materials that are usually removed from the locations where they are made. Many of the wastes people create end up in water. Sometimes wastes are intentionally deposited in water. Other wastes may make it into water from the places they are left. Either way, the wastes end up in water bodies where they are pollutants.

A pollutant is an unwanted substance, and water pollution refers to substances that have contaminated water. Pollutants produce undesired effects in things that live in or use the water.

Types of water pollutants

Consider the following types of pollutants.

TOXIC SUBSTANCES

Toxic substances, or toxins, are poisons. They interfere with processes in living things and can exist in the environment for many years. They may cause cancer or mutations. Toxic substances include radioactive substances and heavy metals, such as lead, mercury and cadmium. Some toxic chemicals are produced by industry. Certain

definition

toxic
poisonous

petroleum derivatives are toxic, as are polychlorinate biphenyls (PCB). Mining and manufacturing may produce cyanide. Agricultural chemicals, such as herbicides and insecticides, are also toxic substances.

Toxic substances are produced by sewage treatment works as well as industry. Toxins may affect living things at very low concentrations. Dioxin is a substance that is toxic at extremely low levels. Some toxic chemicals produce symptoms in organisms through **biological magnification**. In this process the level of toxins builds up along the food chain as the bodies of animals store the toxins from the things they eat until eventually symptoms occur.

A well-known example of biological magnification is the effect of DDT on birds. During the 1940s and 1950s, DDT was sprayed in many areas to kill insects. In the food chain, DDT molecules become concentrated until they reach levels that affect a bird's ability to make healthy eggs. The DDT affects the bird's liver. The liver breaks down the hormone that releases calcium for egg shell production. This results in eggs with thin, fragile shells that are easily broken. As a result, many eggs do not survive and bird populations decline.

ORGANIC MATTER AND PLANT NUTRIENTS

Organic matter is widely produced. Abattoirs, dairies, fish farms and sewage treatment works all produce large amounts of biological material as waste. In water, these wastes provide a rich food source for bacteria. The rapid increase in bacterial numbers reduces oxygen in the water, leading to the death of other aquatic organisms.

Plant nutrients, such as nitrogen and phosphorus, act in a similar way. Run-off from feed lots or land treated with fertilisers can add nutrients to waterways. The process of nutrient enrichment is called **eutrophication**.

Too many nutrients in a body of water can make it unfit to drink by native animals and stock. The nutrient level may result in an explosion of plants and bacteria, particularly blue-green algae, or **cyanophytes**. These organisms are referred to as plankton. They are microscopic producers and, together with microscopic animals, are an important part of the biotic make-up of aquatic ecosystems. Blue-green algae produce toxins that may kill fish and animals that drink the water. When the cyanophytes die, the bacteria that consume them may rob the water of oxygen, killing other organisms in the water. At the very least, algal **blooms** alter feeding relationships in the environment where they occur.

THERMAL WASTE

Hot or cold water can be a pollutant. Hot water is discharged from power plants and industry. Cold water is produced in deep dams. When such water enters a natural system it affects the life cycles of organisms that live there. Fish, for example, spawn at particular temperatures. By changing the water temperature the fish cannot reproduce or grow properly. Some algae, however, thrive in warmer temperatures and if there are excess nutrients in the water they can rapidly increase in number and cause some of the problems mentioned earlier.

OTHER POLLUTANTS

Besides the pollutants listed above we may add infectious agents (such as viruses) and sediment (such as mud). They each have the potential to affect the wellbeing of humans and other living things.

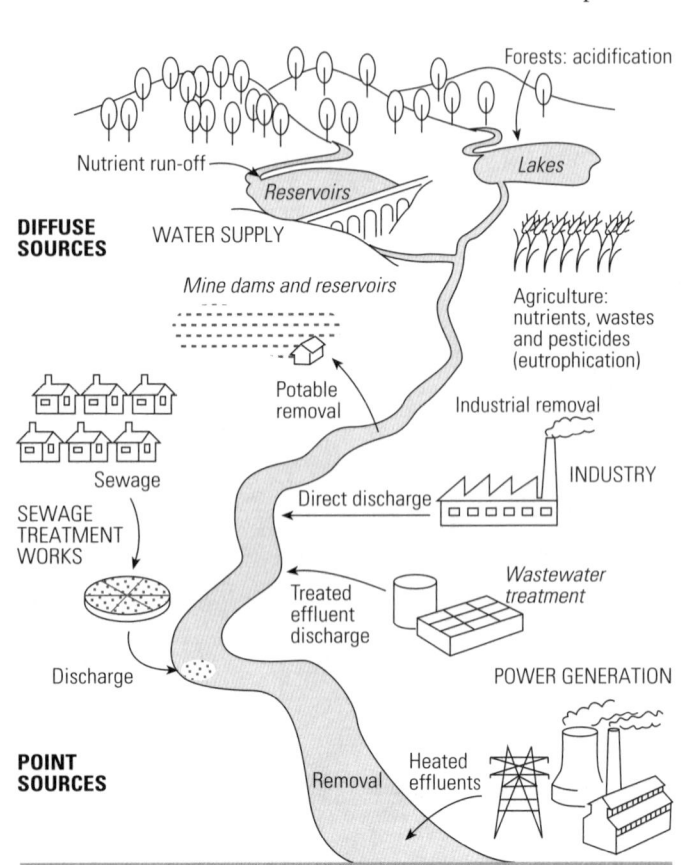

Figure 4.2.3 Sources of pollution in river systems.

Marine pollution

Pollutants not only affect surface waters but ground water and marine waters too. (More will be said about ground water later in the chapter.) Sources of marine pollution include run-off from land; airborne emissions from land; shipping and accidental spills; ocean dumping; and offshore mining. Run-off and discharge from land account for 43% of marine pollution. The type and effects of marine pollutants are similar to those discussed for inland waters.

Preventing pollution

Figure 4.2.3 shows the sources of pollution entering a river system. Some pollutants enter waterways at a particular place. These are called **point sources**. Such sources can be located by systematic testing of water along the river. Point sources can be controlled by regulation with measures such as permits and fines for illegal discharges.

Some pollutants enter water systems over a wide area. These are referred to as diffuse sources. Diffuse sources can be hard to find and control because they originate over a wide area. Some of the methods discussed later in the section on salinity prevent pollution. (See page 241.) A basic rule is that it is much easier to prevent pollution happening than clean up after it has happened.

definitions

biological magnification
the increasing concentration of materials due to selective uptake during feeding in a food chain

eutrophication
nutrient enrichment in a water body

cyanophytes
photosynthetic bacteria; also known as cyanobacteria

blooms
the rapid increase in the number of algae and/or bacteria caused by the presence of high levels of nutrient

point source
a particular place where pollutants enter a waterway

REVIEW ACTIVITIES

1 The quality of water affects the life that lives in it. Identify pollutants in water that affect the growth of plankton.

2 Construct a table to classify types of pollutants and examples of each type.

3 Research the effects of dioxins or radioactive pollutants in the environment.

EXTENSION ACTIVITIES

4 Very strict rules govern mining and mineral exploration within Australia. Locate and summarise information about the guidelines for mining and mineral exploration in environmentally sensitive lands.

5 What are the sources of dioxins or radioactive pollutants and how can their presence be controlled in the environment?

SUMMARY

- Features of an environment can be classified as biotic (living) or abiotic (non-living).

- Water is essential for living things. It is a powerful solvent and carries dissolved gases.

- Oxygen and carbon dioxide are important to living things. The solubility of both is affected by pressure and temperature. However, oxygen behaves differently from carbon dioxide.

- Excessive water evaporation may lead to increased salinity.

- Salinity increases will interfere with biological processes.

- Pollutants, such as heat and nutrients, affect algae growth. Both can cause blooms that affect other types of living things. Toxic substances kill living things, including algae.

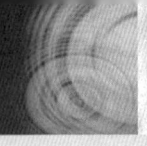

PRACTICAL EXERCISE
Dissolved oxygen and water temperature

This exercise should help you to understand how oxygen solubility varies with water temperature.

Background
The concentration of dissolved oxygen in water is essential for the growth and development of aquatic life. The amount of oxygen that will dissolve in water depends on the water's temperature. In this exercise you will carry out an experiment to measure the oxygen in water at three different temperatures and interpolate the shape of the curve from the data you graph.

Equipment
- A dissolved oxygen test kit
- Three or four large beakers to act as water baths
- A thermometer to measure the temperature of test samples
- Three or four conical flasks with stoppers that fit easily into the large beakers
- Safety equipment: gloves, eye protection and coats

Suggested procedure
1. Half fill each of the conical flasks. Try to ensure that the amount of water in each one is the same.
2. Half fill the beakers with water of different temperatures. Try to ensure that the water in the flasks varies by about 10°C from one to the next.
3. Place the flasks in the water baths and wait until they have reached a steady temperature.
4. Making sure the stoppers are in the flasks, carefully shake each one to mix air with the water. Leave the stoppers in the flasks and return the flasks to the water baths.
5. Wait ten minutes, measure the water temperature in each flask and record the temperature.
6. Test the water in each flask with the dissolved oxygen kit. Try not to shake or stir the water when removing it for testing.
7. Test each water sample three times if you can. Record your results.
8. Clean up your work area and then proceed to analyse your results.

Caution: Some of the chemicals in the kits are hazardous. Wear the safety equipment and read the health and safety sheets that come with the tests before proceeding. Make sure that you follow the instructions carefully.

ANALYSIS

1. Using the data you collected, average the results for each flask.
2. Graph the results using graph paper or a spreadsheet program.
3. Draw a smooth line through your points.
4. Describe the relationship between the temperature and oxygen levels you measured in the flasks.
5. Estimate the sources of error in your experiment.
6. Suggest how the experiment may be improved.

PRACTICAL EXERCISE
Plants and salinity: Experimental design

This exercise should help you to practise the design of a controlled experiment. If you conduct the experiment it should also aid your understanding of the effects of different salt concentrations on seed germination and growth.

Background
Different plants show different sensitivities to salt in their environments. In this exercise you will plan an experiment to test the effect of salt concentration on seed germination. You may carry out the experiment but the exercise is designed to help you focus on the initial design of the procedure.

Suggested procedure
In this exercise you will assume that you have been given a range of salt solutions: 10%, 1%, 0.1%, 0.01% and 0.001% and some distilled water.

Each concentration is made by taking a sample of a known concentration and diluting it so that it makes up one-tenth of the new solution. The procedure is repeated to produce the range of concentrations you need.

Assume you are given seed of four different types, some Petri dishes and cotton wool.

Work through the following questions and steps.

1. What is the question you are seeking to answer? Consider all the things you could study with the materials you have been given.
2. Restate the question as a testable statement—a hypothesis.
3. Identify the things you will measure in the experiment and the things you know.
4. What will you vary in the experiment and what will you keep the same?
5. How long will your experiment run?
6. What will you measure and how will you do it?

7 Decide how you will carry out the experiment and write a draft procedure in point form.

8 What will your results look like? How will they be recorded? How will you ensure that you have repeated measurements?

9 Explain what the results will show if your hypothesis is:
a supported by the results
b not supported by the results.

10 Share your plan with a friend and try and identify the strengths and weaknesses of your approach.

PRACTICAL EXERCISE
Water plants and pollution

The purpose of this exercise is to examine some data and increase your understanding of reed beds and their effect on water quality.

Background
During a study of local water quality some Year 11 students collected water samples from seven locations along a creek near their school. Upstream from the creek is a dam wall, which forms a lake. The first sample was taken from a lake and the others from water in the creek downstream of the dam wall. The samples were collected twice: once after a fortnight of dry weather and again after three days of rain.

Table 4.2.2 Data from tests on samples collected

Date	Test	Results at site A	B	C	D	E	F	G
23 March (dry weather)	Faecal coliform test*	360	65	46	30	8	3	3
	Nitrates**	2.6	1.8	0.5	0.2	0.1	Less than 0.1	Less than 0.1
28 March (wet weather)	Faecal coliform test	280	56	40	20	6	38	29
	Nitrates	2.0	1.1	0.4	0.1	Less than 0.1	1.8	1.1

*Faecal coliform test is measured in colonies per 100 mL of sample **Nitrate test is measured in mg/L

Figure 4.2.4 is a map of the collection sites. The results of the tests are shown in Table 4.2.2.

Faecal coliform counts and nitrate levels are indicators of pollution. The faecal coliform count indicates sewage pollution and the nitrates may be from sewage or other forms of pollution.

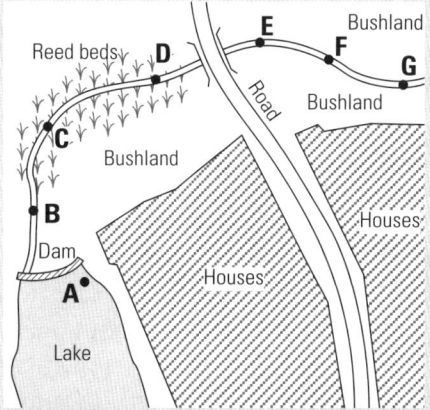

Figure 4.2.4 Map of the area showing collection sites.

ACTIVITIES

1 Graph the coliform count data. Create one graph for the dry weather and another for the wet weather. Describe the shape of the graphs.

2 What effect did the wet weather produce on the pollution in the creek?

3 What can you infer from the graphs about the role of the reed beds?

4 How do you interpret the data from the south side of the bridge?

5 Suggest why the values for the lake are so high.

6 Graph the nitrate results. Again, create one graph for the dry weather and another for the wet weather.
a Do they show a similar pattern to the coliform counts?
b How are the nitrate changes different from the coliform count changes?

7 In dry weather an urban area should have a coliform count under 200 colonies per 100 mL. What are the areas of concern along the area in terms of possible pollution?

8 What plan of action would you suggest to council in order to locate the source of the water pollution around the lake?

WATER ISSUES 225

4.3 Water and weathering

OUTCOMES

At the end of this chapter you should be able to:

- describe the water cycle in terms of the physical processes involved
- distinguish between chemical and mechanical weathering
- identify the role that water plays in breaking down rocks by:
  - abrasion
  - changes in volume of water during freezing
  - dissolving substances
  - acid attack
- identify data, plan and perform an investigation to demonstrate the effects of:
  - abrasion
  - changes of volume of water during freezing
  - dissolving substances
  and analyse information about the impact of these effects on the environment.

THE WATER CYCLE AND ROCK BREAKDOWN

Weathering is the breakdown of rock at, or near, the surface of the Earth. It occurs because rock exposed at the surface adjusts to very different conditions from those where the rock originally formed. As rocks interact with the atmosphere, biosphere and hydrosphere their physical and chemical properties change.

The conditions at the surface of the Earth that cause rocks to weather are related to temperature and the chemical environment. Living things influence the chemical environment of rocks by adding and removing chemicals to and from the environment, as well as allowing air and water to reach minerals deep within rocks.

When a rock is exposed at the surface by weathering and erosion it enters an environment of rapid temperature variation. Within the crust, rocks exist at relatively high temperatures that vary on a scale of thousands or even millions of years. At the surface, a rock body may experience a temperature change of 50°C in one day. The average temperature at the surface will also be very different from that at depth. Even in the hottest places, the average surface temperature will be less than 50°C. Temperature increases with depth in the Earth. Within the upper 20 km of the continental crust, the temperature rises by up to 30°C for every 1 km of depth. This means that in a sedimentary basin a rock at a depth of 4 km may have a temperature over 100°C.

The chemical environment at the Earth's surface contains materials that alter minerals in rocks. Water and the things it carries react with minerals. Oxygen dissolved in water contributes to weathering, as does carbon dioxide. The pH of the water also contributes to mineral breakdown. Salts liberated by weathering may contribute to the weathering of other minerals.

The processes making up the water cycle affect both temperature and the chemical environment. Precipitation and infiltration provide water to minerals deep within the ground. The ability of water to store heat means that the water may raise the temperature at the ground compared to what it may be if it was dry. Evaporation

> **weathering**
> the process in which rocks are altered

and transpiration, together with surface flow, remove water from the surface. Because water is so important in weathering, its removal may slow some types of weathering and allow other processes to dominate.

REVIEW ACTIVITIES

1 Define the term 'weathering'.

2 Compare and contrast the environments of a rock at the surface of the Earth with one 5 km below the surface.

3 Identify the processes that cause a rock to be exposed at the surface of the Earth.

4 Contrast the climate of an area on the far north coast of New South Wales with the Monaro region of southern New South Wales. How would the climate in each area affect weathering?

5 Review the diagram of the water cycle (Figure 4.1.2) on page 208. Tabulate the physical processes that would:
a increase the availability of water for weathering
b decrease the availability of water for weathering.

EXTENSION ACTIVITIES

6 If water is important as a cause of weathering, explain why high latitude areas, such as Siberia or Antarctica, have little chemical weathering but lots of water present.

7 List the features of an environment that allow water to penetrate to sites of weathering in a rock. Explain how the features vary within your local environment.

8 Design an experiment to test the idea that water can raise the temperature of ground compared to when it is dry. This experiment may involve the use of a data logger to measure temperatures over a period of time.

TYPES OF WEATHERING

Weathering is classified into two types: mechanical and chemical. The difference between the two types is their effect on the mineral grains making up the rock. **Mechanical weathering** breaks a rock into smaller pieces but does not alter the chemical composition of the rock's minerals. **Chemical weathering** does alter the minerals in a rock by removing or adding elements to minerals.

Both mechanical and chemical weathering can occur at the same time in a rock. The contribution each makes to a rock's breakdown depends very much on climate, particularly the temperature and rainfall. Chemical weathering dominates in warm wet environments and mechanical weathering is most obvious in places where temperature variations are high, such as deserts, or where water exists as ice.

Mechanical weathering

Mechanical weathering breaks rock bodies into smaller pieces. Some mechanical weathering processes rely on water but others do not.

ICE EXPANSION

Water from rain, or snow, regularly fills the cracks and openings of rocks. When the water freezes the ice produced is capable of splitting the rock into pieces. This happens because water's volume increases. It increases by up to 9% and the pressure of the ice on the surrounding rock may be as high as 110 kg per square centimetre. Repeated freezing and thawing leads to cracks becoming larger and eventually the rock breaks.

definitions

mechanical weathering
breakup of a rock without change in its chemical composition

chemical weathering
weathering involving chemical change within a rock

Ice expansion, or ice wedging, occurs where moisture is available, where fractures or cracks are common and where temperatures regularly fall below freezing. It is important to note that this form of weathering works best when freezing and thawing is repeated.

On steep slopes, ice expansion and gravity act together to form scree, or talus, slopes. When ice expands in soils, pebbles or layers of soil may be raised in a process referred to as frost heaving. A similar process may occur in highly saline areas where salt crystals grow from the evaporation of salt water. This does not, however, produce the amount of weathering that ice expansion does.

ABRASION

Abrasion is a process in which a rock is broken down as the result of something scraping over it. Wind and water both play a role in abrasion. The particles they carry strike or scrape past rocks, breaking pieces off. Good examples of abrasion occur in the cliffs along the New South Wales coast. (See Plate 20.)

In arid areas the lack of vegetation and water to bind soil particles means that winds carry small particles of sand and dust. The sandblasting produced by winds produces polished rock surfaces. Rocks with such surfaces are called vertifacts.

Water and ice are capable of transporting much larger particles than wind. (See Plate 21.) Sediment carried in a stream or river moves along the bottom of the channel by rolling, sliding and bouncing. The sediment particles abrade rocks on the channel floor and also other sediment particles. As a result, the sediment changes along the length of the river, becoming finer and more rounded.

Ice in glaciers can carry enormous boulders. These scrape along the bed of the glacier, leaving lines known as striations. Striations are used to determine the extent and movement of glaciers in the past. The abrasion produced by glaciers creates characteristic landforms and sediments.

OTHER FORMS OF MECHANICAL WEATHERING

Three other types of mechanical weathering that do not involve water are thermal expansion, exfoliation and organic activity. Each of these processes produces weathered material and the importance of the processes varies with location.

Exfoliation

Large rock bodies, such as igneous intrusions, form deep in the crust where pressures are high. The removal of material by erosion releases pressure on the rock bodies and they expand. Doming and uneven expansion cause the rock to fracture, resulting in jointing and sheeting. Joints are particularly important in the weathering process as they allow air and water to gain entry to the inside of these large rock bodies.

Thermal expansion

Cracking and breaking of rocks due to volume changes also occurs due to heating and cooling. In desert areas large variations in temperature between day and night cause rocks to crack.

Organic activity

Living things weather rocks in a number of ways. Plant roots can exert large forces inside cracks in a rock, causing the rock to split. Animals, too, can aid weathering by processes similar to abrasion. Hard hooves and burrowing both weather rocks.

Chemical weathering

Chemical weathering is the breakdown of rocks by a change in the chemical nature of the minerals in the rocks. Water is very important in chemical weathering because:

- water carries dissolved oxygen and carbon dioxide to sites where the chemical reactions occur
- water itself takes part in chemical reactions
- water can remove the products of weathering and allow fresh rock to be weathered.

REACTIONS WITH DISSOLVED SUBSTANCES

Dissolved substances in water include oxygen and carbon dioxide. Oxygen reacts with minerals to form new minerals. An example of this process, known as oxidation, is the reaction between the silicate olivine and oxygen.

Word equation: olivine + oxygen + water → hematite + silicic acid
Chemical equation: $2Fe_2SiO_4 + O_2 + 4H_2O \rightarrow 2Fe_2O_3 + 2H_4SiO_4$

Hematite is an iron oxide. It is deep red in colour. In deserts, hematite forms in thin layers of moisture on rocks, resulting in a characteristic red surface. Note in the example above that water is itself directly involved in the reaction. The yellow iron mineral limonite FeO(OH) is also formed by oxidation and also involves water in the reaction.

Pure water is neutral: it has a pH of 7.0. Rainwater, on the other hand, is acidic: it has a pH of 5.5 to 6.0. Rainwater is acidic because it contains dissolved carbon dioxide. The carbon dioxide and water form carbonic acid, the same acid that is found in soft drinks. Natural waters may be acidic due to materials produced by the decay of rotting plant matter. Pollutants from the atmosphere, such as nitrogen dioxide and sulfur dioxide, also produce acids.

Acids react with minerals to form new substances. The mineral calcite, found in limestones, reacts easily with acid. When a strong acid reacts with calcite it may produce bubbles of carbon dioxide, but in nature the reaction leads to the formation of dissolved bicarbonate. Acidic water in limestones gives rise to caves and to karst topography. Karst topography develops due to rock collapse as the limestone dissolves.

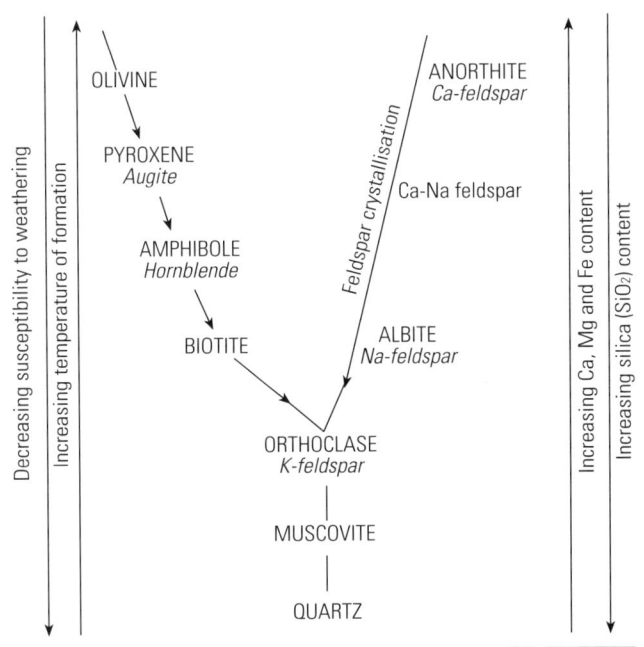

Figure 4.3.1 Weathering and silicate minerals. Minerals that form at high temperatures, such as olivine, weather more rapidly than those that form at low temperatures.

Word equation: calcite + carbonic acid → calcium ions + bicarbonate ions
Chemical equation: $CaCO_3 + H_2CO_3 \rightarrow Ca^{2+} + 2HCO_3$

Other minerals react with acids but they often do so more slowly than calcite.

Water, as we have seen, can take part directly in weathering reactions. It does so because it carries small electrical charges—positive charges on the hydrogen atoms and negative charges on the oxygen atoms—which allows water molecules to disrupt chemical bonds in a mineral.

In silicates, metal atoms often bind silicate tetrahedra to one another. Water molecules react in such a way that the metal ions are removed and replaced with hydrogen ions. Bonds between the tetrahedra are broken and this results in a smaller crystal, together with dissolved substances in the water. This process is called solution.

The tetrahedra combine with hydrogen to form orthosilicic acid $Si(OH)_4$. Some organisms, such as sponges and diatoms, build skeletons using this silica.

There is a relationship between a silicate mineral's rate of chemical weathering and its structure. (See Figure 4.3.1.) Minerals rich in calcium, magnesium and iron (such as olivine and plagioclase) are more prone to solution than minerals consisting mainly of silica (such as quartz).

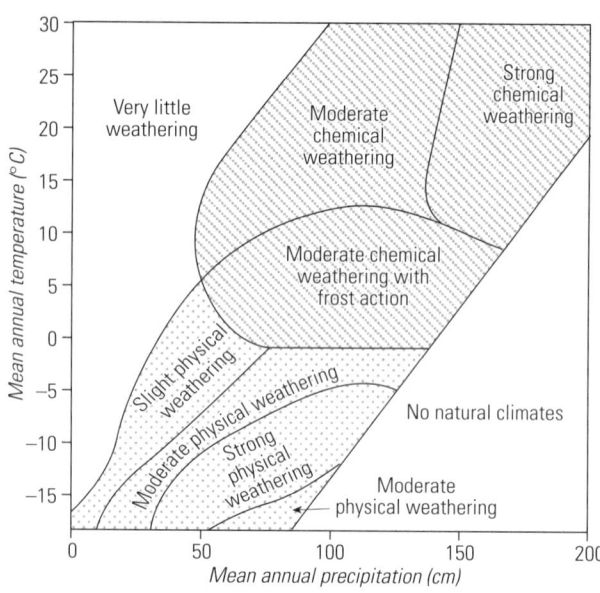

Figure 4.3.2 The relationship between climate and weathering. High temperatures and rainfall (precipitation) favour strong chemical weathering.

This section should suggest to you that there are four factors that determine the way rocks weather. Two factors are climatic: rainfall and temperature. The other two concern the nature of the rock: composition and texture.

Rainfall and temperature act together to determine the availability of water. (See Figure 4.3.2.) The temperature affects the rate of chemical reaction as well as evaporation and the freezing behaviour of the water. Note that only a thin film of water is needed for weathering processes. It may come as condensation from the air rather than as rain.

A rock's composition determines what types of chemical weathering it is prone to. Texture determines the surface of a mineral on which weathering can act. Large minerals have small surface to volume ratios compared with small mineral grains, so they may weather more slowly. On the other hand, large grains removed from a rock will expose more surfaces than a similar number of grains removed from a fine-grained rock. The presence of fractures and joints are other aspects of texture that affect weathering. They do so by allowing water to attack larger surfaces.

REVIEW ACTIVITIES

1 Compare chemical and mechanical weathering processes.

2 Contrast the weathering processes and products you would expect to find in the Snowy Mountains with those you would find in the arid centre of Australia.

3 Summarise the role water plays in the breakdown of rocks.

4 Explain why soils in hot, humid environments are often deeper than soils formed in cold, dry environments.

EXTENSION ACTIVITIES

5 Explain how chemical weathering occurs in deserts.

6 How does a slope affect the rate of weathering in the rocks that make up the slope?

7 Discuss the role of living things in causing weathering.

SUMMARY

 Weathering is the breakdown of rocks at, or near, the Earth's surface.

 Chemical weathering alters the composition of weathered material but mechanical weathering does not.

 Water is an agent of physical weathering when it freezes and thaws and when it carries materials that abrade rocks.

 Water acts as a solvent and transports oxygen in chemical weathering.

 Different rock materials weather in different ways and at different rates.

 The landscape in an area reflects the action of water as a weathering agent and as an agent of erosion.

PRACTICAL EXERCISE
Weathering: A first-hand investigation

In this investigation you will plan and perform an experiment to demonstrate the effects of abrasion or dissolving substances.

Abrasion
The resistance of a rock type to abrasion can be determined by placing eight to ten similar sized pieces in a container and shaking them for a known period of time. Changes in the following can be used to assess the effects of abrasion:
- the shape of the pieces
- the number of pieces
- the volume of abraded fragments.

Dissolving substances
A number of rock fragments can be placed in a warm weak acid solution for three to five days. If a known amount of the acid is removed and evaporated, dissolved salts in the acid can be measured. The evaporation of a fresh acid sample can be used as a control. Limestone is particularly good for such an experiment.

ACTIVITIES

1
Plan and carry out an experiment to assess the abrasion resistance of three local rock types. Ensure you consider:
a any safety issues and how you will deal with them, such as dust and the strength of the container
b what you will measure
c how you will compare results
d a hypothesis to test your results.

2
Assess your results. Comment on:
a whether your results agree with the information in the text
b how your design could be improved.

4.4 Water resources: Past and present

OUTCOMES

At the end of this chapter you should be able to:

- describe evidence in rocks confirming the past presence of large bodies of water in inland Australia (eg limestone, marine fossils, shallow marine and lacustrine sediments) and for each type of evidence, a place (in New South Wales or Australia) where this evidence may be found

- recall pollution as contamination by unwanted substances

- discuss methods used to conserve water including the reuse of water, after treatment

- examine efficiency of water usage in Australia and locally

- outline problems that may occur in ground water systems, such as pollution, salt water intrusion and ground salinity, and give examples of these problems occurring in Australian environments

- outline one State or Federal government policy related to the use of ground water and possible scientific solutions to identified environmental problems associated with the use of ground water

- gather information from secondary sources to summarise landscape features that may identify past aquatic environments

- gather, process and present information as a case study, and use available evidence to illustrate the impact on one or more ecosystems of a change in climate, including a change in water availability

- gather information from secondary sources and use available evidence to present an outline of one environmental problem identified in New South Wales that has arisen from the use of ground water in the past.

EVIDENCE OF PAST WATER BODIES IN AUSTRALIA

Australia is a dry continent. Its centre does not contain the inland seas that early explorers imagined, nor does it contain large bodies of permanent water.

Figure 4.4.1 shows the distribution of wetlands in Australia. Wetlands are areas of land that are naturally flooded on a permanent or intermittent basis. The occurrence of episodic freshwater lakes suggests that, at times, bodies of water are to be found in what is regarded as the dry centre of Australia.

Scientists believe that Australia has not always been dry. The continent has become increasingly arid as continental drift has moved Australia northward. What is the evidence to support the idea of a wetter Australia? If large bodies of water once existed in inland Australia how can we recognise their remains?

A study of modern lakes and shallow marine environments provide scientists with a set of characteristics that might be preserved within rocks. By observing structures,

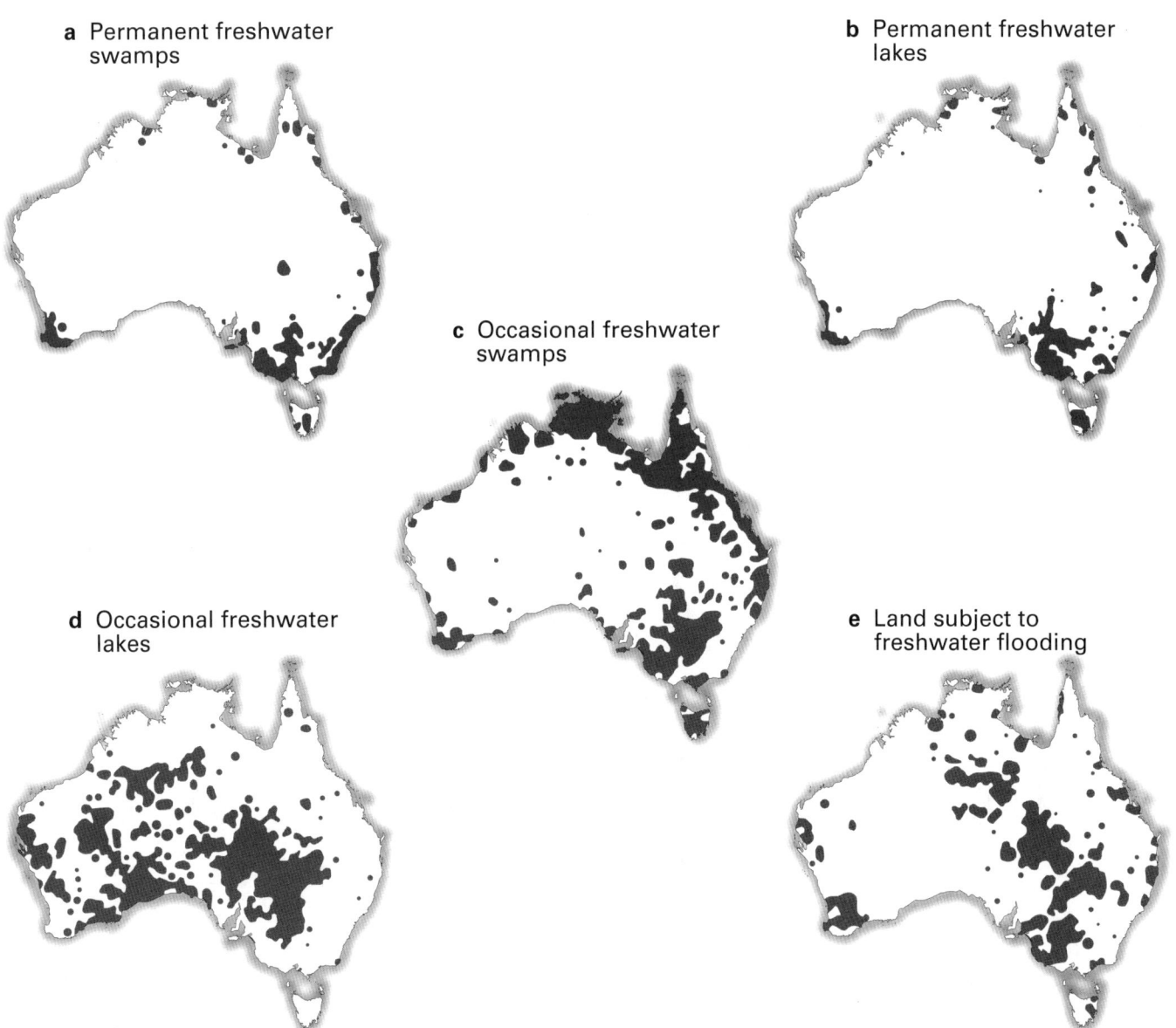

Figure 4.4.1 The distribution of wetlands in Australia. Many areas experience water that persists for short times. Organisms have adapted to these conditions.

fossils and rock types and comparing them with modern environments, scientists can make reconstructions of past environments.

Lakes are characterised by particular types of sediments and by the organisms that live there. Quiet, poorly mixed waters result in dark, layered muds on the lake bottom. Deltas, composed of sands and muds, contain recognisable structures. These can be used to identify the edges of the lake. Shallow water may lead to characteristic structures, such as ripple marks forming in the sediment due to winds acting on the water. Evaporation of water may result in evaporite deposits, such as halite or gypsum.

Terrestrial plants and aquatic organisms may be preserved as fossils in the lake. Worms mix up muds, leaving burrows and other evidence of their presence. The skeletons of diatoms (a type of microscopic life) are preserved in the lake sediments, and water plants often show distributions according to depth.

Shallow marine environments are also characterised by physical and biological factors. The type and amount of sediment may indicate shallow marine conditions. Waves, tides, currents and storms produce recognisable structures in the sediments.

Marine organisms may be preserved in the sediments of shallow seas. If the water is warm and shallow, reefs may occur and leave their mark in the geology of the area.

Sedimentology is the study of the formation and composition of sediments and sedimentary rocks. Together with palaeontology, the study of past life, sedimentology allows us to understand past lakes and seas. We will briefly consider three examples of evidence for large bodies of water in inland Australia.

Jenolan Caves

The Jenolan Caves, west of Sydney, have formed in folded and faulted limestones and shales. Originally, the sediments were about 170 m thick. The sediments were deposited in a shallow marine environment. The environment is known because of marine fossils found in the limestones. The fossils include corals, brachiopods (see Plate 23) and conodonts. The fossils have allowed palaeontologists to date the limestones. They were deposited during the late Silurian period, 414–408 million years ago.

At the same time, rocks near Molong, in the central west of New South Wales, record a shallow marine environment containing coral reefs, mudflats and a lagoon

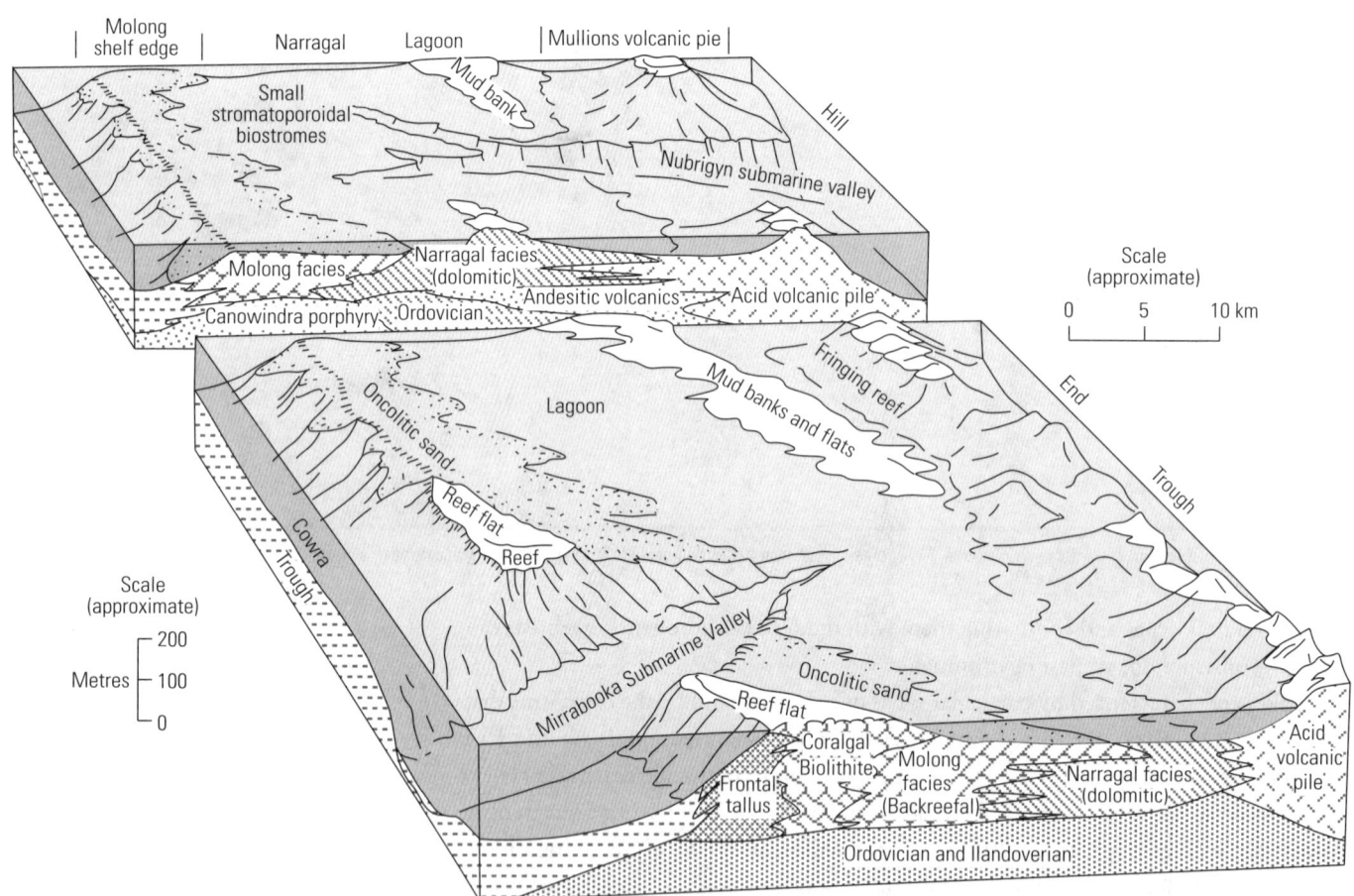

Figure 4.4.2 The types of sediments and environments during the Silurian period in what is now the area near Molong in the central west of New South Wales.

along the western edge of a volcanic island arc. The interpretation is based on a variety of shallow marine fossils, the distribution of sedimentary rock types and structures within the sedimentary rocks that indicate currents and other information. (See Figure 4.4.2.)

Redhead Beach

On the coast just south of Newcastle is Redhead Beach. At low tide you can stand on the rock platform at the northern end of the beach and observe a delta in the cliff.

(See Plate 24.) Giant cross-beds in the upper cliff represent the delta itself. (See Figure 4.4.3.) The delta overlies the Fern Valley coal seam. The rocks are late Permian age and the delta is composed of conglomerates and sand. Fossils in the coal tell us that the environment was a lake or brackish marine area. The climate was cold and a great deal of plant material was being buried as the delta spread into the water. Some of the coal is overlain by material that originated as ash from volcanoes to the east. At this time the formation of the Tasman Ocean by rifting was yet to occur. The conglomerates tell us that the rivers feeding the delta were fast flowing and the conglomerate pebbles are composed of volcanic rocks from the east.

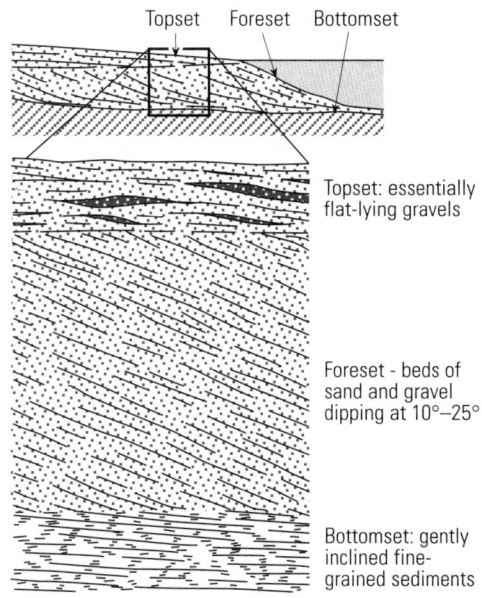

Figure 4.4.3 The structure of a delta and the arrangements of sediments within the delta. The foresets and bottomsets shown can be seen in the cliff at Redhead Beach near Newcastle.

The rapid burial of plant material by the delta as it grew out into the lake is an important factor in the origin of coal near Newcastle. It is possible to use the foreset angles and directions to determine the depth of the lakes and also the direction in which local rivers flowed. Smaller scale cross-beds are common features in many sandstones in New South Wales. More detail on such structures can be found in other textbooks dealing with sedimentology.

Lake Mungo

In south-western New South Wales are the Willandra Lakes. During the last ice age, evaporation rates were less than they are today and the lakes were full from about 55 000 to 30 000 years ago. A short dry period 36 000 years ago saw sandy beaches form on the eastern sides of the lakes, and during dry summers the sands formed dunes. Thin clay layers and salt-rich layers in the sediments indicate other dry periods. Today, the eroded dunes form the Walls of China at Lake Mungo. The history of the lake's changing climate is preserved in the dune's layers of sand and clay.

Aboriginal people lived on the edge of Lake Mungo 40 000 years ago. Artefacts of this age have been dated using radiometric dating. Middens contain the shells of freshwater mussels and the bones of fish. The fossils of other animals have also been found in the lake.

At the start of the Pliocene epoch about 5 million years ago, global sea levels were high and the Murray Basin experienced a return to shallow marine conditions. Sediments deposited near the Southern Ocean included limestones and these are interpreted as shallow marine in origin due to the fossils found there. Marine and freshwater sediments mark a zone where the coastline would have moved backward and forward. Pollen collected from sediments provides information about the plants that lived near the sea. They include chenopods—a group of plants that are found around the world in salt marshes and sand dunes. You may note that this group also contains the saltbushes and bluebushes that characterise arid areas of New South Wales today. The group possessed the adaptations to succeed when the Australian continent began to dry out.

During the Pliocene epoch the changing sea level caused changes in the pattern of sedimentation. In Victoria, deep valleys became choked with sediment and prominent lines of beach ridges formed as the sea level changed.

REVIEW ACTIVITIES

1 Describe the sorts of evidence that would indicate an area was once a lake.

2 Limestones formed in coral reefs indicate warm, shallow seas. Explain why corals are indicators of such environments.

3 For each of the following items, name a place in New South Wales where it may be found:
a marine fossils
b limestones
c shoreline sand dunes
d delta sediments
e fossilised plant material deposited in a lake
f fossils of aquatic organisms that lived in a lake.

4 Compare the characteristics of an ancient shallow marine area and a freshwater lake.

5 Construct a diagram to illustrate the different environments of sediment deposition. Include lakes, shallow marine areas, deltas, organic reefs and lagoons in your diagram. Suggest what common sediment type will occur for each environment.

EXTENSION ACTIVITIES

6 Summarise the changes that have affected drainage patterns in Eastern Australia during the last 100 million years.

7 Hawkesbury Sandstone is common around Sydney. Research the properties of the rock unit and the interpretations of the sediment's origin and depositional environment.

8 Contrast two types of fossils from different marine environments. Describe the information about an environment that can be inferred from the presence of these fossils.

TECTONICS, TOPOGRAPHY AND DRAINAGE

Australia's climate has changed during the last 80 million years as Australia has moved north. It is easy to see that the continent's climate should warm as it moves further from the South Pole. What is less well understood by most people is that tectonic processes, such as rifting and volcanism, also altered the way rivers move across the country and the elevation of parts of New South Wales. Such changes have affected the evolution of Australia's ecology and determined the position of some refugia for native animals and plants.

Today, Australia is characterised by rivers that drain towards the centre of Australia. Rivers within New South Wales either drain to the east if they are on the eastern side of the Great Dividing Range, or toward the west if they are on the western side of the range. But it has not always been this way. Changes in topography due to plate tectonics have changed the way rivers move and the way sediments are distributed.

Using satellite images and studying ancient channel patterns, scientists have been able to determine the history of some modern rivers. During the Jurassic and Early Cretaceous periods the ancestors of many NSW rivers drained towards the west or northward towards the Surat Basin.

During the Mid-Cretaceous period (about 110 million years ago) the uplift of the Great Dividing Range began as the Tasman Ocean started to form by rifting. Approximately 80 million years ago, during the Cretaceous period, rifting between Australia and the New Zealand subcontinent reached a point where the sea flooded the central rift valley. This led to the rivers flowing from the New Zealand subcontinent being cut off.

The formation of the Great Dividing Range caused some of the rivers flowing towards the east to reverse their flow direction. Subsequent erosion of the eastern edge of the Great Dividing Range and volcanic activity has resulted in the complex river patterns along the coast. To the west, the dendritic pattern of rivers has been preserved.

At the beginning of the Tertiary period (about 65 million years ago) the Murray Basin began to subside, or sink, as the southern margin was opened to the sea. This led to a cutting off of drainage to the north and the current drainage within the Murray Basin began to develop, flowing towards the west.

During the Eocene epoch (58–37 million years ago) volcanic activity created a shield volcano at Brown Mountain, damming some of the north-flowing rivers. This caused redirection of the rivers into the Snowy and Murrumbidgee Rivers.

Later, during the Miocene epoch (23.5–5.3 million years ago), volcanism and faulting in south-eastern Australia changed the topography. The Eastern Highlands were elevated as fault blocks hundreds of metres above the surrounding countryside. It was at this time that Lake George, near Canberra, formed in a graben, which is an area between raised fault blocks. Lake George has been a site of deposition ever since, and from its sediments scientists have extracted pollen with which to trace vegetation changes during the Cainozoic era. The movement of the fault blocks also helped change the flow of rivers into the Snowy and Murrumbidgee Rivers.

These changes in topography set the scene for the great evolutionary changes in Australia's vegetation and fauna during the Miocene epoch. During that epoch, Australia's climate changed from being warm and wet to cold and dry. In Antarctica the ice cap began to form and, because more of the Earth's water budget was locked up as ice, Australia became drier. By 6 million years ago there was more ice in Antarctica than there is today and, as a result, the sea level was up to 100 metres lower than it is now.

REVIEW ACTIVITIES

1 Summarise the changes in topography that have occurred in New South Wales during the last 80 million years.

2 Describe why Australia's climate changed during the Miocene epoch.

3 Explain how a lower sea level would affect the Australian coast.

EXTENSION ACTIVITY

4 Thirty million years ago the Nullarbor Plain and the Murray Basin were part of shallow seas. The fossils of Riversleigh in Queensland also date from this period. Describe some of the materials deposited in these places and the effect the deposits have had on human society during the twentieth century.

WATER POLLUTION AND CONSERVATION

Water maintains natural Australian ecosystems and human communities. As we go about our lives we make use of water in many ways. On average, during a year each person in Sydney accounts for 180 tonnes of water entering the city. Some water is used for industry (60% of industrial waste is water) but a lot is processed in our homes. Gardens, showers and baths account for almost 80% of water use in the home. As human development increases the size of our cities and towns, our use of water continues to grow.

definition

resource
something that humans use

Water in the Australian environment is a finite **resource**. We cannot easily increase the amount of water available. By using bores to access water stored in the ground (ground water) we are using more of a fixed amount rather than finding a new source. Ground water is replaced relatively slowly compared to how we use it.

On the coast and in the Murray-Darling Basin our use of surface water has degraded parts of the environment. This has occurred due to irrigation and power generation. These activities change the amount of water available to ecosystems and alter the natural behaviour of rivers and streams. River ecosystems have evolved to make use of the water normally available. By reducing the amount of water available, we change the processes within the environment and cause long-term changes to it.

In order to ensure that adequate water is available to the natural environment and our needs, we must strive to conserve water. Conservation not only involves using less but also maintaining the quality of the water we have. In this section we will consider ways that water can be conserved in both urban and agricultural settings.

Urban water problems include the amount of waste water produced, urban run-off and stormwater and problems related to erosion. While there are a range of problems some of them have common solutions.

The amount of water used in urban areas can be controlled in a number of ways. These include appropriate water-pricing schemes to encourage the conservation and reuse of water. Encouraging the use of water-efficient appliances around the home can significantly reduce water use.

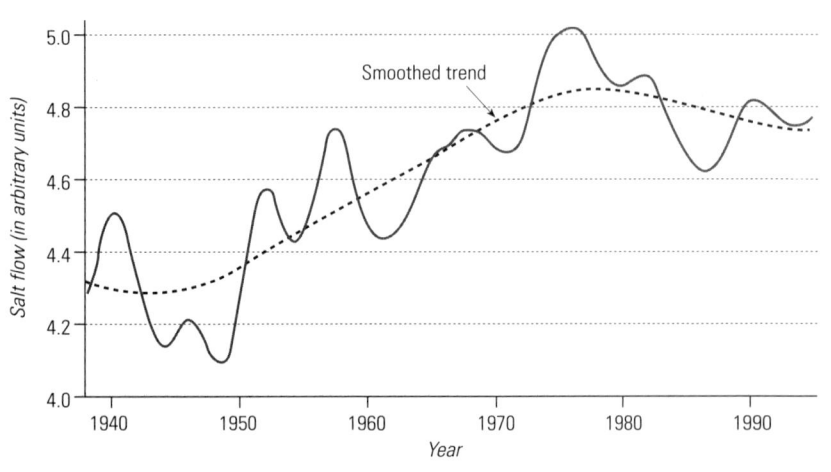

Figure 4.4.4 Salinity in the Murray River at Morgan from 1940 until 1995.

Regulation can also play a role. Town planning can take account of water-sensitive design principles. Current regulations that restrict urban collection and storage of rainwater or on-site use of stormwater inhibit the efficient use of water. Regulations controlling pollution also contribute to the quality of water in urban areas.

Our use of vegetation can also conserve water. Wetlands and reed beds act as efficient filters of nutrients and sediments. Their use can reduce the effect of industrial wastes, sewage and stormwater. Managing vegetation along streams can reduce the amount of water run-off and soil erosion. Encouraging native gardens can also reduce the amount of water used in the urban setting.

In agricultural areas five issues affect water conservation. The first is efficient water use. Irrigation practices that lead to large evaporation losses are inefficient. Salinity is another issue that affects water in a significant way. Overclearing and soil erosion are processes that lead to water pollution. Overclearing increases water run-off and erosion. Erosion adds unwanted sediment to waterways. The fourth issue is farm run-off. Pesticides, organic matter and fertilisers all affect water quality. The last issue is waterway vegetation. Besides reducing run-off and erosion, such vegetation provides habitat and corridors for wildlife. Rich, diverse habitats, such as the reed beds mentioned earlier, also help to maintain water quality.

While many of the measures outlined above conserve water through improving its quality, we can do more to use water more efficiently. Shower water, for example, is disposed of in the same way as sewage, but the quality of the two is very different. Reuse of water that is only slightly altered by use, so-called 'grey water', reduces our use of drinkable, or potable, water. If half the water used in gardens was grey water rather than clean water we could reduce our freshwater demand by one-tenth.

REVIEW ACTIVITIES

1 Summarise the factors that affect water quality in:
a urban areas
b rural areas.

2 Define 'grey water'. Explain how the use of grey water can conserve water.

3 Discuss methods that can be used around the home to reduce water usages.

EXTENSION ACTIVITIES

4 The quality of water affects the life that lives in it. Identify the pollutants in the water that affect the growth of plankton.

5 Explain two ways by which inefficient irrigation practices affect the water available in the local environment.

6 Figure 4.4.4 shows the salinity increase in the Murray River at Morgan from 1940 to 1995. Since 1977 a number of salinity mitigation schemes have been completed upstream of Morgan.
a Describe the trend in salt flow from 1940 to 1970.
b Propose a reason why salinity levels between 1940 and 1950 are low. Do any other parts of the graph support your proposal?
c Discuss the degree to which salinity mitigation schemes have reduced the salinity measured at Morgan.

GROUND WATER: ISSUES, REGULATION AND STRATEGIES

Ground water is water stored in **aquifers**. More than half the population of Australia are totally dependent on ground water and 80% use more ground water than surface water. In addition to humans, ground water sustains many wetlands and other plant communities.

Ground water can be found in sands and gravels as well as rock aquifers. Plate 25 shows the way ground water in aquifers is stored and accessed. Water enters porous rocks in what are called **recharge** areas. Gravity causes the water to move through the aquifer. It cannot escape from the aquifer because the layers around it are non-porous. When a bore drills through the non-porous rocks, the water can rise to the surface. Because the water is under pressure it may rise higher than the watertable where the bore is drilled.

Despite the importance of ground water, our society has not always treated the resource well. When ground water is removed faster than the aquifer can recharge, the bore becomes dry. Water moves through an aquifer very slowly.

Salt water intrusions

The removal of ground water not only changes the amount available, but also can lead to **salt water intrusion**. Plate 26 shows a situation where saline ground water from an ocean lies alongside a freshwater aquifer. This is stable because the freshwater is less dense than the salt water. As the freshwater is used, the saline ground water rises and the watertable falls.

Aquifers are most common in sedimentary basins (such as those shown in Figure 4.4.5, page 240), but they can exist in fractured igneous rocks. Figure 4.4.6 (page 240) shows the underground water and surface water across north-east New South Wales. In some areas, ground water is quite saline because it dissolves salt from the rocks it passes through.

> **ground water**
> water that is stored in rock below the watertable
>
> **aquifer**
> a permeable rock unit saturated with ground water
>
> **recharge**
> the process where surface water seeps into an aquifer
>
> **salt water intrusion**
> the replacement of fresh ground water with saline ground water

1. Great Artesian Basin
2. Gippsland Basin
3. Area of ground water use in Tasmania
4. Otway Basin
5. Murray Basin
6. Cowell Basin
7. Willochra Basin
8. Pirie-Torrens Basin
9. Eucla Basin
10. Officer Basin
11. Albany Basin
12. Perth Coastal Basin
13. Carnarvon Basin
14. Canning Basin
15. Fitzroy Sub-Basin
16. Cambridge Gulf Basin
17. Daly River Basin
18. Wiso Basin
19. Georgina Basin (NW)

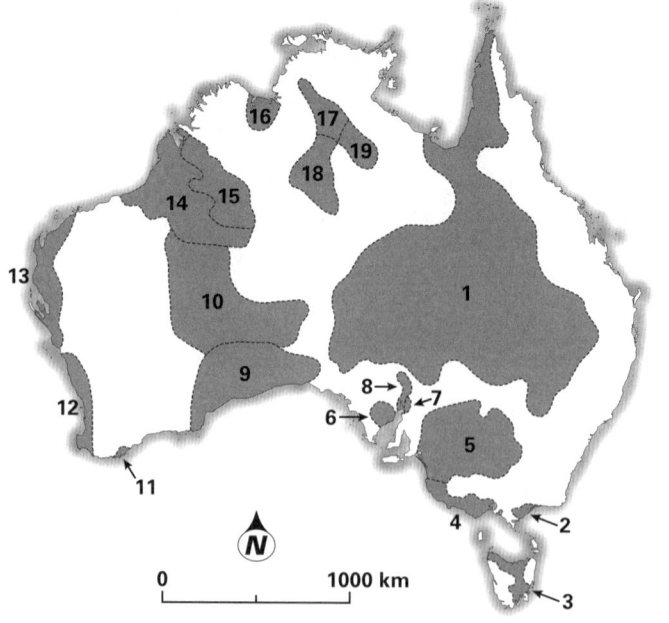

Figure 4.4.5 Ground water basins of Australia. The major basins in New South Wales are the Great Artesian Basin and the Murray Basin.

	Great Artesian Basin		New England Tableland and Slopes		Moreton-Clarence Basin		Coastal sands
	Sedimentary rocks and inland alluvium		Fractured igneous and sedimentary rocks, alluvium restricted to major valleys		Sedimentary rocks and coastal alluvium		Unconsolidated sand
Precipitation	450–630 mm per annum		630–1000 mm per annum		1000–1500 mm per annum		1500 mm per annum
Potential evaporation	1270–1780 mm per annum		1140–1270 mm per annum		1000–1140 mm per annum		1000 mm per annum
Surface water	No major dam sites. Few widely spaced perennial streams. Storage subject to strong evaporation		Good major dam sites and catchments. Numerous perennial streams. Frequent rainfall		Some major dam sites. Numerous perennial streams. Frequent summer rainfall		Poor surface storage
Ground water	Basin: porous sedimentary rocks	Alluvium: unconsolidated sediments	Fractured rocks	Alluvium: unconsolidated sediments	Basin: porous and fractured sedimentary rocks	Alluvium: unconsolidated sediments	
	Bores: depth 15–1200 m; yield 0.5 L/s (pumped) and up to 62.5 L/s (artesian flow); salinity <3000 mg/L	Bores: depth 15–75 m; yield up to 37.5 L/s; salinity <1000 to >14 000 mg/L	Bores: depth 15–90 m; yield 0.1–1.0 L/s, salinity <1000 –3000 mg/L	Bores and wells: depth 5–15 m; yield up to 19 L/s, salinity <1000 mg/L	Bores: depth 15–30 m; yield 0.1–0.6 L/s, salinity <7000 mg/L	Bores and wells: depth 3–30 m; yield up to 6 L/s, salinity <1000 mg/L	Bores: depth 6–30 m; yield 0.4–L9 L/s; salinity <1000 mg/L

Figure 4.4.6 A cross-section across northern New South Wales showing the nature of aquifers and surface waters. Note how the precipitation and potential evaporation change from east to west.

As bores remove freshwater from the ground, the area of freshwater is reduced and salt water moves inland. This may render bores unusable. Some bores draw salt water upward at the same time as they draw the top of the ground water (the ground water table) downwards. Eventually the aquifer becomes so saline that it is unusable.

The control of salt water intrusions is costly. It can involve digging recharge wells to add water to the aquifer, moving wells or reducing their abstraction. One human impact that adds to the problem of salt water intrusions is changes to flooding patterns. By controlling rivers and reducing natural floods that recharge coastal aquifers, we make salt water intrusion more likely.

Pollution

Figure 4.4.7 shows the way urban processes can affect ground water recharge. Not only do impermeable surfaces alter the absorption and recharge rates but pollution can enter the ground water system. Once ground water is contaminated it is a costly and long-term process to clean it up. In parts of Europe, there is widespread contamination of ground water due to agriculture and industry and it has been rendered undrinkable.

Salinity

Salinity in soil and rocks may result from a number of processes. Salt deposited in marine conditions may affect salinity levels. Salt in the marine sediments may dissolve in ground water. During floods, inland rivers carry salts to lakes. Evaporation results in salt crusts on the shores and bed of the lake. Winds can spread this salt to adjacent areas.

Human impacts on the environment are a major cause of salinity in many parts of New South Wales. Salinity reduces the productivity of agriculture and can damage buildings and other forms of infrastructure.

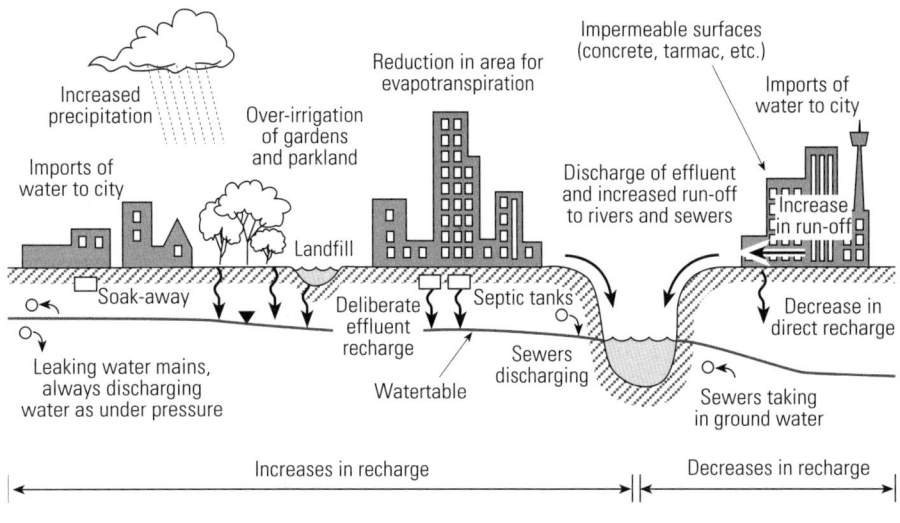

Figure 4.4.7 The effects of urban development on ground water recharge and pollution.

TYPES OF SALINITY

Salinity can be classified into five types: irrigation, dryland, urban, river and industrial salinity.

Irrigation salinity

Irrigation salinity occurs when large volumes of water used in irrigation raise the watertable. As the level of water rises it brings salt into the root zone of plants. Evaporation of irrigation water also concentrates salts at the surface.

Dryland salinity

Land clearing usually leads to deep-rooted plants being replaced by shallow rooted pastures or crops. (See Figure 4.4.8, page 242.) As a result, less water is removed from the soil and it reaches the watertable. The watertable rises, bringing salt with it. The toxic effect of the salt may create less impact than waterlogging and changes to soil structure. The net effects, however, are to kill salt-sensitive plants, create bare areas called salt scalds and increase levels of erosion.

salinity
the amount of salt in a liquid; the problem of high levels of salt in a liquid

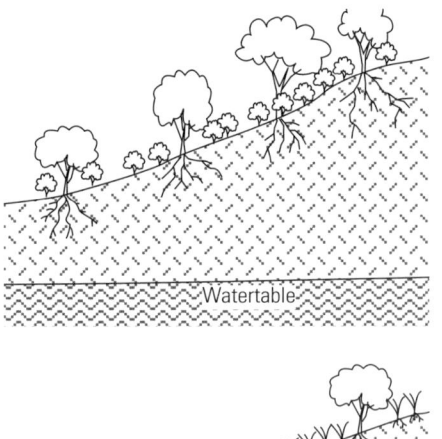

a. Under native vegetation
- Evapotranspiration is high
- Water is drawn out of the soil by deep roots
- Little water seeps deep into the ground
- Salts gradually accumulate in the soil
- The watertable is deep, and possibly saline

b. After trees are cleared
- There is less evapotranspiration
- Crops and pasture have only shallow roots
- Water moves through the soil, flushing out stored salt
- Water penetrates deeply into the soil, raising the watertable close to the ground surface

Figure 4.4.8 The process that creates dryland salinity. Clearing of native vegetation alters the position of the watertable.

Urban salinity

Urban salinity is a combination of the processes that cause dryland and irrigation salinity. Clearing and the over-irrigation of gardens and parks raises the watertable. The salinity that results damages roads and increases building costs in affected areas. Recent studies indicate that at least forty towns in New South Wales are affected by urban salinity.

River salinity

Water run-off from irrigation salinity, dryland salinity and urban salinity adds salt to rivers and creeks. Over time, the salinity of rivers increase. This is influenced by changes to stream flow caused by dams and removal of water for irrigation. Less water in major waterways means less dilution of the salts entering it from affected areas.

Industrial salinity

Wastes from industry and agriculture contain high levels of salt. Industrial processes also concentrate salt or generate saline water. Mines, for example, have to control water generated on mine sites, and it has high salt levels. Coal-fired power stations use water for cooling. Evaporation from cooling towers and adjacent lakes concentrate salt.

REGULATION AND STRATEGIES

Saline water generates costs for people. We need clean water for drinking and manufacturing. Water with high salt concentrations (hard water) damages such things as hot water systems and household appliances. Hard water also leads to people using more soap and detergents, which further degrade waste water quality. Agriculture must deal with land becoming unproductive to sensitive pastures and crops.

Salinity creates problems for natural systems. Changes to salinity alter the types of plants and animals that can live in such water. Blue-green algae can survive in saline conditions and they cause the problems described earlier in the chapter.

State and Federal Governments have had to generate policies to address salinity. Salinity does not act locally. For example problems created in New South Wales can affect areas in Victoria and South Australia via the Murray-Darling system.

Some of the policies that have been generated to help deal with salinity include:
- amendments to the *Soil Conservation Act 1938* (NSW) and the establishment of the Department of Land and Water Conservation
- policies dealing with Total Catchment Management that aim to address issues such as land clearing and erosion
- the *Native Vegetation Conservation Act 1997* (NSW), which controls land clearing
- policies relating to water and native vegetation conservation as well as sustainable agriculture practices and biodiversity.

Salinity is an urgent problem for New South Wales. New initiatives and prospects arise regularly, so you should do some reading of current events in this area.

Science plays a big role in solving the salinity problem. Methods of detecting and monitoring salinity have, and are, being developed. Science is helping us to manage native vegetation and learn about the mix of land uses that will allow less recharge of ground water. Scientists are also engaged in developing ways of using water effectively so that both human activities and natural systems receive the water they need.

REVIEW ACTIVITIES

1 Summarise the problems caused by human impacts on groundwater systems.

2 Explain how a salt water intrusion occurs.

3 Describe the major forms of salinity.

4 List the ways in which salinity affects the environment.

5 List some of the policies that are designed to reduce salinity in the environment.

EXTENSION ACTIVITIES

6 Identify parts of your local area that experience salinity problems.

7 Assess the value of one policy document dealing with salinity that you have studied.

8 Outline ways in which science can assist in dealing with salinity problems.

SUMMARY

- Rock types, structures and formations provide evidence of large bodies of water in areas of Australia that are now dry.

- Changes in tectonic environments have altered topography and drainage systems in New South Wales during the geological past.

- Changes in water availability produce changes in natural ecosystems and human societies.

- Pollution occurs when unwanted substances are added to the environment.

- Ground water systems are of great importance to Australians.

- Ground water systems can be affected by overuse, pollution and saltwater intrusion.

- Water can be conserved by efficient use and treatment.

- Salinity is a major problem in parts of New South Wales. Inappropriate use of ground water can affect salinity levels.

- Salinity strategies are examples of policies that aim to deal with environmental problems stemming from the use of ground water.

PRACTICAL EXERCISE
Past aquatic environments

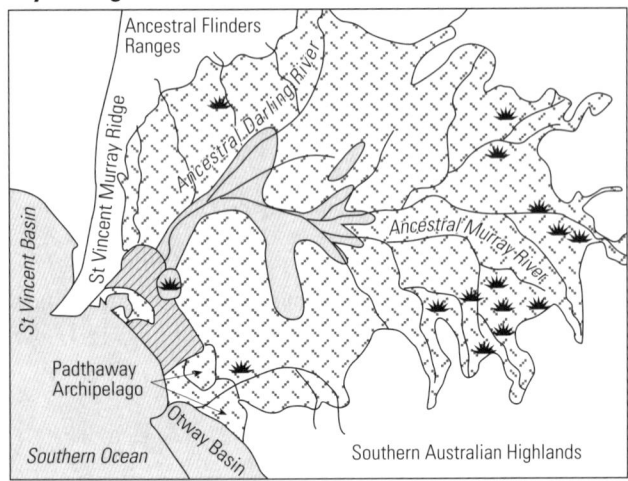

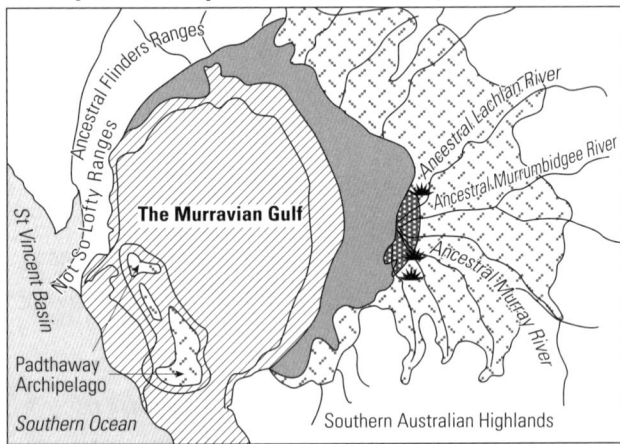

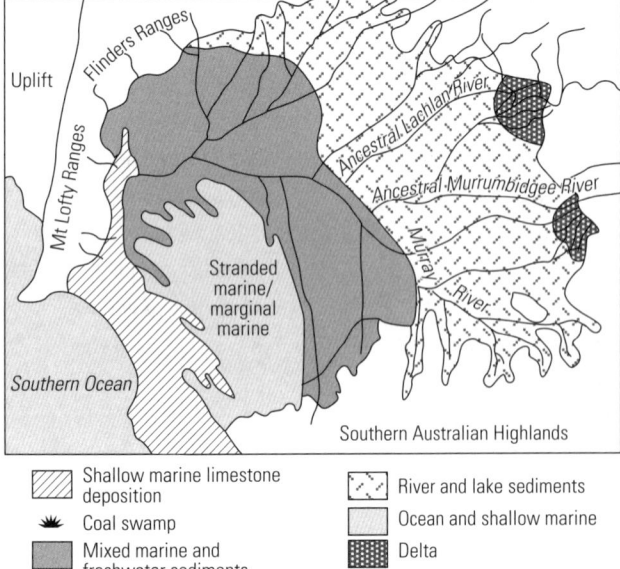

Figure 4.4.9 Evolution of the Murray-Darling Basin from 40 to 5 million years ago.

The purpose of this exercise is to assist you in gathering, processing and presenting information about the affect of climate change on ecosystems.

Background
The Murray-Darling Basin is one of the most intensely studied areas in Australia. The basin provides us with a wide range of resources and, like many intensely managed areas, it contains a number of serious environmental issues.

In this exercise we will look at the Murray-Darling Basin as it was during the period from 40 to 5 million years before the present. You may have to research the procedure activities below to gather the information you need to complete the summary activities.

ACTIVITIES

Procedure

1
Examine Figure 4.4.9. It shows three palaeomaps of the Murray-Darling Basin during the Late Eocene to Early Oligocene epochs, the Early Miocene epoch and the Early Pliocene epoch. Study the maps and try to work out where the present-day coastline is in each map.

2
Describe what was happening to the sea level during the period represented by the three maps. How can you be sure that the sea level was changing rather than the basin rising or sinking?

3
What sorts of climate do coal swamps and shallow marine limestone depositions suggest? Can you confirm what the climate was like during the time in question?

4
Suggest what sort of plants would have lived along the ancestral rivers.

5
For each map, describe the environment and the changes it was undergoing.

6
How were the biotic and abiotic factors within the basin different from the ones that exist there today?

Summary

7
Describe how plant communities within the basin would have coped with the changes shown in the maps.

8
How did the climate differ from the one we have today?

9
What would you propose as the major similarities and differences between the basin in the Miocene epoch and the present?

PRACTICAL EXERCISE
Preparing a case study on environmental change

The purpose of this exercise is to gather information and use available evidence to present an outline of a NSW salinity problem.

Background
In August 2000 the NSW Government released the NSW Salinity Strategy. The document identified that to slow the increase in salinity, the following actions are needed:

- protect and manage native vegetation
- use the land in ways that reduces the amount of water entering the watertable
- use water more effectively and efficiently
- use engineering solutions
- make better use of land affected by salt
- focus the efforts on areas that are most at risk.

ACTIVITIES

1 Prepare a strategy for searching the library about salinity in New South Wales. In selecting search terms consider scientific terms, places such as the Murray-Darling Basin and organisations such as the Department of Land and Water Conservation.

2 Obtain and review a copy of the NSW Salinity Strategy. If you cannot gain access to a hard copy of the report try to find the report at the Department of Land and Water Conservation website (www.dlwc.nsw.gov.au)

3 On the basis of information in the report, identify an area suffering from salinity and find out all you can about the history of salinity in the area and the steps being made to slow the development of salinity.

4 Prepare a short talk on salinity in the area of New South Wales you studied. Prepare five small posters or a PowerPoint presentation to assist you in explaining your topic. In your talk, include information about the steps that have been taken to prevent further salinisation in the area you researched.

RESOURCES

Natural phenomena

Earth's Water Budget: The Numbers
ess.geology.ufl.edu/HTMLpages/ESS/GLY1033_notes/numbers.html

Global Climate Change
jrscience.wcp.muohio.edu/html/globalchange.html

Global Weather Machine
www.pbs.org/wgbh/nova/elnino/anatomy/machine.html

Globe Program *globe.fsl.noaa.gov/*

NASA: Global Change Master Directory *gcmd.nasa.gov/*

NASA: Photo Gallery *www.nasa.gov/gallery/photo/index.html*

Ocean: Virtual Field Trip *www.field-trips.org/sci/oceank/index.htm*

Ocean World *oceanworld.tamu.edu/*

Oceanography from the Space Shuttle
daac.gsfc.nasa.gov/CAMPAIGN_DOCS/OCDST?shuttle_oceanography_web/oss_cover.html

Oceanography Resources on the Internet
www.esdim.noaa.gov/ocean_page.html

The Southern Ocean and Global Climate
www.science.org.au/nova/018/018key.htm

Tornado: A Virtual Field Trip
www.field-trips.org/sci/tornado/index.htm

US Geological Survey: Earthquakes—General Interest Publication *pubs.usgs.gov/gip/earthq1/*

US Geological Survey: This Dynamic Earth
pubs.usgs.gov/publications/text/dynamic.html

Rock weathering

Weathering: Mechanical and Chemical
www.ualr.edu/~saleslie/physical/LWeathering.html

Salinity

1995 Case Study: The Yass Salinity Abatement Demonstration Program (YSADP) *www.epa.nsw.gov.au/soe/95/12_2s1.htm*

Australian Geoscience Portal *www.geoscience.gov.au*

Management and Mitigation of Dryland Salinity
www.brs.gov.au/land&water/drysalinity.html

Murray-Darling Basin Salinity Audit
www.mdbc.gov.au/naturalresources/policies_strategies/projectscreens/Salt_audit/salinity.htm

Murray-Darling Basin: Water and Land Salinity
www.mdbc.gov.au/education/encyclopedia/water_and_land_salinity.htm

Protecting Water: Salinity www.wrc.wa.gov.au/protect/salinity/

Queensland Department of Natural Resources and Mines, Land Fact Sheets: Salinity
www.dnr.qld.gov.au/fact_sheets/landfacts.html

Salinity in New South Wales
www.dlwc.nsw.gov.au/care/salinity/index.html

Water pollution

NSW Environment Protection Authority: Urban Stormwater Program www.epa.nsw.gov.au/stormwater/

Ocean Planet: Pollution
seawifs.gsfc.nasa.gov/OCEAN_PLANET/HTML/peril_pollution1.html

Water resources

Australian Museum: Common Fossils of the Sydney Basin
www.austmus.gov.au/is/sand/fossils.htm

'Better ways to water' article, *Ecos 85*
www.publish.csiro.au/ecos/Ecos85/Ecos85A.htm

Coastal and Marine Pollution
www.environment.gov.au/marine/pollution.html

Inland Waters: Water Quality
www.environment.gov.au/water/quality/

NSW Environment Protection Authority: State of the Environment Report 1997, Chapter 3: Water
www.epa.nsw.gov.au/soe/97/ch3/

River and Estuary Pollution
www.wrc.wa.gov.au/public/waterfacts/3_pollution/index.html

Sustainable Use of Groundwater from Alluvial Aquifers in Northeast New South Wales
www.brs.gov.au/land&water/gab/alluvial.html

US Environmental Protection Agency: Technology Fact Sheets
www.epa.gov/owm/mtbfact.htm

US Environmental Protection Agency: Water Recycling and Reuse—The Environmental Benefits
www.epa.gov/region09/water/recycling/

US Environmental Protection Agency's Office of Water: Constructed Wetlands Case Studies
www.epa.gov/owow/wetlands/construc/

Water Reuse
wwwscience.murdoch.edu.au/teaching/m234/recycle07.htm

Other useful resources

Australian Soil Classification www.cbr.clw.csiro.au/aclep/asc.htm

Ecological Sustainable Development: A Concept Map for Sustainable Development www.origen.com.au/esd.htm

Environment Australia Online www.ea.gov.au

Environment Protection and Biodiversity Conservation Act 1999 (Cwlth): Overview www.environment.gov.au/epbc

Environmental Protection Agency of NSW www.epa.nsw.gov.au

Inland Waters: Wetlands
www.environment.gov.au/water/wetlands/

Murray-Darling Basin Commission www.mdbc.gov.au

NSW Department of Land and Water Conservation
www.dlwc.nsw.gov.au

Publications: Water Facts
www.wrc.wa.gov.au/public/waterfacts/index.html

Salt, Boron and Chloride Plant Tolerance Databases
www.ussl.ars.usda.gov/saltoler.htm

Waterwatch Australia www.waterwatch.org.au/

Buchanan, R. *Bush Regeneration: Recovering Australian Landscapes.* TAFE Student Learning Publications, Sydney, 1989

Clark, I.F. & Cook, B.F. *Perspectives of the Earth.* Australian Academy of Science, Canberra, 1983

Department of the Environment, Sport & Territories. Australian State of the Environment Report, CSIRO Publishing, Melbourne, 1996

Farrier, D. *The Environmental Law Handbook: Planning and Land Use in New South Wales.* 2nd edn, Redfern Legal Centre Publishing, Sydney, 1988

Goudie, A. *The Human Impact on the Natural Environment.* 5th edn. Blackwell Publishers, Oxford, 2000

Johnson, M. & Rix, S. *Water in Australia: Managing Economic, Environmental and Community Reform.* Pluto Press Australia, Sydney, 1993

Scheibner, E. *The Geological Evolution of Australia: A Brief Review.* Geological Survey of New South Wales, Sydney, 1999

White, M. E. *Listen—Our Land is Crying: Australia's Environment; Problems and Solutions.* Reed Books, Sydney, 1999

Young, A. *Environmental Change in Australia Since 1788.* 2nd edn. Oxford University Press, Melbourne, 2000

Effective research

There are two types of research assignment you are likely to be asked to carry out during this course: open-ended investigations and research projects. With an open-ended investigation, students take the initiative to gather information and use it to find an answer to a problem. In the research project, students may be asked to gather, process and present information from secondary sources to fulfil a task. Alternatively, students may need to identify data, choose resources and gather and analyse secondary data.

When you are given a research project or an open-ended investigation, there are a number of steps you need to take:
1 If you are allowed to choose your subject material, select a topic that has a good supply of information.
2 Decide what points need to be covered in your work. As an example, if you are given a research project on salinity legislation, the points to cover could be types and causes of salinity, NSW legislation, comparison with legislation from other states, and federal legislation.
3 Take time to plan out what you are going to do.
4 Decide where you are going to obtain the information from.
5 Don't spend all your time searching for information. Balance it with reading, collating, understanding and writing up your assignment.
6 Make sure you understand the topic before you write up your work.
7 When presenting the information choose the appropriate format. This may be specified or it may be left to the student to decide. Possible formats include posters, oral reports, videos, PowerPoint presentations, pamphlets, written reports or even cartoons.
8 Finally, it is important to evaluate your work.

LOOKING FOR INFORMATION

There are two types of source material: primary and secondary. Primary information is the data you or somebody else has collected (by surveys or experimentation, for example) or original documents. Secondary information includes books or articles written by authors who have interpreted the original material.

Whatever assignment you are carrying out, there are a number of possible information resources that you will be able to access.

The school library

Your school library will not necessarily contain a lot of information on the chosen subject. Also, all the students in your class will make the school library their first port of call for research, so visit early to ensure you can access as much information as possible.

The library will contain a range of sources: books (including those in the reference section), magazines, CD-ROMs, abstracts, indexes and vertical files. Check all possible sources, and remember to consult the librarian if you need assistance.

There are a number of up-to-date textbooks taking a global perspective. However, most have an American or European bias. It is important to check the country of publication. For example a book published in Australia is more likely to contain information on Australian topics than one published in the USA. Very few textbooks are written in Australia, which may make it difficult to find a local perspective. You may need to find other sources, such as specialist magazines or local university Internet sites.

Your school library may have access to SAGE, which is a database of science and geography education containing references to popular science, environmental and geography journals.

The local library

Your local library may be better resourced than your school library and it will have contact with all libraries in New South Wales. If a particular source is not in the library, they should be able to order it from another library. However, this may take time. If it does not have specific magazines in stock, your local library is able to access magazines via the Worldwide Magazine Bank.

Local interest groups

Local interest groups are often valuable sources of information. These groups have expert and detailed information about your local area, or even about your state. Local interest groups include such organisations as:
- Streamwatch
- Landcare
- Bushcare
- Wilderness Society
- The Rare and Endangered Species Group
- Wildlife Information and Rescue Service
- Earth Repair Foundation
- Institute for Earth Education
- Wildplant Rescue and Care Service
- Total Environment Centre
- local natural history societies
- local historical societies.

There are many local interest groups. Some are statewide organisations; others are purely local. Further information about these groups can be sought from either your local library or local council or look in the phone book for contact details.

Government bodies

For a larger perspective and for information about legislation and strategies, for example, it is worthwhile contacting your council regarding the local environment. Each council publishes a state of the environment report for its area. The local council will have documentary information as to how the local environment has changed over the years. They also deal with issues such as noxious weeds, feral animals, local pollution and planning legislation. Most councils also have an environmental education officer who can help you find information.

State bodies may deal with local issues, but they have a statewide viewpoint and deal with state issues, legislation and strategies, for example. These bodies include:
- Environmental Protection Authority
- National Parks and Wildlife Service
- Fisheries New South Wales
- Department of Land and Water Conservation
- Department of Mineral Resources
- Rural Lands Protection Boards
- Agriculture New South Wales
- Waterways Authority of New South Wales.

A number of government bodies in other states can also provide excellent information on a wide range of environmental issues.

Commonwealth government bodies and agencies implement international and federal legislation Australiawide. They include:
- Australian Geological Survey Organisation
- Department of Agriculture, Fisheries and Forestry
- Australian Bureau of Statistics
- Australian Quarantine and Inspection service
- Bureau of Meteorology.

Educational bodies

Many universities and other academic bodies have information that may be very useful to your research project. Examples include the weblinks provided by the School of Geology within the University of New South Wales and the University of Sydney. Some university staff are willing to assist you with your research. It is possible to contact university staff or specialist help by contacting the appropriate university department or by visiting the relevant university's department or school Internet site, which may have a link to the staff or an inquiry section.

A different route would be to contact the education officer at a museum. Museums also have people or exhibits that may help in your research. Tourist centres at caves, radio telescopes and fossil sites, for example, may also be useful sources of information. Information about these sites may be found at a regional tourist information centre, in holiday brochures or in the phone book for the area in which the centre is found.

Examples of educational bodies are:
- University of New South Wales School of Geology
- University of Sydney's Uniserve—a science website that provides links for the stage 6 science syllabuses
- Charles Sturt University website—HSC online pages
- Australian Geological Survey Organisation
- US Geological Survey
- Department of Mineral Resources
- Department of School Education
- The Australian Museum, Sydney
- CSIRO.

The addresses of the above list can be found in a number of places. The contact addresses of a government body will be in the government section at the front of your local phone book or in the phone book of the area where the organisation is based. The Post Office will have copies of the necessary phone books. For non-government bodies, other sources may include the Internet, the phone book, your local council, the library or your teacher.

The Internet

While the Internet is a very useful tool, it is important to check that the information gained is from a reliable source, such as a university or recognised organisation. Researching the Internet is dealt with later in this section.

REQUESTING INFORMATION

Before you write to or phone an organisation with a query, make sure that you have done some basic reading so that you understand what you are asking for. You will get a lot more useful help if you know exactly what you want. Ensure you have specific questions and do not expect the expert to do your project for you. When you request information allow several weeks to receive an answer. The organisation may be able to email what you need, but articles, pamphlets or other documents may have to be sent by normal mail.

COLLATING INFORMATION

Collecting the information may be the easiest part of the assignment; extracting relevant information is possibly the hardest. A template for detailing the extracted information can be found in Table 5.1 (page 250). This template can be used when reading information to record the source, the extracted information and the credibility of that information.

OPEN-ENDED INVESTIGATIONS

There are five steps in carrying out an open-ended investigation.
1. Write a short statement that makes clear what the problem is that you have to solve. Propose a hypothesis that your experiment will test.
2. Write a plan that states what you intend doing. Make sure that the method identifies and controls any factors or variables that may affect the accuracy of the experiment. Explain what measurements you are going to use and how they will be made.

3 Carry out your investigation and record all your observations and measurements. If you have to change your plan, write down what changes you have made and why they were necessary.
4 When you have carried out the investigation, look at your results to see if they reveal any patterns or trends. If there are patterns or trends, try to give an explanation for them. If no trends or patterns are evident, check whether your experimental design has a flaw or problem.
5 Evaluate your investigation. When evaluating your investigation a number of questions need to be answered:
- Did your results support your hypothesis or question?
- Did you successfully control the variables, which may have affected the accuracy of your results?
- Were your results what you expected? If not, why not? Was the experiment inaccurate? Was it testing what you intended?
- Did you repeat the experiment and did you get the same range of results?

The student also needs to ask a number of questions in regard to the work itself. These include:
- Does it address the problem or question?
- Is the information provided relevant?
- Are the sources reliable?
- Does the work display a logical sequence?
- Is the problem or question answered?

If an answer to any of these is in the negative, modifications to the work will need to be made.

USING THE INTERNET AS A RESEARCH TOOL

Key skills you must develop in studying Earth and Environmental Science are the ability to locate, select and analyse information. One source of up-to-date, relevant information is the Internet. Here we will look at what the Internet has to offer and how you can find the information you need in an efficient way. We will also cover how to judge the quality of the information you find.

The Internet and the World Wide Web

The Internet is a network of computer networks. The information stored on one computer, as files or documents, can be shared if another computer can access and view the information. The World Wide Web (WWW) is part of the Internet. Using a program called a browser, you can access documents from computers all over the world. A special feature of the WWW is its use of Hypertext, which is a way of linking information you view on your browser to information on other web pages. By clicking your mouse on a highlighted link the browser can retrieve and display the information at another site. The site can be on a computer in the same room or on the other side of the world.

Search tools

To get the most out of using the Internet you need to know the characteristics of the various search tools available to you. An outline of the five most commonly used search tools follows.

GATEWAY DIRECTORIES

Gateway directories are compiled by experts, academic or otherwise. They are of great value to beginning researchers and you should endeavour to use them as part of your search strategy.

SEARCH ENGINES

A search engine works by searching a database of Internet files. The files are collected by programs called spiders and then indexed. When you use a search engine you provide words and phrases that are looked up in the database. Search engines are useful if you can describe your information well. If the terms you enter are too general, a lot of documents will be documented—many of them not exactly what you want.

META-SEARCH ENGINES

These are search engines that search more than one search engine database. They then consolidate the results and present them in a uniform way. While these tools are good for initial searches, they lack the power of good search engines. At present they capture only about one-tenth of the results in each search engine they visit.

SEARCHABLE DATABASES

These are very useful sources of information. Most of the documents available through the WWW are in such databases, which are sometimes called the 'invisible web' because they are not found using common search engines. To find appropriate databases you may have to use a subject directory or a directory of searchable databases, such as InvisibleWeb.

> Try the InvisibleWeb Catalogue at
> *www.invisibleweb.com*

SUBJECT DIRECTORIES

These are directories where web pages have been organised into categories. They are very useful places to start. When possible, choose academic directories such as Infomine.

Search strategies

Some people have estimated that there are more than a billion separate documents available to users of the Internet. Unless you have a strategy to find the information you need, you may find hundreds of thousands of pages, or 'hits', that are in some way related to the topic you are researching.

A good search strategy will help you find the information you need efficiently. Two critical steps to undertake before you start are:
1 analyse your topic carefully
2 select the best tools for the job.

ANALYSING THE TOPIC

Consider the following:
- List the broad categories your topic might be found under.
- List the unique and distinctive words or phrases that describe your topic. Include abbreviations and acronyms in your list.

Table 5.1 Template for recording sources of information

Resource no.: _____ Assignment headings _____ Page no. _____

				Information found		
				1	2	3
Title						
Author						
Year						
Publisher						
City of publication						
ISBN no.				1		
URL				2		
				3		
		Yes	No			
Is the information believable?				1		
Does it come from a reliable source?				2		
Does it show bias?				3		
Is it accurate?				1		
Do you agree with it?				2		
Other comments				3		

- List organisations, societies and interest groups that may have information about your topic.
- Consider alternative ways of spelling terms (for example 'colour' and 'color' both have the same meaning but the second word is the US spelling) and synonyms (words with similar meaning).
- List the terms you may want to leave out.

SELECTING THE TOOLS

The tools you select will be determined by knowing exactly what you want.

- Are you seeking an overview or are you unsure where to begin? If so, try using subject directories.
- Do you have some distinctive terms or phrases? If so, use a search engine or meta-search engine. If not, try using a search engine with a subject directory or a number of general words dealing with the subject you are searching for. Find out the rules your search engine uses when combining term. For example does 'endangered species' mean 'endangered *and* species' or 'endangered *or* species'?
- Are you looking for specific information about a common topic? Try using a search engine such as Northern Light, which groups documents in categories.

A good site dealing with search strategies and tools is
www.lib.berkeley.edu/TeachingLib/Guides/Internet/Strategies.html

Evaluating web pages

Anyone can produce web pages or documents if they have the right technology and skills. They don't need to know anything about the topic they present. For this reason it is important to evaluate each page or document you read. Be warned: many web pages are essentially advertisements.

Consider the following:

- Who created the site? What are their credentials? Do they have a bias? If so, what is it? Why did they write it?
- Who is the information intended for? If it is for a general audience the words used will be easier to follow than if it is aimed at experts.
- When was the information created or updated? Is it current enough for your purposes?

Where possible, ensure the accuracy of the information by verifying it against other sources.

A good online source to read on website evaluation is Angela Elkordy's tutorial at
www.thelearningsite.net/cyberlibrarian/searching/lesso2a.html

INTERNET EXERCISE

1 Choose a subject that you wish to research and carry out an analysis using the points outlined above.

2
a Choose three tools: two that seem suitable and one that doesn't. Compare the results using the evaluation points above.
b Compare the three search tools you used. What are their strengths and weaknesses?

3 Summarise three pages from your search in a systematic way. A suitable format is shown in Table 5.2.

4 Write a comparison of the three pages you summarised. List the positive, negative and interesting aspects of each page.

Table 5.2 Format for summarising Internet information

Source:	List the URL, author and title of the site and the date you visited it.
Subject:	Summarise what the page is about.
Keywords:	List the terms used to find the site.
Abstract:	Summarise the information in your own words.
Response:	Write *your* thoughts about the information you obtained. Ensure you consider the site's reliability. Consider the evaluation points above.

Answering and asking questions

The ability to answer questions well is important not only for gaining knowledge but also for testing your understanding. In this text, sets of Review Activities have been provided to help you learn the material covered in the text. It is important that you engage with such questions if you are to learn the work well, that is, learn it so that it is not soon forgotten.

Questions can help in three areas of your Higher School Certificate preparation. First, answering questions set by others helps you to develop your basic knowledge, understanding and skills. Second, asking questions of your own about the subjects you study helps to develop the depth and richness of your understanding. Lastly, questions in the formal examinations at the end of Year 12 have a particular form that you need to have practised answering.

In answering questions you should always ask some questions of your own. There are two important questions to ask about questions:
- What is it exactly the question is asking me to do?
- Why am I being asked this question?

The first question is important in a practical sense, because in examinations marks will be awarded according to how well you do what is asked of you. If a question asks you to explain something and you describe it you have not done what the examiner is awarding marks for, that is, your ability to explain a concept.

A set of keywords are used in the Higher School Certificate to tell you what to do. These are shown in the table opposite and you must become familiar with them. Note that they are grouped into three levels of difficulty according to what you have to do. Be aware, however, that a simple process such as 'describe' can be applied to a difficult idea to make a more difficult question.

The second question, 'Why am I being asked this question?', helps you to determine the purpose behind the question. Is the question testing your understanding of part of the syllabus or your skills? In general, looking at the purpose of someone else's question is a powerful way to understand what others see as important and what they expect of you.

Answering and making questions are of central importance in science. The German scientist and philosopher Werner Heisenberg once remarked that nature does not reveal itself to us as it is but only through the questions we put to it. By this he meant that the questions we seek to answer with experiments provide us with knowledge of the world, but only that knowledge the experiment addresses. Similarly, you will only deepen your understanding of concepts by asking and answering questions in a way that builds on what you already know.

So how can you generate questions? One method is to use a definition of something you need to understand. The journalist's questioning of who, what, where, when, why and how will help to generate questions. Students often ask how to apply 'who' and 'why' to some ideas in science; the questions don't appear to apply. This may be true in some cases. If the questions don't apply, don't use them but remember that someone came up with the idea or term and they did so for a purpose. Understanding these things helps you to understand the history of the idea.

Analogies can be used to generate questions and test your understanding. An analogy is a correspondence we draw between things we are familiar with and things with which we are less familiar. We could look at a school and make an analogy with a natural ecosystem. They are both complex systems in which living things have different roles. By looking for similarities, differences and interesting points that arise we may learn new things about both the familiar and unfamiliar parts of our analogy.

Another way to generate useful questions is to ask, 'What if...?' This strategy works very well with processes. For example, knowing that Australia has the highest rate of water storage per person in the world may be of limited use but if we ask the question, 'What if Australia didn't?', we have to examine the reasons behind storage in a way that increases our understanding of the issue.

REVIEW ACTIVITIES

1
Explain why answers to questions starting with 'compare' and 'contrast' are different.

2
Choose three terms at random from the definitions in the text. Generate a list of questions using the journalist's questions listed above.

3
Select a term from the course and research its origin. Who first used it? How was it used? Why was a new term needed?

4
Draw up a table with a list of terms down the first row and who, what, where, when and how across the top row. Use the template to generate questions for the terms and then check you know the answers.

5
Try developing some analogies. In the area of mathematics called topology a doughnut and a coffee mug are classified as the same. Can you see why? Lastly, the plate tectonics cycle moves ocean crust just as the carbon cycle moves carbon. How good is the analogy?

Key words to know

General level of difficulty	Key word	Definition
Low: usually simple and direct questions testing your memory and ability to communicate ideas	Account for	State reasons for, report on
	(Give an) account of	Narrate a series of events or transactions
	Clarify	Make clear, make plain
	Define	State meaning and identify essential qualities
	Describe	Provide characteristics and features
	Identify	Recognise and name
	Outline	Indicate the main features of; sketch in general terms
	Recall	Present remembered facts, ideas and experiences
	Recount	Retell a series of events
	Summarise	Express relevant details concisely
Medium: these require thought about the subject and the use of knowledge when answering outcomes information	Account for	Make clear; make plain
	Analyse	Identify parts and the relationships between them; identify and relate implications
	Apply	Use or employ in a certain situation
	Assess	Make a judgment about the value, size, quality or outcomes
	Calculate	Determine from the given facts, figures or other information
	Classify	Arrange into categories
	Compare	Show how things are similar and/or different
	Contrast	Show how things are different
	Deduce	Draw conclusions
	Demonstrate	Show by example
	Discuss	Identify issues and provide points for and against
	Examine	Inquire into
	Explain	Relate cause and effect; make relationships between things plain; provide the 'why' and/or 'how'
	Interpret	Draw meaning from
	Predict	Suggest what may happen based on information available
	Recommend	Provide reasons in favour
Difficult: these questions are often complex and frequently require planning and careful thought	Assess	Make a judgment about the value, size, quality or outcomes
	Construct	Make, build or put together items or arguments
	Critically (analyse/evaluate)	Add a level of depth, knowledge, understanding, logic, questioning or reflection to the task
	Evaluate	Make a judgment based on criteria; determine the value of
	Justify	Support an idea or conclusion using reasons
	Propose	Put forward for consideration or action
	Recommend	Provide reasons in favour of

Glossary

abiotic concerning non-living things

absolute dating gives the age of the rocks in years based on atomic breakdown of unstable elements

absorption spectrum when light from a source that has a continuous spectrum is shone through a gas with a lower temperature and pressure, dark lines are formed on the spectrum. The lines are at the wavelengths of the atoms in the gas.

accretion when material collides and sticks together

accretionary wedge the material scraped off the top of a subducting plate, forming a ridge

active continental margin a continental margin located within the zone of seismic activity

aeration the amount of air present in the soil

aerobic respiration release of energy using oxygen

aerosols liquid droplets or solid particles that remain suspended in the air

affluence having material wealth

albedo the reflectivity of a surface

alluvial horizon a zone of mineral accumulation in a soil

amino acid an organic compound, required in the formation of proteins

amplitude the height of a wave, from the midpoint of the wave to the top of the peak. The maximum displacement of the medium from its equilibrium position.

anaerobic the absence of free oxygen

anaerobic respiration less efficient form of respiration by which energy is released in the absence of oxygen

antimatter matter composed of the counterparts of ordinary matter, such as antiprotons instead of protons and positrons instead of electrons

apparent polar wandering the appearance that the positions of the poles moved

aquifer a permeable rock unit saturated with ground water

archaeobacteria a very primitive type of bacteria

asteroids this word means 'starlike'. When viewed through a telescope, asteroids look like faint stars. They are made mostly of rock and maybe ice and are usually less than 2 km across but they can be over 100 km across.

asthenosphere the layer or shell of the Earth that lies directly below the lithosphere and behaves like a semisolid

atmosphere the gaseous envelope surrounding the Earth

atom the smallest part of an element that can exist as a stable entity. An atom is composed of a central nucleus containing protons possessing a positive charge and neutrons possessing no charge. Surrounding the nucleus are layers of electrons. Each layer possesses a different energy level.

basalt a dark-coloured, fine-grained, basic volcanic rock

benioff zone a zone of grinding between two plates, where one is subducting

big bang theory the universe started with an explosive event, which created matter. The matter is expanding outwards from the point of the original explosion.

biodiversity diversity of life forms, genes and biological systems

biological magnification the increasing concentration of materials due to selective uptake during feeding in a food chain

bioregion an area defined on the basis of the characteristic plant communities found there

biosphere the sum total of the Earth's organisms and organic matter

biota the animals and plants living in a particular ecosystem

biotic concerning living things

blooms the rapid increase in the number of algae and/or bacteria caused by the presence of high levels of nutrient

blue giants massive stars with high temperatures, high luminosities and diameters ten to one hundred times that of the Sun

caldera a very large bowl-shaped volcanic depression, formed by the combination of the explosion and collapse of the top of a volcanic cone or groups of cones

chemical weathering weathering involving chemical change within a rock

clastic sedimentary rock a rock formed of fragments cemented together

cleavage flat, sheet-like structures along which a mineral tends to break easily

closure temperature the temperature at which the radio-isotope is sealed in a crystal

cohesion the property of something that makes its parts hold together

comets celestial bodies from space, usually of small mass that circle in elliptical orbits around the Sun. As a comet approaches the Sun it becomes visible because the surface of the centre, or nucleus, begins to warm and volatile gases evaporate. The evaporated molecules boil off and carry small solid particles with them, forming the comet's tail, or coma, of gas and dust. The coma absorbs ultraviolet radiation and begins to fluoresce. When the nucleus is frozen it can only be seen by reflected sunlight.

composition a description of the things making up an object. A chemical composition describes the elements or compounds in a substance and their relative abundances.

condensation the process of forming a liquid from a vapour

continental drift hypothesis the theory that the continents have undergone movement

continental rise a slight slope in the oceanic crust leading up to a continental slope

continental shelf a wide, shallow active or passive continental boundary made up from sediments

continental slope the slope leading to deep water at the edge of a continental shelf

continuous spectrum a hot, opaque gas, solid or liquid that under high pressure will produce a broad band of wavelengths of light, forming a continuous spectrum

convection current a pattern of mass movement of mantle material in which the outer area is downflowing and the central area is uprising due to heat differences

core the innermost shell composed of a liquid outer layer and a solid inner layer

Coriolis effect the tendency of fluids to be deflected from straight-line flow; caused by the Earth's rotation

crust the crust forms most of the lithosphere and is made up of solid rock

crystal a solid formed of atoms arranged in a regular manner

curie point the temperature below which a mineral can become a permanent magnet

cyanophytes photosynthetic bacteria; also known as cyanobacteria

deep sea trench the point where two converging plates meet

density the mass per unit of volume of an object. Density is calculated as follows:

$$\text{Density} = \frac{\text{Mass}}{\text{Volume}}$$

deoxyribonucleic acid (DNA) a nucleic acid which holds the blueprints of life, responsible for inheritance

dieback the long-term decline in the health of rural gum trees

dissolve to become part of a liquid

eccentricity the shape of the Earth's orbit as it changes from an elliptical orbit to a circular orbit

ecology study of organisms and where they live

ecosphere the zone in which life exists

effervescence bubbles of gas being released from something

effluent liquid discharged as waste

electron an elementary particle that is a constituent of all atoms and has a minute mass of approximately 9.1×10^{-31} kg. It has a negative charge.

electron configuration the arrangement of electrons in the different shells

element a substance that cannot be further divided by chemical methods. They are the basic substances that build up chemical compounds.

emission spectrum a hot gas under low pressure will emit individual wavelengths of light. These form an emission spectrum, which is a series of bright lines on a dark background.

energy the quality something has that allows it to cause change

ENSO El Niño-Southern Oscillation

eukaryote advanced cells that have membrane-bound organelles. Examples are animal and plant cells.

eutrophication nutrient enrichment in a water body

extrusive rocks magma that has been extruded onto the Earth's surface, cooled and formed rock

faeces the undigested part of food together with bacteria and chemicals from the gut

fault a fracture in a rock along which movement occurs

feldspars the most common group of rock-forming minerals that make up 60% of the crust. They are made up of silicates.

felsic a type of rock containing light-coloured aluminium and silicon minerals. Examples are quartz and feldspars.

felsic igneous rock an igneous rock rich in feldspars and quartz

ferromagnesian minerals minerals containing iron and magnesium

fission track method particles from uranium decay leave tracks in glass or mica

flux the rate at which material or energy flows

foliated having a sheet-like texture due to the growth of minerals

frequency the number of peaks or troughs to pass a particular point every second

fumarole deep sea volcanic vent

galaxies large collections of stars, dust and gas in space; systems of millions or billions of stars held together by gravitation

ground water water that is stored in rock below the watertable

gyres systems of wind-driven ocean surface currents

heat capacity the amount of heat needed to raise the temperature of a substance by 1°C

heterotrophs organisms that are unable to make their own food

hydrosphere the totality of the Earth's water, including oceans, lakes, rivers, ice, snow and underground water

hydrothermal hot fluids created by igneous activity. The fluids carry dissolved minerals and gases.

igneous material formed by cooling and freezing of molten material

inorganic materials not containing carbon compounds other than carbon oxides and similar simple compounds

intrusions the emplacement of magma into cracks in a pre-existing rock

intrusive igneous rock a rock that solidified from magma within the crust

island arc a curved chain of volcanic islands, formed on the landward side of a trench

isotope elements that have the same number of protons in their nucleus (atomic number) and similar chemical properties, but different atomic weights (their mass relative to ^{12}C)

jointing the formation of cracks in a rock where, unlike a fault, no major movement across the cracks occurs

Jovian planets the outer planets composed mainly of gas: Jupiter, Saturn, Uranus and Neptune

law of horizontality sediments are deposited in layers parallel to the Earth's surface

law of superposition each rock layer is younger than the layer below it if strata have not been overly disturbed

leaching a process in which soluble material is carried from a material by water flowing through it

light year the distance travelled by light in one Earth year (1 light year = 9 500 000 000 000 km)

liquefaction when soil behaves as if it is liquid

lithosphere the solid outer shell of the Earth

lustre the appearance of a mineral in reflected light

mafic a group of rocks containing dark minerals composed of magnesium and iron

mafic igneous rock an igneous rock containing minerals rich in iron and magnesium but no quartz

magma molten rock containing some crystals and dissolved gases

main sequence star first and longest stage of a star's life

mantle the semiliquid layer below the crust

mass the quantity of matter or atoms found in an object when the force of gravity is not acting upon it.

matter the material that objects are made of. Matter is composed of atoms or the parts of atoms.

matter particles fundamental particles. They have no known smaller parts. They are also called quarks.

mechanical weathering breakup of a rock without change in its chemical composition

mesosphere the shell below the asthenosphere composed of rocks at a very high temperature which are very strong and highly compressed

metamorphic rock a rock formed from other rocks by crystal growth

methanogenic organisms that use methane as a source of food

microsphere a stage before the formation of primitive cells

mid-ocean ridge occurs when two crustal plates separate in the mid ocean and basaltic material wells up through the spreading centre

mineralogy the study of minerals, including their formation, composition, properties and classification

molecule the smallest particle of matter; composed of two or more atoms

mor a layer of acidic organic matter with leaf litter on top and fermenting plant remains below

natural selection a process that changes the frequency of genes and traits in a population

nebulae gas and/or dust clouds

neutrino an elementary particle with zero electrical charge and a mass of zero when at rest

neutron an elementary particle that is a constituent of all atomic nuclei except normal hydrogen. It has a zero electrical charge and approximately the same mass as a proton.

nitrogen-fixing bacteria a type of bacterium that is able to convert gaseous nitrogen to a more useable nitrate form

nitrogenous bases part of the DNA and RNA molecules

nuclear decay the emission of particles from the nuclei of an element that results in a change from one isotopic form of an element to another or from one element to another

nucleotides part of the nucleic acid molecule, formed from a ribose sugar and a nitrogenous base

ophiolite suites rock types that appear to have been formed on the sea floor and are now emplaced on land

orogenesis the process of mountain formation

outgassing the escape of gases from volcanic vents

oxides compounds containing metal or non-metals combined with oxygen

ozone layer a layer within the stratosphere containing ozone. It absorbs harmful ultraviolet radiation.

Pangea a supercontinent that existed about 200–300 million years ago

parallax the relative positions of stars change when viewed from different positions of the Earth's orbit

passive continental margin a continental margin that is located far from a seismically active mid-ocean ridge

pause a thermal boundary between each atmospheric layer

perihelion the point on the Earth's orbit when it is closest to the Sun

periodic table a table of elements arranged in order of atomic number, the arrangement emphasising the chemical relationship between the elements

permeability the ability of a fluid to pass through a material

pH a measure of the amount of hydrogen ions present. A measure of how acid or alkaline something is.

photon a small bundle of electromagnetic energy

photosynthesis the process by which plants make their own food from inorganic compounds

planetesimal an early stage in the formation of a planet

plankton very small or microscopic animals and plants living in the oceans

plastic changes shape when a force is applied to it

point source a particular place where pollutants enter a waterway

polar describes a molecule that has an uneven distribution of electrical charges

polymer formation of large molecules consisting of repeated structures

porosity the amount of pore space in a material

precession the wobble that the planet has as it spins on its axis

precipitation a process in which an insoluble material is formed in a liquid

principle of faunal succession each rock formation contains a unique assemblage of fossils and each assemblage succeeds one another in an orderly and predictable way

prokaryote a primitive type of cell, lacking membrane-bound organelles (structures with a specialised function). Examples are bacteria and blue-green algae.

proteins very complex organic compounds, required for numerous processes in living things

proton an elementary particle present in every atomic nucleus, the number of protons being different for each element. A proton has an electric charge equal in magnitude to that of an electron but of opposite charge and has a mass of 1.7×10^{-27} kg.

protoplanetary disc an early stage in a planet's formation

protostar a young star that is still forming ('proto' means 'before' or 'early')

pulsating universe theory starting with the big bang, the universe expands. This is followed by a contraction of matter, called the 'big crunch'. This expansion and contraction occurs continuously.

pyroclastic flow a hot mixture of gas and particles that moves away from a vent at high speed

quarks *see* matter particles

radiation in order to become stable some elements decay, releasing radiation. There are three types of radiation: alpha, beta and gamma radiation.

radioactivity spontaneous emission of energy from unstable atom nuclei

radio-isotopes elements that have the same atomic number and similar chemical properties but different atomic weights

radio-isotopic ratio the ratio of parent to daughter radio-isotopes

rain shadow the sheltered side, or downwind flank, of a mountain range, which receives less rainfall than the upwind flank

recharge the process where surface water seeps into an aquifer

red giant a late stage in a Sun-sized star's life. The outer layers expand and cool, and nuclear fusion is replaced with nuclear fission.

red star a small star of less than 0.1 solar masses

relative dating the dating of rocks by working out their place in a sequence of rocks in one locality and then comparing or correlating these rocks to those in other regions

remnant communities small areas containing what remains of once widespread communities

residence time the average time a molecule exists in a certain place

resistant resisting weathering and erosion in the environment where it is found

resource something that humans use

ribonucleic acid (RNA) a nucleic acid needed for making proteins

ribose a sugar required in the structure of nucleic acids

ribosome a cell organelle involved in the production of proteins

rift valley occurs when continental crust is splitting apart. The new material forms the valley bottom.

rock a material made up of minerals bound together

salinity the amount of salt in a liquid; the problem of high levels of salt in a liquid

salt water intrusion the replacement of fresh ground water with saline ground water

scleromorphy having the anatomy and characteristics of scerophyllous plants, that is plants with tough leaves that help to reduce water loss

sediment material that is transported and deposited by wind, water, ice or gravity

sedimentary rock a rock formed from accumulated and consolidated sediment

shells the energy levels surrounding the nucleus of an atom. The shells are occupied by electrons.

soil a material formed by the weathering of surface materials

soil profile a vertical section through a soil showing differentiation and structures

solar nebular hypothesis the solar system is formed from a solar nebula

solar system a group of nine planets orbiting a main sequence star called the Sun

solubility the amount of a material that will dissolve in another material

solute the material that dissolves in something

solvent the substance into which something dissolves

steady state theory a steady state universe has no beginning or end of time. The average density and arrangement of galaxies does not change.

stratigraphy the study of rock layers or strata

stratosphere the second of four layers of the atmosphere; found above the troposphere

stromatolites cyanobacteria that form columns in shallow, warm seas

structure the way in which soil particles form units, called peds

subduction the movement of one crustal plate under another so that the descending one is 'consumed' by the mantle

subduction zone the area where one plate goes under another plate

sulfaphilic organisms that use sulfur as a source of food

supernova a red supergiant star whose outer layers are blown off in a massive explosion. The remaining core collapses and forms a neutron star.

system something that takes in matter and/or energy, processes it in some way and then gives out matter and/or energy

technology materials, techniques and knowledge arising from the application of science and art

tectonic plate theory states that the lithosphere has fragmented into several large plates, which move. As the continents are part of these plates, they move accordingly.

terrestrial planets the inner planets formed from solid material by accretion: Mercury, Venus, Earth and Mars

Tethys Sea an ancient sea that was located between the two landmasses of India and Asia

tetrahedron (plural: tetrahedra) the basic building block of silicate minerals

texture the size, shape and arrangement of particles in a rock

thermosphere the outermost layer of the atmosphere

topographic relief variation in the shape of the Earth's surface; the set of landforms found in an area

toxic poisonous

troposphere the first layer of the atmosphere above the Earth's surface

true star a star in which nuclear fusion reactions are taking place

tsunami a tidal wave

urbanisation the building of towns and cities, replacing agricultural land and native bushland

viscosity the property of a substance that prevents it from flowing

visible spectrum when white light can be split into the range of its constituent colours. The constituent colours are red, orange, yellow, green, blue, indigo and violet.

wavelength the distance between two adjacent troughs or crests on a wave

weathering the process in which rocks are altered

weight the force with which an object is attracted to the Earth. This attraction will vary depending on the gravitational pull on the object.

white dwarf a late stage in a star's life. It is the result of the outer layers of the star dispersing and the core collapsing to form an extremely dense, small star. It may have half the mass of the Sun but it is only the size of the Earth.

Index

Page references followed by *f* indicate figures; those followed by *t* indicate tables.

a
Aboriginal concept of the universe, 16–17
absorption spectrum, 9
adaptation, 169–70
albedo, 52, 210–11
amino acids, 58, 60–1
amplitude, 6
apparent polar wandering, 97–8
aquifers, 185, 240*f*, plate 25
archaeobacteria, 61–3
asteroids, 2, 26–7
asthenosphere, 38–9, 87–8, 89, 207
astronomy, 3–6, 17–18. *See also* cosmology
atmosphere, 38–9, 43; convection and circulation, 210–11; effect of photosynthesis, 64; gases in, 45–7; layers, 47–8, 206–7
atoms, 8–9; arrangement and structure, 39–40, 74
Australia: aquifers and groundwater basins, 233*f*, 240*f*; atmospheric circulation, 211*f*; climate and climate zones, 211*f*, 212*f*, 213*f*, plate 18; droughts, 181*t*; habitat disturbance and destruction, 179–84; hot spot volcanoes, 116*f*; past water bodies, 232–5; structural units (NSW), 140; topography and tectonics, 236–7; vegetation, plate 19

b
banded iron formations, 45
Bessel, Friedrich, 15
big bang theory, 15, 17–20
biodiversity, 198–201
biological communities, 162–5
bioregions, 179–80, plate 19
biosphere, 37, 207, plate 6
blue giants (stars), 26
Blue Mountains, 162–4
Bondi, Hermann, 15, 21
brachiopods, plate 23
bushfires, 183

c
calcium cycle, 152*f*
carbon 14 dating, 77–8
carbon cycle, 51, 152*f*
carbonates, 129
cells: origins and evolution, 60–3; types, 62
chemosynthetic origin of life, 59
clastic sedimentary rocks, 136
climate change, 48–54, 53–4, 181, 210–15, 230*f*
comets, 2, 26–7
continental drift, 50–1, 83–4, 85*f*, plate 10
continental drift hypothesis, 83–4, 90–3
continental margins, 104–5
continuous spectrum, 8, 9
convection currents, 86, 87–8
convergent plate boundaries, 107–9
Copernicus, 14
Coriolis effect, 210–11
cosmology, 14–17
crust (Earth), 38–9; structure and movement, 87–93. *See also* plates (tectonic)
cyanobacteria (cyanophytes), 63–4, 222–3

d
dams, 187–8
dating methods: age of universe, 23–4; animal and plant remains, 77–9; rocks, 72–3, 77–8
deep sea trenches, 103
deltas, 233–5, Plate 24
dendrochronology, 54, 78–9
density particle model, 39–40
divergent plate boundaries, 104
DNA (deoxyribonucleic acid), 58, 199
Doppler effect, 17–18, plate 4
drainage systems, 186–7
droughts, 181*t*

e
Earth: age, 42, 79; albedo, 52, 210–11; atmosphere, 43, 45–8, 206–7; early character, 42–3, 57; hydrosphere and biosphere, 37, plate 6; magnetic field, 96–9; orbit, 49–50; structure and crustal movement, 38–9, 87–93, 129*t*, plate 8; subsystems, 206–7; water distribution and movement, 208–11. *See also* oceans; plates (tectonic)
earthquakes, 110–11
ecologically sustainable development, 193
ecosphere, 207
ecosystems, 172; diversity, 199–200; human impacts on, 184. *See also* biological communities
Einstein, Albert, 15, 18
electromagnetic radiation, 2–3, 6–7
electrons, 20, 74
elements, 8, 9, 32, 129*t*, 154–5
emission spectrum, 8, 9
ENSO (El Nino-Southern Oscillation), 181, 212, 213*f*
environmental change, 177–88; as problem, 191
environmental regulation, 192–5, 238, 242
eukaryote cells, 62
extrusive igneous rocks, 112–13, 134–5

f
faults and faulting, 109, 141–2
faunal succession, 73
feldspars, 87
felsic rocks, 87, 141
feral animals, 182–3
fission track method (of rock dating), 78
Fleming, Willamina, 9
foliated rocks, 138–9
frequency, 6
Friedman, Aleksandr, 15
fumaroles, 61–2, plate 11

g
galaxies, 2, 15, 18, 24–5
Galileo, 15
gamma rays, 7
genetic diversity, 199
geological time scales, 70–1
geology, effect on landscape, 141–2. *See also* minerals; plates (tectonic); rocks
global warming, 52
Gold, Thomas, 15, 21
Gondwanaland, 85*f*, 86, 91–2
grazing, 181–2
greenhouse effect, 44
greenhouse gases, 44, 47, 51
groundwater, 239–41
gyres, 212–15

h
half-life, 23, 75, 77
Hess, Harry, 84, 86
heterotrophs, 63
hot spot volcanoes, 115–16
Hoyle, Sir Fred, 15, 18, 21–2
Hubble Space Telescope, 4–5
Hubble, Edwin, 15, 18
hydrosphere, 37, 207
hydroxides, 130

i
ice ages, 48, 53
igneous rocks, 112, 133–5
infra-red radiation, 7
infra-red telescopes, 5
Internet, as research tool, 249–51
intrusions (igneous), 106
intrusive igneous rocks, 112, 134–5
island arcs, 107
isotopes, 23

j
Jeans, Sir James, 15, 21
Jenolan caves, 234
jointing, 142
Jovian planets, 34–5

k
Kepler, Johannes, 15

l
Lake Mungo, 235
land clearing, 179–80, 186, 238, 242
landscape: effect of geology on, 141–2
leaching, 153, 159
Lemaitre, George, 17
life: definition, 57; origins and preconditions, 57–63
light, 8
light years, 15
liquefaction, 111
lithosphere, 38–9, 87–8, 207

m

mafic rocks, 87, 129, 141
magma, 112, 133–4
magnetism, 96–9
main sequence stars, 25
matter particles, 20
mesosphere, 38–9, 47–8, 206–7
metamorphic rocks, 137–9
meteorites, 26–7, plate 7
microspheres, 60
microwave astronomy, 5
microwaves, 7
mid-ocean ridges, 86, 103–4, 105f, 112
Milankovitch cycles, 49–50, 54
Miller, Stanley, 59
mineralogy, 54
minerals, 127–32
molecules, 8–9
mor, 153
mountain formation, 53
Mt Narryer zircons, 79, 86

n

natural selection, 169–70
nebulae, 2, 26
neutrino, 20
neutrons, 20, 74
nitrogen-fixing bacteria, 45
nuclear decay, 23
nucleotides, 59

o

ocean-continent convergence, 108
ocean-ocean convergence, 107–8
oceans: currents and gyres, 212–15; formation, 44; sea floor spreading, 92–3; sea level changes, 52–3; subduction, 86
Oparin, A. I., 59
ophiolite suites, 106
optical telescopes, 3–4
orogenesis, 52–3
outgassing, 43
oxides, 130
ozone layer, 47, 64

p

palaeomagnetic banding, 98–9
palaeomagnetism, 96–7
Pangea, 84, 90
panspermia theory, 58
Parkes telescope, plate 2
pauses, 47
perihelion, 50
periodic table, 32
pH (of soils), 153–4, 158–9
photons, 6, 20
photosynthesis, 61, 64
Pickering, Edward, 9
planets, 34–7
plankton, 54, 222
plant communities, 172–5
plate tectonics theory, 84–7, 92
plates (tectonic): boundary movements, 88, 103–10; movements, 50–1, 83–93
plutonic rocks, 112
polarity reversals, 98, 99f
pollution (water), 221–3, 237–8, 241

polymers, 58
precession, 49
prokaryote cells, 62–3
proteins, 58
protons, 20, 32, 74
protostars, 25
Ptolemy, 14
pulsating universe theory, 22
pyroclasts and pyroclastic flows, 113–15

q

quarks, 20
quasars, 22

r

radar astronomy, 5
radiation, 2–3, 6–7, 18, 23
radio astronomy, 5
radio telescopes, 4–5, plate 2
radio waves, 7
radio-isotopes, 74–8
radioactivity, 74–6
radiometric clock, 76
Redhead Beach, 234–5, Plate 24
refugia, 201
research methods, 247–53
resources management, 192–3
rift valleys, 104
RNA (ribonucleic acid), 58
rocks: and landscape structure, 141–2; composition, 127–130; identification, 139–41; magnetic inclination, 96; methods of dating, 72–3, 77–8; types, 87, 105–6, 112, 133–44; weathering, 159, 161, 226–30, plate 20, plate 22

s

salinity, 220–1, 241–3
salt water intrusion, 239–41, plate 26
satellites, 4, 6
scleromorphy, 169
sea floor spreading, 92–3, 99f
sea level changes, 52–3
Secchi, Angelo, 8, 9
sedimentary rocks, 135–7
silicates, 127–9, 229
soil erosion, 161, 238
soils: and biological communities, 162–5; chemical properties, 151–5; composition, 146–7, 151–2; fertility, 158–9; formation, 147f, 159–62; New South Wales, 164f; organic content, 151–3; physical properties, 148–51; soil profiles, 156–8
solar nebular hypothesis, 31–2
solar system formation, 31–5
species diversity, 198–9
spectra, 9–10, plate 3
stars: classes, 26–7; evolution, plate 5; life cycle, 25–6
steady state theory, 15, 21–2
stellar spectra classification, 9–10
stellar spectroscopy, 9
Stonehenge, 3
stratigraphy, 72
stratosphere, 47, 206–7
stromatolites, 64, plate 12
structural units, 140

subduction, 86, 88
subduction zones, 86, 103
successions (ecological), 170–1
Suess, Edward, 86
sulfates and sulfides, 129
Sun, 37; formation, 32–3; output and energy, 49, 209–10
sunspots, 49
supernova, 26, 32
systems, 124–5

t

tectonic forces, 103–16, 236–7
tectonic plate theory, 84–7, 92
telescopes, 3–6, plate 2
terrestrial planets, 35–7
Tethys Sea, 109
thermosphere, 48, 206–7
topographic relief, 160–1
topography: tectonic effects on, 236–7
transform plate boundaries, 109–10
tree ring dating, 54, 78–9
trees: dieback, 180–1
troposphere, 47, 206–7
true stars, 25
tsunami, 111, 115

u

ultraviolet radiation, 7, 64
universe, 2, 14; components, 24–7; dating the age of, 23–4; observing and exploring, 2–4; theories of origin, 14–22
Urey, Harold, 59
Urey-Miller experiment, 59

v

visible light, 7
visible spectrum, 8
volcanic rocks, 112
volcanism, 36–7
volcanoes, 52; effects, 114–15, 142, 207, 237; eruption types, 113–14, plate 13; types, 112, 115, plate 13

w

water: and rock weathering, 226–30, plate 20, plate 22; evidence of past bodies of, 232–5; global distribution and movement, 208–11; pollution, 221–3, 237–8, 241; properties, 218–20; use, conservation and regulation, 237–8, 242
water cycle, 208–9, 226–7
water erosion, plate 21
water systems: human impacts on, 185–8
weathering, 159–60, 226–30, plate 20, plate 22
weeds, 182–3
Wegener, Alfred, 83–4, 86, 90, 92
wetlands, 233f
white dwarfs (stars), 25

xyz

X-ray telescopy, 6
X-rays, 7
zircons, 79, 86